Architecting the Internet of Things

Dieter Uckelmann • Mark Harrison
Florian Michahelles
Editors

Architecting the Internet of Things

With a foreword by Bernd Scholz-Reiter

Springer

Editors
Dieter Uckelmann
University of Bremen
Hochschulring 20
28359 Bremen, Germany
dieter@web-of-things.com

Mark Harrison
University of Cambridge
Institute for Manufacturing
17 Charles Babbage Road,
Cambridge, CB3 0FS, UK
mark.harrison@cantab.net

Florian Michahelles
ETH Zürich
Information Management
ETH Zentrum SEC E4
Scheuchzerstrasse 7
8092 Zürich, Switzerland
fmichahelles@ethz.ch

ISBN 978-3-642-19156-5 e-ISBN 978-3-642-19157-2
DOI 10.1007/978-3-642-19157-2
Springer Heidelberg Dordrecht London New York

Library of Congress Control Number: 2011925652

Cover design: eStudio Calamar S.L.

Printed on acid-free paper

Springer is part of Springer Science+Business Media (www.springer.com)

Foreword

The Internet of Things – Threats and Opportunities of Improved Visibility

The Internet has changed our business and private lives in the past years and continues to do so. The Web 2.0, social networks and mobile Internet access are just some of the current developments in this context. Ubiquitous computing and ambient intelligence have been fields of research where changes of computing in everyday situations have been examined. Today, the Internet of Things is a foundation for connecting things, sensors, actuators, and other smart technologies, thus enabling person-to-object and object-to-object communications.

The development of the Internet of Things is aligned with ongoing changes in information technology, logistics and electronic (e-)business. The significant reduction of message exchange times from analogue to digital messaging has led to reduced message sizes while increasing the number of message transactions. Additionally, there is a shift from mass broadcast to mass customisation and user-specified subscription to content tailored to an individual's interests. We expect to retrieve personalised information, as needed to cope with the growing information overflow. These changes are not limited to the Internet. We see similar changes in logistics, for example the increasing number of smaller deliveries, which has been influenced by e-business and improved material handling in the past years. The Internet of Things will bridge the gap between information technology and objects. The automatic identification of things and improved data handling capabilities allow individual product identification where we have previously been limited to types of products or batch identification. Large product recalls, which have led to severe financial and brand reputation losses, may be replaced by individual selective product recalls in direct business-to-consumer communication. E-business has changed our shopping habits. We retrieve information from the Internet, buy products online and contribute with information through product ratings. The speed of change in doing business has increased, thus requiring a higher level of agility. Some catalogue-based retailers that were previously very successful have been among the first victims because they were not prepared for the digital age.

The Internet of Things may just prove to be the missing link between logistics and information. However, there is still no clear understanding of how the Internet of Things will change our lives. First visions of smart fridges being able to automatically send replenishment orders have not yet become a reality. We might argue that consumers as well as businesses are not prepared for this yet or that this scenario is too complex – but is it? Printer manufacturing companies have integrated automatic identification for print cartridges, sensors to measure the ink fill levels, user interfaces to inform the consumer about the current status, instant online ordering of replenishment cartridges through corresponding software utilities, e-business and e-fulfilment, e-servicing and, last but not least, e-billing and e-payment.

Nevertheless, this only represents one stand-alone solution dominated by a single large business company. We would not want isolated "business tunnels" for every Internet of Things application. One key to success is the freedom of choice! We want to choose between different manufacturers, suppliers, service providers, delivery options, and payment services without the need for proprietary technologies. For this, we need to cope with the heterogeneity of the involved technologies and architectures. Interoperability across businesses, service providers and consumers will only be achieved if standardised interfaces can be provided.

Additionally, we need to overcome the structural shortcomings of IT investments in businesses. So far, Small and Medium Enterprises (SME) have been burdened by large key-players through mandates to invest in new technologies that rarely provide substantial benefits for the SME themselves. Cost benefit sharing and other compensation approaches need to be researched to make the Internet of Things a solution that is not limited to large companies.

We will need different human interfaces as well as machine interfaces to release the full potential of the Internet of Things. While we see Barcode and 2D-reader software being installed on mobile phones to identify objects, only few users are using this functionality to link to Internet-based information. Near-Field Communication (NFC) seems to be the next technology to enable unique identification and linking automatically to Internet services. Billing and payment services operated through mobile providers will be in the forefront to exploit the business opportunities of NFC. Radio frequency (RF) SIM cards provide another option that may enable non-NFC mobile phones to participate in mobile business and product related information access. In addition to multi-purpose devices we may see dedicated personal identification gadgets that are simpler to operate. USB-sticks have been more successful than mobile phones for portable data-storage. A small and easy to use identification device may be just as beneficial to link objects to their virtual representations in the Internet.

Will the Internet of Things make our lives easier? Or will it just be another component in a world of information overflow? Currently, the Internet of Things is all about information visibility – it is not about autonomous decision-making. To relieve us from everyday decision tasks and to avoid delays between information availability and decisions, new methods and technologies need to be integrated. In logistics, autonomous cooperating logistic processes are being researched. The main idea of this concept is to use decentralised and hierarchical planning and control methods. The combination of autonomous control and the Internet of Things would provide a higher level of infrastructural robustness, scalability and agility.

However, the integration of autonomous concepts in the Internet of Things is not limited to logistics. Personalised software agents will cater for our needs in private life, including shopping, smart home and public environments. Bidding agents are already quite common in the Internet. Nonetheless, software agents need to go beyond simple if-then algorithms, integrate sensor data to perceive the nature of their environment, communicate with other agents, learn from experience, and allow human intervention. Nevertheless, they need to be easy to use and configure to reach a higher level of user acceptance among the general public. Current developments in the Internet, enabling end-user participation through mash-ups and other user-friendly do-it-yourself software tools, are leading in a similar direction of leaving the developer community and reaching out to the end-users.

However, technology can only provide us with new opportunities. It is up to us to use these for holistic innovation approaches. We need to rethink traditional business setups. Other research disciplines need to integrate the Internet of Things into their every-day thinking. Civil architecture needs to develop RF friendly factories to avoid reflections and interferences. Industry designers need to develop forklift trucks with information technology ergonomically integrated, instead of bulky attachments. Public infrastructures, such as toll systems, need to be extended to support additional services for and through the Internet of Things. Objects, such as cars, need to be able to communicate with each other and with their own environment to exploit a limited infrastructure and enable new sustainable sharing models. Wearable computing needs to be enhanced to "sleek fashion computing" – where stylishness, usability, intelligence, connectivity and mobility are integrated to produce superior end-user friendly devices. Smart phones, personal data terminals, and other mobile computing devices are still far away from what a future Internet of Things will require to connect people and things.

The advantages of the Internet of Things are obvious. Improved efficiency, effectiveness, and new business opportunities may be achieved. Nevertheless, there are also certain threats and issues of governance, security, and privacy that need to be considered. Open governance in an Internet of Things remains an important issue. However, it may be assumed that the ongoing discussions between different regions and countries will lead to a federated structure in the longer term, similar to the domain structures we know from the Internet today. Anyway, proprietary industrial approaches ignoring international standardisation approaches as well as political discussion will try to set their own de-facto-standards. A recent malware attack (Stuxnet), aiming to spy on and reprogram Supervisory Control And Data Acquisition (SCADA) systems, has revealed once more the need for security in a future Internet of Things. The Internet has been misused to manipulate the virtual world, such as stock markets; and the Internet of Things will have direct implications on the physical world. In relation to privacy, it is important that personal data should be treated as such. New legislation is being proposed to deal with the misuse of personal data by employers. According to current political discussions in Germany, secret video surveillance of employees shall be banned and social websites shall not be used for research in the employment process. The Internet of Things enables further surveillance possibilities concerning employees and consumers. Again, it is up to us to use the advantages of the Internet of Things while promoting a responsible usage of the newly achieved visibility. Improved laws and regulations will help, but self-regulating control mechanisms will be even more important. Responsible usage will be rewarded in a world that is more and more influenced by social and sustainable management. Businesses have already seen boycott requests in the Internet of Things that have forced them to rapidly change their strategy. It will be important for enterprises to understand that these self-regulating mechanisms are extremely powerful and can change their business for better or for worse. We should always remember the power of a webcam showing an oil stream from a broken oil pipeline.

The Internet of Things provides far more visibility than a webcam – yet, it also enables faster exception handling and agility, which may help to save money, the environment or even lives.

Prof. Dr.-Ing. Bernd Scholz-Reiter

Contents

Figures

Tables

Abbreviations

4PL – Fourth Party Logistics

6LoWPAN – IPv6 Low Power Wireless Personal Area Networks

ACA – Assisting Cargo Agent

ACEA – European Automobile Manufacturers Association

ACID – Atomicity, Consistency, Isolation, Durability

AJAX – Asynchronous JAvaScript and XML

ALE – Application Level Events

API – Application Programming Interface

AR – Augmented Reality

ASAM – Associsation for Standardisation of Automation and Measurement Systems

B2B – Business-to-Business

B2C – Business-to-Consumer

BIBA – Bremer Institut für Produktion und Logistik GmbH

BMS – Building Management Systems

BOL – Beginning of Life

BPEL – Business Process Execution Language

BPMN – Business Process Modelling Notation

CAN – Controller Area Network

CBS – Cost Benefit Sharing

CERP – Cluster of European Research Projects on the Internet of Things

CH – Cluster Head

COBRA – Common Object Request Broker Architecture

CO-LLABS – Community-Based Living Labs

COM – Component Object Model

CPG – Consumer Packaged Goods

CPU – Central Processing Unit

CRC – Collaborative Research Centre

CRM – Customer Relationship Management

DCOM – Distributed Component Object Model

DFG – German Research Foundation

DiY – Do-it-Yourself

DiYSE – DiY Smart Experiences

DLNA – Digital Living Network Alliance

DNS – Domain Name Service

DoD – Department of Defense

DOGMA – Developing Ontology Grounded Methods and Applications framework

DOGMA-MESS – DOGMA Meaning Evolution Support System

DPWS – Device Profile for Web Services

DSRC – Dedicated Short Range Communications

EAN – Electronic Article Number

EANCOM – EAN Communication

EASE – Ecological Approach to Smart Environments

ebXML – Electronic Business Extensible Markup Language

ECM – Enterprise Content Management

ECR – Efficient Consumer Response

EDI – Electronic Data Interchange

EDIFACT – Electronic Data Interchange For Administration, Commerce and Transport

EEML – Extended Environments Markup Language

EnoLL – European Network of Living Labs

EOL – End of Life

EPC – Electronic Product Code

EPCIS – Electronic Product Code Information Service

ERP – Enterprise Resource Planning

ESSI – European Semantic Systems Initiative

ETSI – European Telecommunications Standards Institute

EURIDICE – European Inter-Disciplinary Research on Intelligent Cargo for Efficient, safe and environment-friendly logistics

FIFO – First In, First Out

FIPA – Foundation for Intelligent Physical Agents

FOSSTRAK – Free and Open Source Software for Track and Trace

FSF – Free Software Foundation

GEF – Graphical Editing Framework

GNSS – Global Navigation Satellite System

GPRS – General Packet Radio Service

GPS – Global Positioning System

GRAI – Global Returnable Asset Identifier

GSM – Global System for Mobile Communications

GUI – Graphical User Interface

HAL – Hardware Abstraction Layer

HF – High Frequency

HTML – HyperText Markup Language

HTTP – HyperText Transfer Protocol

I/O – Input/Output

IC – Integrated Circuit

ICT – Information and Communication Technology

ID – Identifier

IDE – Integrated Development Environment

IERC – Internet of Things Research Cluster

IETF – Internet Engineering Task Force

IFC – Industry Foundation Classes

IFTF – Institute for the Future

IMSAS – Institute for Microsensors, -actuators and -systems

IOT – Internet of Things

IoT IS – Internet of Things Information Service

IP – Internet Protocol

IPv6 – Internet Protocol version 6

IRTF – Internet Research Task Force

ISO – International Organization for Standardization

IT – Information Technology

ITEA2 – Information Technology for European Advancement, period 2

ITS – Intelligent Transport Systems

ITU-T – International Telecommunication Union - Telecommunication Standardisation Sector

IWT – Agency for Innovation by Science and Technology (Belgium)

J2SE – Java 2 Platform, Standard Edition

JADE – JavaAgentDEvelopment framework

JSON – JavaScript Object Notation

KB – Knowledge Base

LCD – Liquid Cristal Display

LED – Light-Emitting Diode

LOD – Linked Open Data

LTE – Long Term Evolution

M2M – Machine-to-Machine

MAC – Medium Access Control

MANET – Mobile Ad-hoc NETworks

MAS – Multi-Agent Systems

MIDI – Musical Instrument Digital Interface

MIT – Massachusetts Institute of Technology

MOL – Middle of Life

NFC – Near Field Communication

NIP – Non-Internet Protocol

NJMF – Norwegian Iron and Metal Workers Union

OBU – On-Board Unit

OCA – Operational Cargo Agent

ONS – Object Name Service

OOS – Out of Stock

ORiN – Open Robot Resource Interface for the Network

OSGi – Open Services Gateway initiative

ORPHEUS – Object Recognition and Positioning Hosted European Service

OS – Open Source

OS – Operating System

OSGi – Open Service Gateway initiative

OWL – Web Ontology Language

OWL-S – Web Ontology Language for Web Services

P2P – Peer-to-Peer

PaaS – Product as a Service

PbH – Power by the Hour

PBL – Performance-based Logistics

PD – Participatory Design

PDT – Personal Data Terminals

PEID – Product Embedded Information Device

PLCS – Product Life Cycle Support

PLM – Product Lifecycle Management

PMI – PROMISE Messaging Interface

PuSH – PubSubHubbub

QoS – Quality of Service

R&D – Research and Development

RDF – Resource Description Framework

RDFa – Resource Description Framework in attributes

REST – Representational State Transfer

RFC – Remote Function Call

RFD – Reduced Functionality Devices

RFID – Radio Frequency Identification

ROI – Return on Investment

ROLL – Routing Over Low power and Lossy networks

RTI – Returnable Transport Items

RTP – Real Time Protocol

RTSP – Real Time Streaming Protocol

SAC – Social Access Controller

SCM – Supply Chain Management

SDO – Sensor Data Ontology

SGTIN – Serialised Global Trade Identification Number

SHO – Sensor Hierarchy Ontology

SHOE – Simple HTML Ontology Extension

SMD – Service Mapping Description

SME – Small and Medium-sized Enterprises

SOA – Service Oriented Architecture

SOAP – Simple Object Access Protocol

SOC – Service-Oriented Computing

SPARQL – SPARQL Protocol and RDF Query Language

SPI – Serial Peripheral Interface

SQL – Structured Query Language

SSCC – Serial Shipping Container Code

SUMO – Suggested Upper Merged Ontology

SWS – Semantic Web Services

SWSF – Semantic Web Services Framework

TCP – Transmission Control Protocol

TDS – Tag Data Standard

TDT – Tag Data Translation

UCD – User-centered Design

UCSD – University of California San Diego

UDDI – Universal Description, Discovery and Integration

UHF – Ultra High Frequency

UI – User Interface

UML – Unified Modelling Language

UMTS – Universal Mobile Telecommunications System

UPnP – Universal Plug-and-Play

URI – Uniform Resource Identifier

URL – Uniform Resource Locator

URN – Uniform Resource Name

USB – Universal Serial Bus

UWB – Ultra Wide Band

VSP – Virtual Service Point

W3C – World Wide Web Consortium

WIMP – Window, Icon, Menu, Pointing device

WMS – Warehouse Management System

WoT – Web of Things

WS – Web Services

WSAN – Wireless Sensor and Actuator Network

WSDL – Web Service Definition Language

WSMO – Web Service Modeling Ontology

WSMX – Web Service Execution Environment

WSN – Wireless Sensor Network

WWAI – World Wide Article Information

WWW – World Wide Web

XML – Extensible Markup Language

XMPP – Extensible Messaging and Presence Protocol

XOL – Ontology Exchange Language

1 An Architectural Approach Towards the Future Internet of Things

Dieter Uckelmann[1], Mark Harrison[2], Florian Michahelles[3]

[1]LogDynamics Lab, University of Bremen, Germany

[2]Institute for Manufacturing, University of Cambridge, United Kingdom

[3]Information Management, ETH Zurich, Switzerland

Abstract Many of the initial developments towards the Internet of Things have focused on the combination of Auto-ID and networked infrastructures in business-to-business logistics and product life cycle applications. However, a future Internet of Things can provide a broader vision and also enable everyone to access and contribute rich information about things and locations. The success of social networks to share experience and personalised insights shows also great potential for integration with business-centric applications. The integration and interoperability with mainstream business software platforms can be enhanced and extended by real-time analytics, business intelligence and agent-based autonomous services. Information sharing may be rewarded through incentives, thus transforming the Internet of Things from a cost-focused experiment to a revenue-generating infrastructure to enable trading of enriched information and accelerate business innovation. Mash-ups and end-user programming will enable people to contribute to the Internet of Things with data, presentation and functionality. Things-generated physical world content and events from Auto-ID, sensors, actuators or meshed networks will be aggregated and combined with information from virtual worlds, such as business databases and Web 2.0 applications, and processed based on new business intelligence concepts. Direct action on the physical world will be supported through machine-interfaces and introduction of agile strategies. This chapter aims to provide a concept for a future architecture of the Internet of Things, including a definition, a review of developments, a list of key requirements and a technical design for possible implementation of the future Internet of Things. As open issues, the evaluation of usability by stakeholders in user-centric as well as business-centric scenarios is discussed and the need for quantifying costs and benefits for businesses, consumers, society and the environment is emphasised. Finally, guidelines are derived, for use by researchers as well as practitioners.

1.1 Introduction, Background and Initial Visions

The term *Internet of Things* first came to attention when the Auto-ID Center launched their initial vision of the EPC network for automatically identifying and tracing the flow of goods in supply-chains, in Chicago in September 2003 (EPC Symposium 2003). Whereas the first mention of 'Internet of Things' appears in an Auto-ID Center paper about the Electronic Product Code by David Brock in 2001 (Brock 2001), increasing numbers of researchers and practitioners have followed this vision, as it is documented by books, conferences and symposia having Internet of Things in their titles.

The Internet of Things is a concept in which the virtual world of information technology integrates seamlessly with the real world of things. The real world becomes more accessible through computers and networked devices in business as well as everyday scenarios. With access to fine-grained information, management can start to move freely from macro to micro levels and will be able to measure, plan and act accordingly. However, the Internet of Things is more than a business tool for managing business processes more efficiently and more effectively – it will also enable a more convenient way of life.

Since the founders of the Auto-ID Center coined the term 'Internet of Things' (Santucci 2010), it has widely been used by researchers and practitioners to describe the combination of the real world with the virtual world of information technology (Fleisch and Mattern 2005, Bullinger and ten Hompel 2007, Floerkemeier et al. 2008) by means of automatic identification technologies, real-time locating systems, sensors and actuators.

Thanks to the recent advances of miniaturisation and the falling costs for RFID, sensor networks, NFC, wireless communication, technologies and applications, the Internet of Things suddenly became relevant for industry and end-users. Detection of the physical status of things through sensors, together with collection and processing of detailed data, allows immediate response to changes in the real world. This fully interactive and responsive network yields immense potential for citizens, consumers and business.

RFID is increasingly being deployed in applications across supply chains with readers that are distributed across factories, warehouses, and retail stores. Sensor technology is also being adopted in manufacturing and logistics in order to control processes and the quality of goods. In traditional RFID applications, such as access control and production automation, tags moved in closed-loop processes, and the RFID data was consumed only by a single client system. Accordingly, there was little need for exchange of data across organisational boundaries. In the same way that monolithic business information systems of the past have evolved into highly networked systems that use the Internet extensively, open-loop RFID applications in networked environments represent a challenge that various stakeholders from industry are facing and partly solving.

Accessing real-time information through Information and Communication Technology (ICT) usage in the 'anytime, anywhere' manner, as suggested by the paradigm of the Internet of Things, calls for open, scalable, secure and standardised infrastructures which do not fully exist today. These have been developed and continue to be developed for example in working groups within the EPCglobal community in order to gather user requirements and business cases to develop open global technical standards for improved visibility. Similarly, members of the Open Geospatial Consortium (OGC) are building a framework of open standards for exploiting Web-connected sensors and sensor systems of all types, including flood gauges, air pollution monitors, stress gauges on bridges, mobile heart monitors, webcams and satellite-borne earth imaging devices. Today's technology-centric instead of user-centric developments are some of the problems that hinder a broader and faster adoption. The arrival of NFC and RFID technology in the consumer market (e.g., Nabaztag.com, Touchatag.com) together with the availability of mobile Internet (e.g., Apple iPhone, HTC Touch) and scalable information sharing infrastructures (e.g., Twitter.com) opens an enormous space for end-user innovation and user-centric developments. People and things are getting closer. An open and holistic approach of a network of products and people has yet to be developed.

Most existing RFID-installations in production and logistics today can be considered as an Intranet of Things or Extranet of Things. Traditional communication means, such as EDIFACT, are used to communicate with a limited number of preferred partners. These early approaches need to be extended to support open Internet architectures.

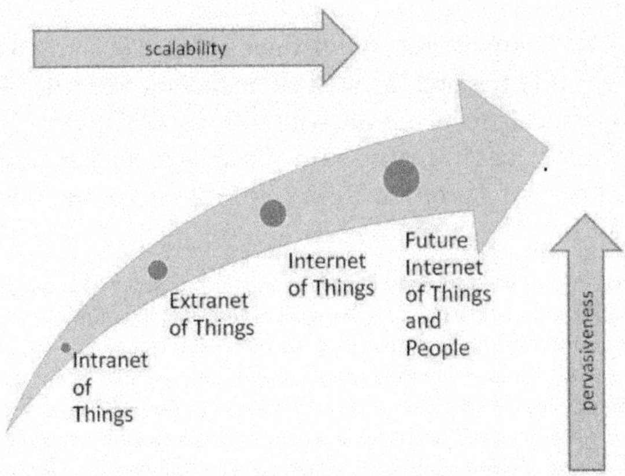

Fig. 1.1 A Phased Approach from the Intranet of Things to a Future Vision on the Internet of Things

Figure 1.1 shows a phased approach from the current Intranet / Extranet of Things to a future Internet of Things and People. While pervasiveness increases through new applications and wider adoption, the scalability requirements of the Internet of Things have to be met.

Additionally, a solid business case and flexible mechanisms for balancing costs and benefits are missing in many of today's early implementations. The usability needs to be improved by providing flexible but simple devices and services to connect things and people. The Internet of Things can benefit from the latest developments and functionalities commonly referred to as Web 2.0 through provision of new intuitive user-centred and individually configurable and self-adapting smart products and services for the benefit of businesses and society. Whereas the successful examples of Web 2.0, such as Facebook or Twitter, connect people with data, this is achieved by proprietary Application Programming Interfaces (APIs) that do not provide powerful data-sharing models capable of Business-to-Business (B2B) requirements, such as data management and analysis.

This chapter will focus on providing an overview of the Internet of Things and its future requirements. In section 1.2 we will provide a definition of the Internet of Things. Section 1.3 will provide a broad review of development projects and initiatives, whereas in section 1.4 we will highlight ten key requirements for the future Internet of Things. Section 1.5 will explain a holistic architectural approach and, finally, in section 1.6 we will provide a conclusion and a further outlook towards future developments.

1.2 Definitions and Functional Requirements

The term Internet of Things is not well defined and has been used and misused as a buzzword in scientific research as well as marketing and sales strategies. Until today it remains difficult to come up with a clear definition of the Internet of Things. One definition has recently been formulated in the Strategic Research Agenda of the Cluster of European Research Projects on the Internet of Things (CERP-IoT 2009):

> "Internet of Things (IoT) is an integrated part of Future Internet and could be defined as a dynamic global network infrastructure with self configuring capabilities based on standard and interoperable communication protocols where physical and virtual 'things' have identities, physical attributes, and virtual personalities and use intelligent interfaces, and are seamlessly integrated into the information network. In the IoT, 'things' are expected to become active participants in business, information and social processes where they are enabled to interact and communicate among themselves and with the environment by exchanging data and information 'sensed' about the environment, while reacting autonomously to the 'real/physical world' events and influencing it by running processes that trigger actions and create services with or without direct human intervention. Interfaces in the form of services facilitate interactions with these 'smart things' over the Internet, query and change their state and any information associated with them, taking into account security and privacy issues."

While this definition lists the possible technical components of the Internet of Things, it still has three major shortcomings. Firstly, it lists components that have been mentioned before in relation to other visions such as pervasive or ubiquitous computing and therefore it is difficult to distinguish from these concepts. Secondly, it misses wider consideration of current developments and user-interactions in the Internet commonly referred to as Web 2.0. Similar to the relationship between the World Wide Web (WWW) and the Internet, the addition of Web 2.0 functionality may be seen as a user-centric extension to the Internet of Things rather than an integral part of it. However, whereas the development of the Internet began more than thirty years before the realisation of the WWW in the early 1990s, the Internet of Things is already being influenced by Web 2.0 functionality right from the beginning. Both technology developments have been happening in parallel rather than consecutively. Thirdly, it does not provide a reason why or how the Internet of Things will be a self-sustainable and successful concept for the future. Self-sustainability encompasses viability, including a dynamic global network infrastructure with self-configuring capabilities based on standards and interoperable communication protocols as well as openness for future extensions, ideas, and technologies. Economic success may never have been a part of a definition for the Internet or other technical network infrastructures. Nevertheless, we consider it a valid consideration within a holistic definition approach as economic success and adoption is just as important as technical sustainability in a forward-looking statement.

For the purposes of differentiation it may be best to consider what the Internet of Things is not – or at least not exclusively. A corresponding blog discussion has been started by Tomas Sánchez López (Sánchez López 2010). He considers that the Internet of Things is not only:

- *ubiquitous / pervasive computing*, which does not imply the usage of objects nor does it require a global Internet infrastructure
- the *Internet Protocol (IP)*, as many objects in the Internet of Things will not be able to run an Internet Protocol
- a *communication technology*, as this represents only a partial functional requirement in the Internet of Things similar to the role of communication technology in the Internet
- an *embedded device*, as RFID tags or Wireless Sensor Networks (WSN) may be part of the Internet of Things, but stand-alone they miss the back-end information infrastructures and in the case of WSN the standards to relate to 'things'
- the *application*, just as Google or Facebook could not be used in the early 90's to describe the possibilities offered by Internet or WWW

With these negations in mind it is easier to differentiate the Internet of Things. Consequently, this implies that most publications claiming to address the Internet of Things are not really covering the real essence of the Internet of Things. We suggest two more negations. The Internet of Things is *not the Internet of People* (although we believe that the Internet of People will link to the Internet of Things) and it is *not the Intranet or Extranet of Things*. Therefore, applications that provide only access to a small group of stakeholders (e.g., few companies) should not be considered to represent the full scope of the Internet of Things. However, all fields of research that have been mentioned above overlap partially with the Internet of Things (Figure 1.2).

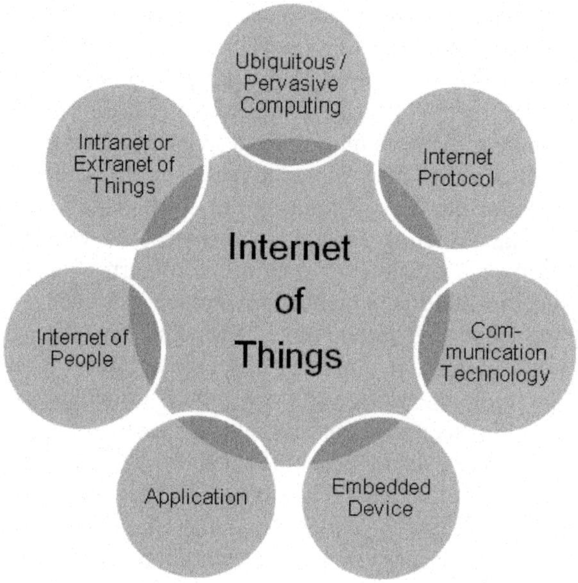

Fig. 1.2 Overlaps of the Internet of Things with Other Fields of Research

The second problem in the CERP-IoT definition is the missing Web 2.0 integration. One could argue that the Web 2.0 is exemplified only by certain types of applications in the Internet of People, which again is not equal to the Internet of Things. However, the Web 2.0 has changed usage of the WWW by providing more intuitive interfaces for user interaction, social networking and publication of user-generated content, without requiring fundamental changes to the design and existing standards of the internet. The primary advantage of Web 2.0 technology has been the use of intuitive interfaces to enable web contributions by end-users irrespective of their technical expertise. The interaction between things and people will be one core issue in the future Web of Things. End-user product ratings and usage instructions provide a valuable set of information on things. Unfortunately today this information is very much scattered across the WWW and there is no direct link to a product identifier.

Thirdly, the reason for success is missing in the above CERP-IoT definition. Maybe a definition on the Internet of Things does not require a benefit statement – the Internet of Things itself surely does, if it is ever to become a reality. Initially, most applications of Auto-ID technologies were internal or closed-loop applications rather than applications across company boundaries. The main reason is the missing benefit for the individual participants. While benefits can be easily calculated across supply chains or product life cycles, input data to cost-benefitanalysis is most often based on "educated guessing" (Gille and Strüker 2008, Laubacher et al. 2006) rather than on hard facts.

Another approach towards a definition of the Internet of Things can be derived from logistics where it is common to ask for the *right product* in the *right quantity* at the *right time* at the *right place* in the *right condition* and at the *right price*. In this analogy the *right product* relates to accurate and appropriate information about a uniquely identifiable physical object as well as its form, fit and function. This includes the usage of Auto-ID and appropriate sensor information or any other kind of linked information to the object that can be accessed through the Internet of Things. The *right quantity* can be achieved through high granularity of information combined with filtering and intelligent processing. The *right time* does not necessarily mean anytime, but more precisely 'when needed'. It may be sufficient to receive information about an object only once a day or only in the case of a status change. Consequently, right-time does not equal real-time, a term that is mentioned quite often in relation to the Internet of Things.

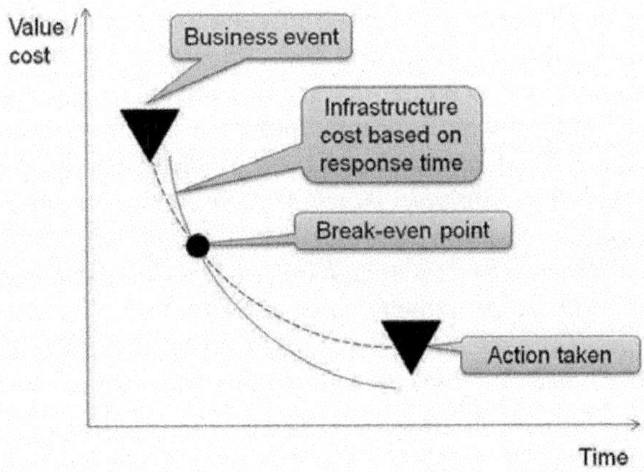

Fig. 1.3 Infrastructure cost vs. response time (based on Hackathorn 2004)

In general, real-time access to data is desirable to reduce the latency between a business event and a corresponding action; the ability to achieve such a reduction is also referred to as agility. Unfortunately, real-time capability is linked to high infrastructure cost (Figure 1.3).

Similarly, the information availability at *right place* does not imply any place - but rather, where the information is needed or consumed (which may not necessarily be the same place as where it is generated). If information is not generated and consumed in the same place and if either of these places have unreliable or intermittent network connectivity, then effective data synchronisation protocols and caching techniques may be necessary to ensure availability of information at the right place. Again, the cost of any place availability has to be seen in relation to its profit potential. But as mobile devices are more and more ubiquitous, there will evidently be an opportunity to access information in the Internet of Things at any place at a reasonable price. The *right information* condition is met if it can be utilised with a minimum effort. This includes human readable information for human interaction as well as semantically and syntactically enriched machine-readable information, which may in turn require transformation of low-level raw data (possibly from multiple sources) into meaningful information and may even require some pattern recognition and further analysis to identify correlations and trends in the generated data. The *right price* is not automatically the lowest price, but instead it is a price between the costs for information provisioning and the achievable market price. Information provisioning costs include labour costs as well as infrastructure costs.

A minimalist approach towards a definition may include nothing more than *things*, the *Internet* and *a connection in between. Things* are any identifiable physical object independent of the technology that is used for identification or providing status information of the objects and its surroundings. *Internet* in this case refers to everything that goes beyond an extranet, thus requiring access to information for more than a small group of people or businesses. A closed loop application consequently has to be regarded as an *Extranet of Things*. The *Internet* acts as a storage and communication infrastructure that holds a virtual representation of *things* linking relevant information with the object.

Combining the different approaches we can conclude that *the future Internet of Things links uniquely identifiable things to their virtual representations in the Internet containing or linking to additional information on their identity, status, location or any other business, social or privately relevant information at a financial or non-financial pay-off that exceeds the efforts of information provisioning and offers information access to non-predefined participants. The provided accurate and appropriate information may be accessed in the right quantity and condition, at the right time and place at the right price. The Internet of Things is not synonymous with ubiquitous / pervasive computing, the Internet Protocol (IP), communication technology, embedded devices, its applications, the Internet of People or the Intranet / Extranet of Things, yet it combines aspects and technologies of all of these approaches.*

1.3 A European Perspective on Funded Projects, Technologies and State of the Art in Relation to the Internet of Things

Several projects related to the Internet of Things have been carried out and have contributed to the current state of the art. Especially in Europe, numerous projects have been funded to research certain aspects of the Internet of Things.

EPoSS[1] brings together European private and public stakeholders to create an enduring basis for structuring initiatives, for co-ordinating and bundling efforts and for establishing sustainable structures of a European Research Area on Smart Systems Integration. EPoSS has published the 'Internet of Things in 2020' (EPoSS 2008) report, which elaborates on what the Internet of Things might become in future. In particular, governance, standardisation and interoperability are named as absolute necessities on the path towards the vision of things that are able to communicate with each other. Furthermore, the report states that the real advantages of the Internet of Things have to be shown convincingly, addressing and considering all citizens' concerns when developing innovative solutions and proposals. The objective of the *BRIDGE*[2] project was to research, develop and implement tools to enable the deployment of Radio Frequency Identification (RFID) and EPCglobal Network applications. Based on an initial vision by the *Auto-ID Center*, the architecture of the *EPCglobal Network* (2007) has developed to become an architecture of industry-driven open standards based on unique item identification via the Electronic Product Code (EPC) encoded on data carriers, such as RFID. It defines standards for capturing, filtering, storing and querying EPC data and includes layered standards spanning the whole architecture range from RFID tag memory layout and air interfaces to look-up services that return pointers to data repositories given a particular identifier. The BRIDGE project was dedicated to the development of easy-to-use technological solutions for the European business community including small and medium sized enterprises (SME), ensuring a basis for collaborative EPCglobal systems for efficient, effective and secure supply chains. The technical work in BRIDGE made significant progress on some required services for the Internet of Things, such as discovery services. The ITEA 2[3] funded *Do-it-Yourself Smart Experiences* project (*DiYSE*)[4] has just recently started and aims to enable ordinary people to easily create, setup and control applications in their smart living environments as well as in the public Internet of Things space, allowing them to leverage aware services and smart objects for obtaining highly personalised, social, interactive, seamless experiences at home and in the city. DiYSE is not looking at business-to-business communication. A single architecture that addresses both business and public

---

[1] www.smart-systems-integration.org

[2] www.bridge-project.eu

[3] www.itea2.org

[4] www.dyse.org

applications based on a standardised infrastructure would be beneficial to bridge the gap.

In 2010, further projects funded by the EU such as *Internet of Things – Architecture* (IoT-A)[5], *Enabling the business-based Internet of Things and Services* (ebbits)[6], *The Network is the Business* (NISB)[7], *Software Platform for Integration of Engineering and Things* (SPRINT), *Experiential Living Labs for the Internet Of Things* (ELLIOT), *Networked Enterprise transFormation and resource management in Future internet enabled Innovation CloudS* (NEFFICS), *Internet of Things Initative* (IOT-i)[8], and *Internet of Things at Work* (IoT@work) have started their work and will contribute to the ongoing research concerning the Internet of Things in Europe.

There are several projects and standardisation initiatives on sensor networks, which may eventually converge with the Internet of Things. The core objective of the *COBIS*[9] project was to provide the technical foundation for embedded and wireless sensor network technology in industrial environments. *SENSEI*[10] creates an open, business-driven architecture that fundamentally addresses the scalability problems for a large number of globally distributed wireless sensors and actuator devices. It provides network and information management services to enable reliable and accurate contextual information retrieval and interaction with the physical environment. Likewise, other smaller research projects exist, such as *GSN* (Aberer et al. 2006), *SARIF* (Shim et al. 2007), and *MoCoSo* (Sánchez López et al. 2009), that combine concepts of object identification, sensor data and the Internet. Sensor networks can be integrated in the Internet of Things for example, by integration with the EPCglobal Architecture Framework. Although the EPCglobal Network does not yet provide adequate support for the inclusion of sensor values in the streams of data, the Action Groups inside the GS1/EPCglobal community are actively researching issues such as 'Active Tagging' and 'Sensor and Battery Assisted Passive Tags'. The EPC Sensor Network (Sung et al. 2007) is an effort of the Auto-ID Lab Korea to incorporate Wireless Sensor Networks (WSN) and sensor data into the EPCglobal Network architecture and standards. The Open Geospatial Consortium (OGC) Sensor Web Enablement (SWE) initiatives are establishing the interfaces and protocols that will enable a 'Sensor Web' through which applications and services will be able to access sensors of all types over the Web. The OGC SWE defines standards for modelling, encoding, transporting, querying and discovering sensor data (Botts et al. 2006). Valuable lessons can be learned from this work and from other standardisation initiatives (e.g., IEEE 1451, ISO/ICE 24753) for incorporation of sensor support into the

---

[5] www.iot-a.eu

[6] www.ebbits-project.eu

[7] www.nisb-project.eu

[8] www.iot-i.eu

[9] www.cobis-online.de

[10] www.sensei-project.eu

Internet of Things architecture. A public deliverable from the BRIDGE project provides a detailed survey of standards relevant for integration of sensor information (BRIDGE 2009). While most of the sensor network standardisation activities are still in an early stage, there are already established industry standards promoted through the OPC Foundation[11] and the Association for Standardisation of Automation and Measuring Systems[12] with a focus on industry automation. It should be possible to achieve synergies between these different approaches in an open Internet of Things architecture.

While identification, sensing and actuator integration are core functionalities in an Internet of Things, there are further requirements such as scalability and robustness that need to be addressed. Again, there are numerous existing research activities to build on. Clustering of resources seems to be one valid approach to address this issue. Much work on clustering has been done for MANETs (Mobile Ad-hoc Networks) with little regard to strongly constrained devices, such as those most common in the Internet of Things (e.g., wireless sensor networks). Even so, specialised protocols exist for certain desirable features , for example *energy-efficiency:* EECS (Ye et al. 2005), EDAC (Wang et al. 2004) and HEED (Younis and Fahmy 2004), *mobility:* DMAC (Basagni 1999), *heterogeneity*: GESC (Dimokas et al. 2007), but there is no unified work. Additionally, research on autonomous concepts will influence the further development of the Internet of Things.

Technical projects are supplemented by research and coordination activities on standards and privacy. The *GRIFS*[13] project, seeks to identify all relevant standards for the operating characteristics of physical things (readers, tags, and sensors), infrastructure standards for defining the communications, addressing and structures, and data exchange standards. *CASAGRAS2*[14] looks at global standards, regulatory and other issues concerning RFID and its role in the Internet of Things. *PRIME* [15] focussed on privacy and identity management for private consumers but this proposal did not consider how to empower users e.g. to make informed and balanced choices in their purchasing decisions, supported by the Internet of Things. The *European Research Cluster on the Internet of Things* (IERC[16] finally aims to achieve a consensus on how to realise the vision of the Internet of Things in Europe.

[11] www.opcfoundation.org

[12] www.asam.net

[13] www.grifs-project.eu

[14] www.iot-casagras.org

[15] www.prime-project.eu

[16] www.internet-of-things-research.eu

1.4 Opportunities and Motivation

Even though there are numerous projects and developments concerning certain aspects of the Internet of Things, an open and accessible infrastructure for a wider adoption of the Internet of Things is missing. A more generic approach towards a future development schedule is needed. While technologies are important building blocks, they are not enough to embrace the large research spectrum that needs to be addressed. The following five subject guidelines may be used to trigger successful and sustainable contributions to the Internet of Things.

1. *Envision* – A vision of the Internet of Things needs to provide holistic scenarios focusing on private, social and business benefits. Experimentally-driven, participative research approaches will be needed to allow involvement of different stakeholders for identification of requirements, usability testing, evaluation and active participation. Mechanisms are needed for empowering citizens to fully participate and innovate in the Internet of Things, in order to provide a new multi-directional communication infrastructure for researchers, industries and citizens. This user-centric concept maybe referred to as the 'Web of Things' as it provides intuitive graphical user interfaces that include functionalities familiar to Web 2.0 applications.

2. *Extend* – To leverage state-of-the-art developments and accepted technologies, existing architectures, such as the EPCglobal Network, should be utilised and extended by adding new functionalities to support diverse means of identification (RFID, barcode, 2D-code), sensors, actuators, intelligent devices and other information sources (e.g. user-generated content, commercial databases) within an open framework. The value of product-related data needs to be increased through semantic enrichment. Extending existing approaches will allow utilisation of prior efforts and investments and allow a phased approach towards the Internet of Things. Disruptive new approaches should be avoided unless they provide substantial new benefits or build on existing work. It should be noted that this approach does not exclude integration of other heterogeneous technologies, but it promotes the usage of a single core architecture.

3. *Enable* – It is crucial to solve today's adoption challenges. There is still a lot of research needed on technical challenges that too often are considered to be solved (especially by researchers and practitioners lacking the technical knowledge). Privacy, security and confidentiality are key factors to provide a trustworthy Internet of Things. New mechanisms for sharing costs and benefits to enable the creation of opportunities for new market entrants are needed.

4. *Excite* – New stakeholders need to be excited to contribute to the future Internet of Things. Ease of participation, collaboration and generation of benefits are major requirements to excite new entrants to the Internet of Things. Open frameworks and end-user programming environments may empower citizens to create cost-free as well as billable micro services, such as a product guides and reviews.
5. *Evaluate* – New approaches need to be discussed with a large variety of stakeholders and verified in industry pilots and user-centric environments. A good example for the future Internet of Things is the informed and ethical consumer who requires product-related data (e.g., country of origin, ingredients, dynamic best-before date, carbon-footprint) and who is willing to add information to the Internet of Things. Other popular examples include public user-centric scenarios that build on the concept of Smart Cities and Smart Homes. Furthermore, we need to evaluate the Internet of Things in a philosophical context as things will become social actors in a networked environment.

1.5 Outlook to Future Developments

Based on the development schedule described above, we see a list of key requirements that need to be considered in the Internet of Things:

1. *Meet key societal needs for the Internet of Things including open governance, security, privacy and trustworthiness.* The Internet of Things should not be owned by single interest groups, as it should be an open global infrastructure as the Internet and WWW are today. One of the key issues in Europe and Asia in the past years has been the predominance of VeriSign, an American company operating the Object Name Service (ONS) under contract for the EPCglobal Network (Clendenin 2006, Heise online 2008). Federated structures are needed to provide a power balance. Security, privacy and trustworthiness need to be considered, but are in most aspects not specific to the Internet of Things. The same technologies that have been successfully used in the Internet can be utilised in the Internet of Things as well, although there are some specific challenges due to characteristics of the Internet of Things application scenarios, which often include mobile or portable objects that change custody or ownership during their lifetimes. However, there is a difference in the Auto-ID, sensor and actuator part, where different attacks on the network are possible. Nevertheless, it has to be remembered that the highest achievable security level is not always required. There are for example different levels of security required for passports or logistic applications.

2. *Bridge the gap between B2B, business-to-consumer (B2C) and machine-to-machine (M2M) requirements through a generic and open Internet of Things infrastructure.* While there has been a clear focus on B2B requirements in the last years, B2C and M2M will gain importance in the future Internet of Things. While in B2C ease of use as well as human readable data are important, in M2M communications, the data should be machine-readable structured and semantically well-defined.
3. *Design an open, scalable, flexible and sustainable infrastructure for the Internet of Things.* The Internet of Things has to be open by definition. Open standards are required to use and extend its functionality. It will be a huge network, considering that every object has its virtual representation. Therefore, scalability is required. The Internet of Things will need to be flexible enough to adapt to changing requirements and technological developments. Its development can be accelerated through the availability of open source software, such as Fosstrak[17] to allow anyone to implement and test new functionalities. Another opportunity to experiment and test new functionalities are living lab initiatives, where service providers and users participate in a collaborative environment. Finally, it needs a sustainable infrastructure to provide a basis for the necessary investments.
4. *Develop migration paths for disruptive technological developments to the Internet of Things.* Rather than requiring disruptive new and parallel approaches, there have to be means of integrating new developments into the fundamental infrastructure, otherwise there can be no guarantee of sustainability or enduring value. Examples include autonomous objects that do not essentially require a networked infrastructure. Nevertheless, providing a migration path for autonomous control in the Internet of Things would broaden its usage and provide a solid networked infrastructure for autonomous objects (Uckelmann et al. 2010).
5. *Excite and enable businesses and people to contribute to the Internet of Things.* If stakeholders cannot benefit from the Internet of Things, they will not participate. In contrast, any user benefiting from the Internet of Things will attract and excite more participants. Research on how to benefit from the Internet of Things is needed. Business needs to see a clear business case. End-users need to find a personal benefit. Funded research, such as that described in section 1.3, can trigger new ideas and stakeholders, but in a longer view benefits have to be generated from within the network and not through external funds.
6. *Enable businesses across different industries to develop high added value products and services.* New business models (both industry-specific and cross-sector) are required based on retrieving and contributing information to/from the Internet of Things. Researchers can help to identify new

[17] www.fosstrak.org

potentials but business entrepreneurs are needed to actually raise the potential of the Internet of Things.

7. *Encourage new market entrants, such as third party service and information providers, to enter the Internet of Things.* Information in the Internet of Things can be accumulated, processed and sold independently of owning the physical product. Service providers should be encouraged for example to provide access to multiple sources of information about things and adding technical billing capabilities for information access.
8. *Provide an open solution for sharing costs, benefits and revenue generation in the Internet of Things.* Information should be freely tradable, irrespective of the physical product. Today, wider usage of the Internet of Things is most often hindered by missing concepts on human, organisational and technical shortcomings to share cost and benefits, or even generate revenue from the Internet of Things.
9. *Public initiatives to support the usage of the Internet of Things for social relevant topics.* Legislation has always been a push mechanism for adoption of new technologies. While it is obvious that the Internet of Things can be used to provide society with relevant data, some legislative requirements on topics such as carbon footprint, green logistics, and animal welfare would help to show the utility of the Internet of Things for society.
10. *Enable people to seamlessly identify things to access as well as contribute related information.* How many people carry an Auto-ID reader all day to identify objects and access corresponding information? Mobile phones today already include a camera that can scan barcodes and 2D matrix symbologies. Near Field Communication (NFC) is expected to be the next logical step for user interaction with the Internet of Things. However, it is questionable how many mobile phone owners will use these technologies. Besides mobile phones, there may be cheap dedicated devices. Nabaztag[18] provides a set including reader, tags and internet-based applications for about 40 Euro. Mobile barcode scanners and RFID readers that can be attached to a key chain and that are as easy to operate as a USB-stick are yet another opportunity to enable mass participation in the Internet of Things.

These ten key requirements are not intended to provide a complete set of requirements. They are meant to focus on certain aspects of the Internet of Things to start a rethinking process for future developments.

[18] www.nabaztag.com

1.6 A Possible Architecture for the Future Internet of Things

While it is quite obvious that there are and will be numerous approaches towards the Internet of Things, thus leading to a creative variety of applications in the Internet of Things, we favour an architectural approach that is based on extensions to a successful standardised open architecture – the EPCglobal Network. The EPCglobal Network is widely accepted and has gained the biggest support from IT companies that have adopted the standardised interfaces into their own applications. Numerous products have been developed and certified (EPCglobal 2010). Therefore, the EPCglobal Network provides a solid foundation, despite the fact that it is still under development.

However, the Internet of Things requires a more holistic architecture as described before. This can build on the same design principles as the EPCglobal Architecture Framework (EPCglobal 2007). These include layering of standards, separation of data models and interfaces, provision of extension mechanisms, specification of data models and interfaces, initially in a neutral abstract manner (e.g., using UML), then with provision of specific transport bindings (e.g., web services) and schema bindings (e.g., XML).

A future Internet of Things has to integrate stakeholders who will be affected by the Internet of Things, such as citizens, small and medium enterprises, governmental institutions and policy makers, to meet and match key societal and economic needs. Applications that recognise and improve the fundamental qualities of life for users, businesses, society and the environment are needed.

The foundation will need to provide open architectures, protocols and technologies for new classes of smart Internet-/Web-based public and business applications. Social platforms to share experience and personalised insights will be integrated with business-centric applications. Discovery and retrieval of useful and relevant information beyond personal expectations will be achieved though engineering for serendipity. Users shall be empowered to access more information about things (e.g., Where has an item been produced? – Who owned it previously? - What was it used for?) instantly at their fingertips, subject to compliance with privacy regulations. Mash-ups and end-user programming will enable people to contribute to the Internet of Things with data, presentation and functionality. Things-generated 'physical world' content from Auto-ID, sensors, actuators or meshed networks shall be aggregated and combined with information and events from 'virtual worlds', such as business databases and social platforms, and processed based on new business intelligence concepts. Results will be displayed in a user-centred design, including intuitive interfaces and Web 2.0 functionalities. Direct action on the physical world will be supported through Internet of Things machine-interfaces and introduction of agile strategies. Buying decisions will be supported through the access to relevant information as needed. Agile strategies in this context refer to real-time management and execution capability under consideration of conflicting optimisation values (e.g., shipment size). Information sharing will be rewarded through incentives, including transparent, open billing interfaces between numerous stakeholders, thus transforming the Internet of Things from a cost-focused infrastructure to a benefit-focused infrastructure to accelerate business innovation. Distributed data ownership across the object life cycle will be addressed by integrated billing. Information will be as easily tradable as products and services. The gap between distributed intelligence concepts (e.g., autonomous logistics) and the Internet of Things will be overcome through integration of open interfaces, protocols and lookup services as well as information services on mobile devices, acting as a mediator among decentralised information systems. Openness, scalability and security will be addressed as an integral part of the core architecture. Openness includes social (e.g., governance, privacy), organisational (e.g., industries) and technical (e.g., infrastructures, identifiers) dimensions. The integration and interoperability with mainstream business software platforms will be enhanced and its functionality will be extended through real-time analytics and business intelligence.

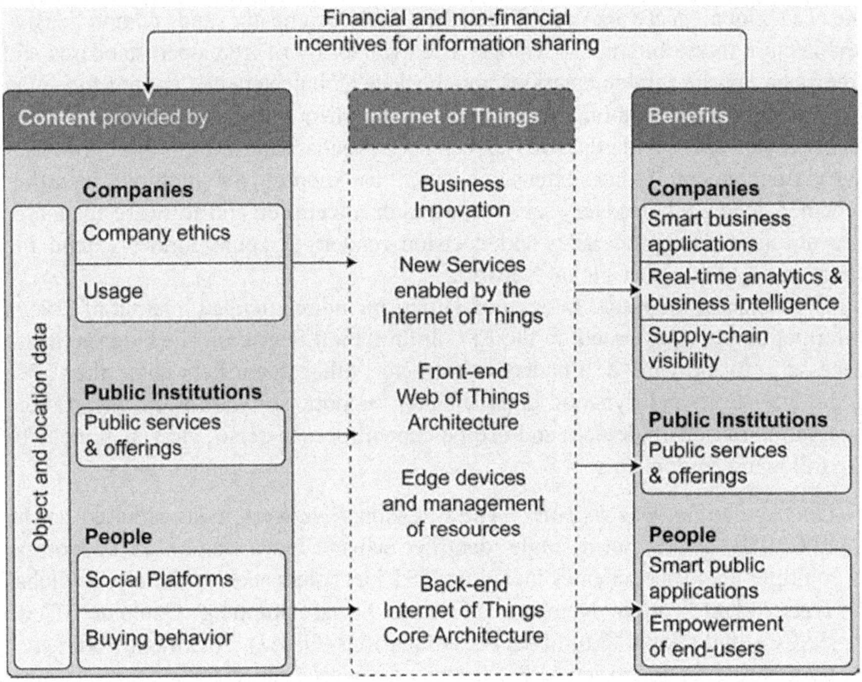

Fig. 1.4 A Holistic Internet of Things Scenario Including Companies, Public Institutions and People

Figure 1.4 shows one possible scenario that includes content providers (producers) and content users (consumers) that utilise the Internet of Things and share benefits. Company data includes for example product and usage data as well as company ethics that may influence buying behaviour. Public institutions as well as people will be able to contribute content. New services and business innovation will be enabled by an enhanced Internet of Things infrastructure including edge devices and back-end services as well as front-end user-interfaces. Companies, public institutions and people will be able to access data for their own benefits and financial as well as non-financial benefit compensation will further add to a fast adoption process of the Internet of Things.

Key goals for a future Internet of Things architecture to achieve are:

- An open, scalable, flexible and secure infrastructure for the Internet of Things and People
- A user-centric, customisable 'Web of Things' including interaction possibilities for the benefit of society
- New dynamic business concepts for the Internet of Things including flexible billing and incentive capabilities to promote information sharing

The EPCglobal Network architecture is currently only one aspect of the broader Internet of Things. However, if openness, scalability and security can be assured, the EPCglobal Network could be the most promising and comprehensive architecture in the Internet of Things. The availability of free, open standards and free open source implementations for the EPCglobal Network architecture may play a significant enabling role in its development, alongside complementary technologies and standards, such as Open Geospatial Consortium (OGC) Sensor Web Enablement. Other extensions, such as support for multiple identifier schemes, federated discovery services, actuator integration and software agents for decentralised data processing and decision rendering, could further extend the functionality of the EPCglobal Network.

The vision of the future Internet of Things includes extended Internet of Things Information Services based on the EPC Information Services. The extensions are necessary to provide a broader support for other identifiers than the EPC, additional static and dynamic data, actuator support, software agent integration, integration of non-IP devices and offline-capabilities. In detail, the vision includes the following components:

- *Extended static data support* – The EPCglobal Network today is based on the EPC. The EPC is not a single identifier scheme but a framework supporting multiple identifier schemes including GS1 identifiers such as Serialised Global Trade Identification Number (SGTIN), Serial Shipping Container Code (SSCC), and Global Returnable Asset Identifier (GRAI). This framework is not limited to GS1 identifiers; EPC formats are also defined for unique identifier constructs specified by the US Department of Defense. In principle, other approaches such as the Uniform Resource Names (URNs) could be used to support identifiers based on ISO 15962 and even identifiers based on Uniform Resource Locators (URLs) could be included, since they are a subset of Uniform Resource Identifiers (URIs). There is a need to support all things that carry a unique ID, because changing an established identifier scheme in an industry can cost millions of Euro and should be compared to the efforts involved for changing databases in the last millennium to make them year 2000 compliant. There have been and continue to be approaches to transform existing established identification schemes into a format that is compatible with the EPCglobal Network, as well as EPCglobal standards such as Tag Data Standard (TDS) and Tag Data Translation (TDT) that enable two-way translation between an EPC representation and an existing legacy representation. Additional structured data in barcodes (e.g., for best-before-date) may need to be supported to fully integrate existing optical identification techniques and to exploit the user memory capabilities of RFID tags, as well as facilitating stock rotation, product recalls, etc. An open, universal identifier translation framework would enable all things that carry a unique ID to be part of the Internet of Things. However, until everything carries a unique ID, the Internet of Things may also need to support objects identified by a classID (productID) and attributes.

- *Integration of dynamic data* – In order to bring the real and the virtual world closer together there is a need to sense environmental conditions as well as the status of devices. A standardized sensor interface to the Internet of Things would help to minimise costs and foster implementation. Sensors are key components of the next generation of internet services because they empower bottom-up interaction with things by enabling the gathering of information about their state or condition within the real world. The state of the things can be used to feed services at the infrastructure layer, transforming everyday things into true enablers of the Internet of Things.
- *Support for non-IP devices* – Non-IP devices offer only limited capability. They can be integrated in the Internet of Things through gateways that take care of the computational overhead required to share physical devices over the Internet, while also providing advanced functionality that are not available on the devices themselves.
- *Integration of an actuator interface* – Actuator integration into the Internet of Things will allow standardised communication with machines executing decisions either rendered by humans or software-agents on their behalf. Actuators complement bidirectional interaction processes by providing the means for services and users to influence the state of things. The combination of sensors and actuators and their integration in the core Internet of Things infrastructure is an indispensable feature and needs to be considered at all layers of the architecture.
- *Optional integration of software agents* – The complexity of global supply networks will require more decentralised and automated decision making. Software-agents have been researched broadly but have not yet gained considerable acceptance in industries. The reason for this may be the lack of standardisation. A standardised interface in the Internet of Things would help to boost the usage of software agents. Smart objects in the Internet of Things need to execute intelligent algorithms to be able to discard irrelevant data, interact with other things in an efficient way, raise warnings about their state or the

state of their environment, and take informed decisions and actions on behalf of human end-users to eliminate or assist control / management activities by humans. Additionally, software agents may help to increase scalability and robustness in the Internet of Things (Uckelmann et al. 2010). In a holistic scenario we imagine things to host a certain infrastructure subset of the Internet of Things. These things may not always be connected to the Internet. Therefore, we envision a certain degree of smart characteristics and autonomy.

- *Extended, federated discovery services* – The EPCglobal Network today does not yet provide ratified standards for federated discovery services, although a technical standard for discovery services is currently under development. At the time of writing, the only lookup service currently provided by EPCglobal is the ONS, which only holds class-level records pointing to authoritative information. This is currently operated under contract by VeriSign Corp. under the onsepc.com domain. The existing ONS implementation is distributed across multiple servers globally. Nevertheless, there are political concerns that the ONS is defined under the .com Top-Level-Domain, which is under the authority of the US Department of Commerce and that the ONS service is operated only by one American company. This has led to political discussions on governance in the Internet of Things, resulting in national focused approaches in China and Europe (Muguet 2009). Federated discovery services are needed to enable open governance, scalability and choice of lookup service in the Internet of Things.
- *Data-synchronisation for offline support* – The EPCglobal Network requires online connection to access data related to the identified product. In certain cases online-connectivity cannot be assured. Data-synchronisation is needed to support mobile scenarios and decentralised decision making.
- *Interface to federated billing services* – In order to enable competition between billing service providers, a standardised interface to these services is needed. This billing interface will enable balancing of costs and benefits as well as new business models and revenue generation opportunities for business and citizens based on micro-trading of information in the Internet of Things.

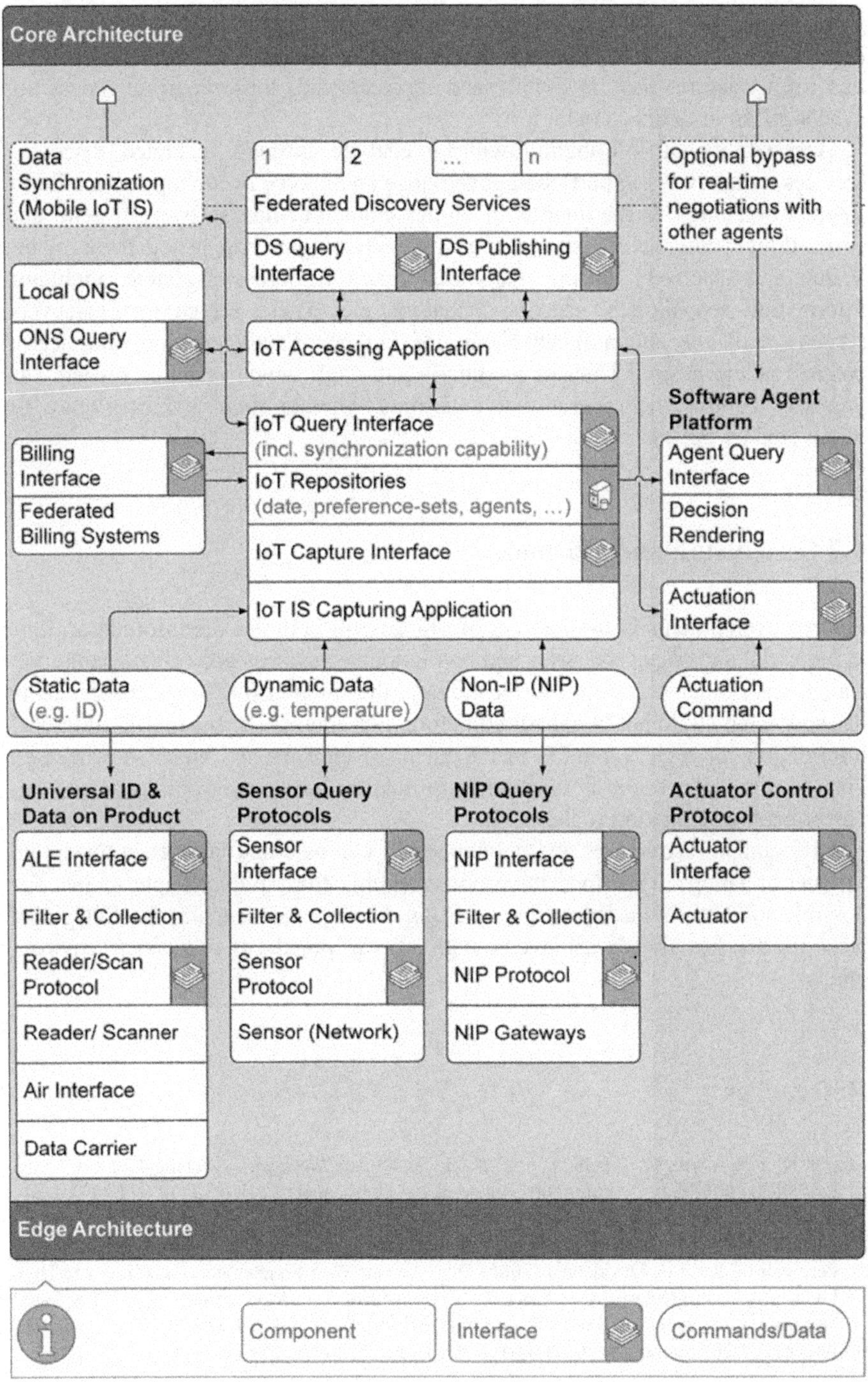

Fig. 1.5 An Extended EPCglobal Architecture Towards a Future Internet of Things

In Figure 1.5 the integration of sensors, actuators and software agents connected to the Internet of Things Information Service (IoT IS) is shown. Parts of this infrastructure may be mobile and disconnected, thus requiring means for synchronisation of data and logic.

Accessibility of information will be enabled through federated discovery services, which will support open governance and choice of lookup service in the Internet of Things. In the Internet of Things, human beings, software systems and smart things will have a strong need for technologies supporting them in the search and discovery of the many distributed resources available, including information repositories, sensors, actuators, etc. These search and discovery services will rely upon mechanisms for universal authentication and access control, at the desired level of granularity, through which resource owners can precisely control the criteria that determine whether their resources may be discovered by others.

1.7 Conclusion and Outlook

Future developments in the Internet of Things will optimise the information flow in industrial and social scenarios and revolutionise business and private communication. Like other milestones in technology, the Internet of Things enables us to measure what could not be measured before. For companies this means additional information for high resolution management of industry and business processes. For citizens the possible implications are manifold, ranging from consumer empowerment to rethinking society.

Different infrastructures and networks will compete and interact in the future Internet of Things. Therefore, the proposed architecture in this chapter is just one possible solution, but it is based on existing developments such as the EPCglobal Network that has already achieved a high level of popularity in business environments.

References

Aberer K, Hauswirth M, Salehi A (2006) Global Sensor Networks, Technical report LSIR-REPORT-2006-001. http://lsirpeople.epfl.ch/salehi/papers/LSIR-REPORT-2006-001.pdf. Accessed 1 May 2010

Basagni S (1999) Distributed Clustering for Ad Hoc Networks. Proc. ISPAN'99 Botts M, Percivall G, Reed C, Davidson J (2006) OGC Sensor Web Enablement: Overview and High Level Architecture. Open Geospatial Consortium Whitepaper. http://portal.opengeospatial.org/files/?artifact_id=25562. Accessed 1 May 2010

Bullinger H-J, ten Hompel M (2007) Internet der Dinge. Springer, Berlin

BRIDGE (2009) Sensor-based Condition Monitoring. http://www.bridge-project.eu/data/File/BRIDGE_WP03_sensor_based_condition_monitoring.pdf. Accessed 5 July 2010

Brock L (2001) The Electronic Product Code (EPC) – A naming Scheme for Physical Objects. http://autoid.mit.edu/whitepapers/MIT-AUTOID-WH-002.PDF. Accessed 5 July 2010

CERP-IoT (2009) Internet of Things Strategic Research Roadmap, http://www.grifs-project.eu/data/File/CERP-IoT%20SRA_IoT_v11.pdf. Accessed 1 May 2010

Clendenin M, (2006) China aims for homegrown RFID spec by '07. http://www.embedded.com/news/embeddedindustry/191600488?_requestid=355245. Accessed 1 May 2010

Dimokas N, Katsaros D, Manolopoulos, Y (2007) Node Clustering in Wireless Sensor Networks by Considering Structural Characteristics of the Network Graph. Proc. ITNG'07,IEEE Computer Society, USA.

EPCglobal (2007) The EPCglobal Architecture Framework, Standard Specification. www.epcglobalinc.org/standards/architecture/architecture_1_2-framework-20070910.pdf. Accessed 1 Mai 2010

EPCglobal (2010) EPCglobal Certification Program. http://www.epcglobalinc.org/certification/. Accessed 7 July 2010

EPC Symposium (2003) Inaugural EPC Executive Symposium. http://xml.coverpages.org/EPC-Symposium200309.html. Accessed 5 July 2010

EPoSS (2008) Internet of Things in 2020 – A roadmap for the future. http://old.smart-systems-integration.org/internet-of-things/Internet-of-Things_in_2020_EC-EPoSS_Workshop_Report_2008_v3.pdf. Accessed 1 May 2010

Fleisch E, Mattern F (2005) Das Internet der Dinge: Ubiquitous Computing und RFID in der Praxis: Visionen, Technologien, Anwendungen, Handlungsanleitungen. Springer, Berlin

Floerkemeier C, Fleisch E, Langheinrich M, Mattern, F (2008). The Internet of Things: First International Conference, IOT 2008. Springer, Berlin

Gille D, Strücker J (2008) Into the Unknown – Measuring the Business Performance of RFID Applications. 16th European Conference on Information Systems (ECIS 2008). http://is2.lse.ac.uk/asp/aspecis/20080218.pdf. Accessed 1 May 2010

Hackathorn R (2004) The BI Watch: Real-Time to Real-Value. http://www.bolder.com/pubs/DMR200401-Real-Time%20to%20Real-Value.pdf. Accessed 1 May 2010

Heise online (2008) Frankreich schlägt europäische Root für das "Internet der Dinge" vor. http://www.heise.de/newsticker/meldung/Frankreich-schlaegt-europaeische-Root-fuer-das-Internet-der-Dinge-vor-209807.html. Accessed 1 Mai 2010

Laubacher R, Kothari S, Malone TW, Subirana B (2006). What is RFID worth to your company? Measuring performance at the activity level. ebusiness.mit.edu/research/papers/223%20 Laubacher_%20APBM.pdf. Accessed 1 Mai 2010

Muguet F (2009) A written statements on the subject of theHearing on future Internet Governance arrangements – Competitive Governance Arrangements for Namespace Services. http://ec.europa.eu/information_society/policy/internet_gov/docs/muguet_eu_internet_hearing.pdf. Accessed 27 October 2010

Sánchez López T (2010) What the Internet of Things is NOT. http://technicaltoplus.blogspot.com/2010/03/what-internet-of-things-is-not.html. Accessed 1 May 2010

Sánchez López T, Kim D, Canepa GH, Koumadi K (2009) Integrating Wireless Sensors and RFID Tags into Energy-Efficient and Dynamic Context Networks. Comput J 52:240-267. doi:10.1093/comjnl/bxn036

Santucci Gérald (2010) The Internet of Things: Between the Revolution of the Internet and the Metamorphosis of Objects. http://ec.europa.eu/information_society/policy/rfid/documents/iotrevolution.pdf. Accessed 18 October 2010

Shim Y, Kwon T, Choi Y (2007) SARIF: A novel framework for integrating wireless sensors and RFID networks, IEEE Wirel Commun14: 50-56. doi:10.1109/MWC.2007.4407227

Sung J, Sánchez López T, Kim D (2007) The EPC Sensor Network for RFID and WSN Integration Infrastructure. Pervasive Computing and Communications Workshops, Fifth IEEE International Conference on Pervasive Computing and Communications Workshops (PerComW'07)

Uckelmann D, Isenberg MA, Teucke M, Halfar H, Scholz-Reiter B (2010) An integrative approach on Autonomous Control and the Internet of Things. In: Ranasinghe DC, Sheng QZ, Zeadally S (eds) Unique Radio Innovation for the 21st Century: Building Scalable and Global RFID Networks. Springer, Berlin

Wang Y, Zhao Q, Zheng D (2004) Energy-Driven Adaptive Clustering Data Collection Protocol in Wireless Sensor Networks. Proc. ICMA'04. http://ieeexplore.ieee.org/stamp/stamp.jsp?arnumber=01384266. Accessed 1 May 2010

Ye M, Li C, Chen G, Wu J (2005) An Energy Efficient Clustering Scheme in Wireless Sensor Networks. Proc. IPCCC'05, Phoenix, USA

Younis O, Fahmy S (2004) HEED: A Hybrid, Energy-Efficient, Distributed Clustering Approach for Ad Hoc Sensor Networks. IEEE Trans Mob Comput 3:366-378. doi:10.1109/TMC.2004.41

2 About the "Idea of Man" in System Design – An Enlightened Version of the Internet of Things?

Sarah Spiekermann

Institute for Management Information Systems, Vienna University of Economics and Business, Austria

Abstract This article aims to argue that, as we move into an era of ubiquitous computing, where the traditional Internet evolves to embrace an Internet of Things, it may be beneficial to embed an "Idea of Man" into its systems design. The "Idea of Man" is a holistic philosophical concept that considers what Man is, what Man should be, and how Man lives with others in society. The article provides arguments for the relevance of the Idea of Man in system design in general. I argue that the Idea of Man influences the power relationship between men and computer systems as well as the values that we build into these systems. Furthermore, I argue that programmers' Idea of Man influences the values which embed systems. Finally, I highlight future challenges involved in integrating an Idea of Man into systems. The article is a viewpoint and its arguments are purely deductive. Its contribution is that it shows how the Idea of Man could serve as a foundation for a variety of considerations relating to computer ethics. If we take today's Idea of Man in the Western world, which views men as responsible and mature, able to act rationally, and capable of defining themselves through moral autonomy and freedom of choice, we establish high level guidance for how systems should be built and what an Internet of Things could, or should not, do for us.

2.1 Introduction

In the early 1990s, a vision was born that deeply influenced the discipline of computer science. Researchers from Xerox Park claimed that computing in the 21st century would become a ubiquitous service resource that would weave itself "into the fabric of everyday life" (Weiser 1991). 20 years later, we see this vision come true, thanks to giant technical and scientific leaps in data processing, -storage and -transfer capabilities, miniaturisation, material sciences, and energy harvesting. Sensors, identification technologies, video systems, and online tracking and location technology systems constantly observe the environment, detect the people within it, and help those people accomplish tasks. Systems carry out bookings, coordinate dates, open and close doors, make sure we drive "correctly", remind us about important events, tell us to buckle our seatbelts, etc. They carry out private tasks that human beings either performed themselves or had others perform for them in the past. Suddenly, machines are acting as "agents" of human principals in everyday situations. As teachers, guardians, servants, playmates, and private secretaries, they become "social actors" in a networked environment. Some scholars have started to call this networked environment the Internet of Things (Fleisch and Mattern 2005).

With the rapid shift from a solely industrial and corporate Internet of Things to a more holistic approach, including everyday personal computing, ethical questions are beginning to arise. To what extent can surveillance be accepted? How much control should be delegated to machines? How much transparency is needed for machine operations? How should content be shared through systems? Designers are encountering an incredibly lengthy list of issues around how systems should behave, be used, and be deployed. Many IT companies are unsure of how to embed all relevant ethical standards into their IT solutions systematically. As a result, these companies are left alone in a trial-and-error game of what is feasible and acceptable and what is not, often at the cost of consumer trust and brand equity. Yet, even if companies wanted to build systems to meet ethical expectations, they would face the major challenge of determining which expectations are important. Does sensitivity to privacy drive ethical acceptability?, Security?, Universal usability?, Control?, What standards need to be met?, What goal are we striving for when we debate what is ethical and what is not?

In this article, I explore to what extent the philosophical construct of the Idea of Man may promote ethical system design. To date, philosophy seems to have rarely inspired the Internet of Things or even computer science in general. All the more, I believe the reflection to be worthwhile. If we take today's Idea of Man in the Western world, which views men as responsible and mature (German: "mündig"), able to act rationally, and capable of defining themselves through moral autonomy and freedom of choice (Kant 1784/1983), we already establish some high level guidance for how systems should be built. At the very least, we create a counterbalance for the only Idea of Man that is currently accepted in computer science, that of the "Dumbest Assumable User".

The current article is not intended to "operationalise" the Idea of Man for system design or for the design of the Internet of Things. This task would probably require long-term interdisciplinary effort. Instead, this article introduces the concept of the Idea of Man to the technical community. It shows what the Idea of Man is, how it relates to technology, how important programmers' Idea of Man is for system design, and what challenges must be overcome to embed an Idea of Man into our work as technology designers.

2.2 About the Idea of Man: Definition and Relation to System Design

The Idea of Man is an ambiguous concept that has been debated in philosophy for decades (if not centuries) (Fahrenberg 2007). As a term, translated from the German "Menschenbild", it may alternatively be referred to as an "Image of Humanity" or "Conception of Humanity". But the best translation for the way the German language conceives of the term may be Idea of Man. According to Diemer (1978), "Menschenbild" contains a double meaning: On the one hand, speaking about a "Bild" (=Image, Picture of Man) implies the existence of an effigy of Man. What does he or she look like? On the other hand – and viewed holistically – the term implies an object of aspiration, an ideology or pedagogic idea of what is desired from mankind. What IS Man?

Both conceptions of the term have considerable impact on technological design. Scientists in robotics and software agent development, for example, experiment with humanoid representations of technical systems. These scientists want to create an effigy of Man. Humanoid robots like the Japanese Geminoid[19] or embodied interface avatars are examples of how the idea of human appearance can be translated into technology.

Yet, according to the vision of Ubiquitous Computing, a majority of systems that interact with human beings might not be represented as humanoid artifacts. Instead, they may be integrated into objects and infrastructures, such as the Internet of Things that will surround us in our everyday life. The present essay therefore focuses less on questions of human effigy and more on the question of what role the Idea of Man can play as an ethical concept for technological design in an Internet of Things.

According to Fahrenberg, the Idea of Man (in the sense of a "role model") contains the sum of all assumptions and opinions about what human beings by nature are, the way they live in their social and material surroundings, and the values and objectives their lives should embrace (Fahrenberg 2007). This definition integrates two main dimensions of the Idea of Man that could be significant for technological design: First, the Idea of Man involves assumptions about the nature of mankind, our individual existence, and our individual abilities. These abilities relate in very specific ways to the nature and abilities of computer systems. And second, the Idea of Man comprises assumptions and opinions about social interaction and society at large. How should people live with each other? And when computer systems (such as smart "things") mediate these social interactions, do they become subject to the same assumptions and opinions?

The next two sections will discuss these two dimensions of the Idea of Man and their significance for the design of everyday computing systems.

---

[19] Geminoid is a man-size robot that is an effigy of his creator Prof. Hiroshi Ishiguro (for more information and pictures see: http://www.irc.atr.jp/Geminoid)

2.3 The Idea of Man as Opposed to the Nature of a Computer System

The Idea of Man differs between cultures and is subject to change over time. When I refer to the Idea of Man hereafter, I rely on only a single conception, namely the one particular Idea of Man shared by many in Western civilisation. In this conception, the medieval Idea of Man, which is marked by a fatalistic belief in destiny and a God-given disparity between men, is supplanted by a humanistic view of an enlightened mankind. In this view, men act rationally and define themselves through moral autonomy and freedom of choice (Kant 1784/1983). Moreover, the era of postmodernism considers men to be "constructors of their own selves" (Eickelpasch and Rademacher 2004). Some modern sociologists use metaphors of decline to characterise the postmodernist being, describing individuals as uprooted and "mentally homeless" (Baumann 1995), even "released" (Beck 1986). However, sociology is still based on the notion that all human beings can live a self-determined life. This view of men is regarded as an achievement of our Western civilisation and a fundamental prerequisite for democracy.

In a highly automated and networked environment, how will people retain their self-determination, their ability to make decisions for themselves? Could Ubiquitous Computing undermine the autonomy and choices that are said to characterise mankind?

A fundamental step towards answering this key question is to define the power relationship between human beings and computer systems. When we discuss peoples' power in contrast to computer systems, unfortunately, we frequently encounter an Idea of Man that is subject to a deep uncertainty. When we compare ourselves to computer systems, we tend to question our human skills and capacities: Who makes faster and better decisions? Who do we trust to tell the truth? Who will evolve more rapidly? All too often, we display a latent disposition to trust the power of the machine more than the human subject. But what do such views imply? Do we risk slipping into a perspective that views man as inferior to computer systems and that questions human power and decision-making? If we adopt such a perspective of human inferiority, do we risk reentering, as Kant calls it, a stage of "self-inflicted immaturity"? Do we give up the autonomy and freedom of choice that we are so proud of?

In contrast to what many science fiction novels tell us, automation scholars regularly show that the overall superiority of machines is not a given (Sheridan 2002). Technical systems exist to "assist" (Wandke 2005) human beings in areas where humans need support. In his acclaimed 1951 essay, Fitts tries to objectify the competence relationship between men and machines for the engineering sciences (Fitts 1951). He states that machines outperform humans in terms of fast reaction time, the use of strong power in a soft and precise way, the complete deletion of information, or deductive argumentation. However, men are superior to machines when it comes to improving present processes, judging, or arguing inductively. Despite considerable advancements in computing since the 1950s, this fundamental view of the man-machine relationship still has virtue. Machines may be getting better at making complex decisions, but complexity also adds cost and risk to machine operations.

The tradeoff between these risks and costs and the efficiency of control delegation, this fundamental decision, regarding the distribution of work between men and machines, remains a grey area (Sheridan 2000). Sheridan, one of the leading automation scholars, describes the control allocation problem between men and machines as "algorithm, alchemy or apostasy" (Sheridan 2000). Are fully automated airplane cockpits safer than human pilots? Are electronic voting machines better at counting ballots than electoral staff? Do video control systems prevent crime more efficiently than human guards?

It is within this alchemistic grey area of control-allocation decisions that the Idea of Man comes into play. Do we opt for men or for computer systems? Wherever an objectively detectable superiority of people over machines is not a given, the Idea of Man can help people decide whether human beings will be allowed to maintain control. Because system developers, operators, and manufacturers make such "grey-area-decisions" based on some intuition, it is their particular Idea of Man that influences control-allocation decisions in an important way.

2.4 Social Interaction and Norms at the Human/Machine Interface

According to Fahrenberg, the Idea of Man is not simply a construct of individuals' identity, ability, and nature; it also relates to individuals' interaction in society and society's conception of how men should treat each other (Fahrenberg 2007). Hence, the Idea of Man manifests itself in both, behavioral rules of social interaction and values that underlie positive cooperation between humans. Values and behavioral rules define how Man should be.

Today, this idea of how Men should be is undeniably affected by pluralism: in a rapidly changing global society, no value monopoly exists. However, we still share ethical norms that are widely accepted; such norms are reflected, for example, in international agreements like the Universal Declaration of Human Rights of the United Nations.

These values and norms are also meaningful for technical design. When computer systems become social actors, interact with human beings in their everyday lives, and handle tasks for them, people expect them to act like people. Socially developed norms of interaction and behavior are conferred upon machines (Reeves and Nass 1996). But which norms will help us uphold our Idea of Man in an evolving Internet of Things landscape? Using the term "Value Sensitive Design", Friedman and Kahn propose a number of ideas that should constitute Man in relation to the machine. These include the right to privacy; the right to be calm when we require it; the right to make autonomous decisions and control our surrounding electronic environment; machines' accountability for actions made in the name of men or vis-à-vis men; man's freedom from machine bias and right to be treated impartially by machines; the right to respectful interaction; the ability to trust in the machine; the right to make informed decisions when machines ask us to make them and the right to be the master of our own identity in machine systems (Friedman and Kahn 2003).

If systems are to become social actors in an Internet of Things that reflect the "Idea of Man" and serve as trustworthy extensions of the self, they must act in a way that is consistent with all of these aspects. Current exemplary discussions about electronic privacy demonstrate what happens when machines ignore our "ideas of them as social actors": Governments' surveillance practices are overruled by supreme courts (for example see: (Bundessverfassungsgericht 2010) and companies need to change technology they just launched (Claburn 2010). Over 80% of consumers claim that they would stop doing business with a company if they learned about improper data practices (Ernst & Young LLP 2002) and express their expectations in public hearings, such as the EU's public consultation processes (Article 29 Working Party 2005).

2.5 The Impact of the Programmer's Idea of Man

What machines are allowed or forbidden to do, and how they behave towards people, depends on how the machines are programmed. Machine developers therefore have a tremendous influence on the Idea of Man that is embedded in machines.

To ensure the "usability" of a system, developers adapt systems to peoples' abilities through "physical and cognitive engineering", a standard stage in system development lifecycles today (Nielsen 1993; Norman 2007; Te'eni et al. 2007). The question of how machines treat people and how people deal with machines, however, is an issue that extends beyond the traditional notion of "usability". On a macro level, system developers make fundamental decisions about the role a machine is allowed to play with regard to people. A system may excel on the physical and cognitive level of engineering, but nevertheless "betray" its users at the back end. For example, privacy policies may ostensibly be in place, while, at the back end, their technical implementation is neither supervised nor permanently adhered to.

Developers choose the values a system lives up to (see above). On a micro level, these decisions are translated into concrete machine actions. On a macro level, the values that Friedmann summarises could inform developers about key points of system development (Friedman and Kahn 2003). However, these macro level principles must be translated into concrete micro level system design guidelines.

To put the Idea of Man into practice, three areas of system design can be differentiated:

First, system designers determine how people interact with machines and influence machine actions (manipulation). Second, they design the way machines treat humans (contact). These two areas are frontend design decisions. Third, engineers determine the way machines act at the back end and to what extent such actions are transparent and subject to influence by users. Consequently, at the micro level, the Idea of Man manifests itself in how programmers design frontend interaction and backend behavior. Figure 2.1 depicts these interdependencies.

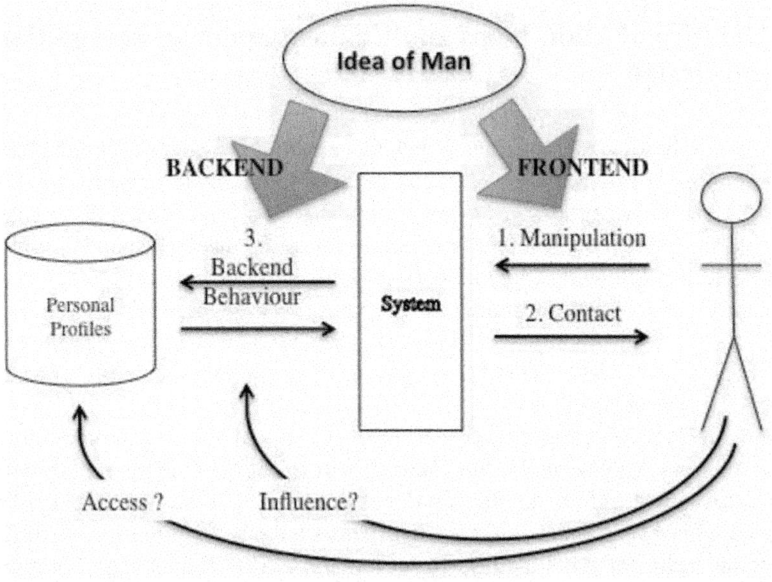

Fig. 2.1 How The 'Idea of Man' Influences System Design

As an example, consider system control in the context of intelligent cars: On a macro level, we know that being able to exercise control, especially over our possessions, is highly valued. For instance, although car owners enjoy "intelligent vehicles", they may still want to control the vehicles' operation. But can they? Let's continue the example for the seatbelt warning feature. Law regulates that every vehicle must be equipped with a seatbelt warning system. However, it is the manufacturer or vehicle developer who, on the micro level, determines its concrete design, making decisions, such as: Can drivers manipulate the system by turning off the warning signal? (Manipulation) How does the vehicle (at the frontend) warn its drivers: by means of a drown-out, shrill acoustic signal that forces them to perform the desired action? Or does it discreetly remind them that it would be wiser to fasten their seatbelts the moment they start the engine? (Contact) And finally, how does the vehicle behave at the back end? Will it register and save data about the drivers' behavior? Will it make that information available to insurance companies in case of an accident? And do drivers have the right to access and delete this information? (Backend Behavior)

This example shows the tremendous impact, a single component of the Idea of Man, namely control, can have on concrete micro decisions in technological design. It also illustrates the broad margin that designers enjoy on the micro level, one that allows them to develop a system in a variety of ways.

2.6 The Idea of Man: Steps and Challenges for its Recognition in System Design

A series of interviews with software developers about privacy in technological design produced a provocative result that may be transferable to many social and ethical issues in system design. When asked how data protection is taken into account during prototype development, nearly all interviewees responded with one or more of the following arguments (Lahlou et al. 2005): privacy is an abstract problem; privacy is not an immediate problem, because firewalls and cryptography take care of it; privacy is no problem at all; privacy is not their problem, but one for politicians, lawmakers, or society at large; privacy is simply not part of their project deliverables.

These responses indicate that even privacy, one of the most commonly discussed societal "values" and a fundamental part of our Idea of Man, does not figure into the concerns of technological development. The reason for this lag of consciousness may be that the engineering sciences have been primarily interested in enhancing technical functionality. However, the era of "Functional Computing" may be slowly replaced by (or at least overlapped with) an era best described as "Human Centric Computing". Mass market technologies, such as home IT, mobile communications, video games, or navigational systems are key drivers of technological progress today. And since their market success depends on the usability and consumer-friendliness of products, the "human factor" in engineering has gained in importance. "Human Centric Computing" considers how users can manipulate machines and how the contact is designed (Zhang 2005). Yet, less emphasis has been put on how to respect ethical system behavior systematically when designing backends.

To embrace "Ethical Computing" and the Idea of Man, we must face a number of challenges: First, we need to embrace the Idea of Man and the integration of human values into technology as fundamentally important for the engineering sciences. While a few scientists have tried to raise awareness of ethical issues in computer science for decades (e.g., Weizenbaum 1977), they are often marginalised. Consequently, we lack knowledge about what constitutes socially acceptable technology. The "Systems Development Life Cycle" and its manifold variations

(Kurbel 2008), as well as the “Human Centric Systems Development Life Cycle” (Te'eni et al. 2007), do not incorporate processes for the consideration of human values or an Idea of Man in machine construction at the micro level. Such processes should be developed. At the same time, reference models, used as blueprints for system concepts, could incorporate mechanisms of “value management”. Finally, we need to investigate how modeling languages could systematically consider the immaterial aspect of “value”. Some are already taking first steps to develop micro level mechanisms for ethical engineering (i.e. those researchers developing privacy-enhanced technologies). But few of their concepts and approaches are integrated into the teaching of computer and engineering sciences.

Commitment from practitioners of the computer and engineering sciences, however, will ultimately not be enough to craft technology that is more socially compliant. After all, developers in a given company use technology to implement the demands of product management departments (“requirement”). Therefore, management must emphasise socially acceptable technical design. Yet, because companies are driven by profitability, they limit themselves to meeting basic legal regulations in an effort to make the development of systems as cost-effective as possible and maximise the potential use of their technical systems (such as data collection). Companies who take the time to reflect upon different options of system design, with all its varying ethical pros and cons, might boost development costs and restrain the company’s scope of strategic action (Spiekermann and Cranor 2009). As a result, developers often try to avoid the topic until they are forced to confront it by the market or regulators.

The reaction time of lawmakers, however, is often too slow to affect rapidly developing technical markets. Especially in Europe, there is a latent fear of overregulation; politicians want to reduce the risk of stifling the innovative spirit of technology markets by limiting ethical regulations. Some experts argue that market mechanisms should be responsible for sanctioning socially incompatible technological designs and rewarding socially compatible ones.

Would economic incentives justify private investment in socially compatible technologies? The development of social networking websites, such as Facebook in recent years, shows that companies have become more aware of their clients’ wishes and expectations. As a result of strong pressure and negative reactions from clients, social network operators now allow clients to adjust privacy settings for their data.

Another possible scenario is that clients, who begin to value sustainable, ethical technological designs, might be willing to pay more for them than for traditional designs. In some cases, such as organic food, consumer markets have developed in this direction. However, it remains unclear, whether markets that are less transparent and more technically complex, like IT services, can become clear enough for clients to understand the added value of socially compliant services. IT services’ operations are prone to information asymmetry, particularly in view of their operating modes and backend functions. Many clients use socially risky services (such as privacy invasive customer loyalty cards), because they lack information about backend practices (Bizer et al. 2006). If transparency increases and more information about common backend operations become available, markets might be forced to change completely.

Even if clients recognise one technology as more value-sensitive than another, they might not pay more to avoid the risk of long-term damages. For example, Internet users do not pay much attention to their privacy on the World Wide Web. They seem to value the short-term advantages of Internet services more than they fear the long-term potential loss of their privacy. People exhibit such lax behaviors because they have difficulty evaluating risks. They often underestimate (discount) long-term risks and overvalue short-term benefits (Acquisti and Grossklags 2005).

2.7 Conclusion

Questions about the social impact of technical designs have been asked for many years; as far back as 1980, an intense discussion was held about the potential threat artificial intelligence might pose to human beings. Time and again, governments have launched research programs to analyse the ethical aspects of computerisation, while impact assessment studies have addressed the social implications of technology. With the increasing popularity of the Internet of Things and its implications for society the relation of "things" and "people" has to be redefined.

This essay is only a small contribution to the endeavor of making our technical environment more humane. It deals with the specific notion of the Idea of Man and its potential value for the technical design of systems and networked environments, such as an Internet of Things. It shows that the Idea of Man can act on three levels: first, it enables us to reflect upon the power relationship between human beings and machines on a higher level; second, a decomposition of the Idea of Man helps us to identify concrete values that should impact technical design at a macro level; and third, a conscious sharing of an Idea of Man supports the respect of values at a micro level, where developers make daily decisions about how to structure interactions between men and machines.

References

Article 29 Working Party (2005) Results of the Public Consultation on Article 29 Working Document 105 on Data Protection Issues Related to RFID Technology. http://ec.europa.eu/justice/policies/privacy/docs/wpdocs/2005/wp111_en.pdf. Accessed 7 December 2010

Acquisti A, Grossklags J (2005) Privacy and Rationality in Individual Decision Making. IEEE Secur Priv 3:26-33

Baumann Z (1995) Ansichten der Postmoderne. Argument Verlag, Hamburg

Beck U (1986) Risikogesellschaft. Auf dem Weg in eine andere Moderne. Suhrkamp, Frankfurt am Main

Bizer J, Günther O, Spiekermann S (2006) TAUCIS - Technikfolgenabschätzungsstudie Ubiquitäres Computing und Informationelle Selbstbestimmung. Humboldt University Berlin, Unabhängiges Landeszentrum für Datenschutz Schleswig-Holstein (ULD), Berlin, Germany

Bundessverfassungsgericht (2010) Konkrete Ausgestaltung der Vorratsdatenspeicherung nicht verfassungsgemäß. 1 BvR 256/08, 1 BvR 263/08, 1 BvR 586/08, Karlsruhe

Claburn T (2010) Google Sorry About Buzz Privacy. InformationWeek. http://www.informationweek.com/news/windows/security/showArticle.jhtml?articleID=222900563. Accessed 7 December 2010

Diemer A (1978) Elementarkurs Philosophie - Philosophische Antropologie. Econ Verlag, Düsseldorf

Eickelpasch R, Rademacher C (2004) Identität. transcript Verlag, Bielefeld

Ernst & Young LLP (2002) Privacy: What Consumers Want

Fahrenberg J (2007) Menschenbilder - Psychologische, biologische, interkulturelle und religiöse Ansichten. Institut für Psychologie, Universität Freiburg, Freiburg

Fleisch E, Mattern F (2005) Das Internet der Dinge - Ubiquitous Computing und RFID in der Praxis, Springer Verlag, Berlin, Heidelberg, New York

Fitts PM (1951) Human Engineering for an Effective Air-Navigation and Traffic-Control System. Columbus, Ohio, USA

Friedman B, Kahn P (2003) Human values, ethics, and design. In: Jacko J, Sears A (eds) The Human-Computer Interaction Handbook. Lawrence Erlbaum Associates, Mahwah, NY, USA

Kant I (1784/1983) Was ist Aufklärung?. Wissenschaftliche Buchgesellschaft, Darmstadt

Kurbel K (2008) System Analysis and Design. Springer Verlag, Heidelberg

Lahlou S, Langheinrich M, Röcker C (2005) Privacy and Trust Issues with Invisible Computers. Commun ACM 48:59-60

Nielsen J (1993) Usability Engineering. Morgan Kaufman, Mountain View, CA, USA

Norman D (2007) The Design of Future Things. Basic Books, New York, USA

Reeves B, Nass C (1996) The Media Equation: How People Treat Computers, Television, and New Media Like Real People and Places. Cambridge University Press, New York, USA

Sheridan TB (2000) Function allocation: algorithm, alchemy or apostasy?. Int J Hum-Comput Stud 52:203-216

Sheridan TB (2002) Humans and Automation: System Design and Research Issues. John Wiley & Sons, Santa Monica, USA

Spiekermann S, Cranor LF (2009) Engineering Privacy. IEEE Trans Softw Eng 35:67-82

Te'eni D, Carey J, Zhang P (2007) Human Computer Interaction - Developing Effective Organizational Information Systems. John Wiley & Sons, Inc, New York, USA

Wandke H (2005) Assistance in human-machine interaction: a conceptual framework and a proposal for a taxonomy. Theor Issues Ergon Sci 6:129-155

Weiser M (1991) The Computer of the 21st Century. Sci Am 265:94-104

Weizenbaum J (1977) Die Macht der Computer und die Ohnmacht der Vernunft. Suhrkamp Verlag, Frankfurt

Zhang P (2005) The importance of affective quality. Commun ACM 48:105-108

3 Enabling the Masses to Become Creative in Smart Spaces

Orienting User Creation in the Internet of Things in the Context of the ITEA2 DiYSE Project

Marc Roelands, Laurence Claeys, Marc Godon, Marjan Geerts, Mohamed Ali Feki, Lieven Trappeniers

Alcatel-Lucent Bell Labs, Antwerp, Belgium

Abstract In this chapter we present a first analysis towards the enablement of mass creativity in the Internet of Things, potentially leading to a wide range of new tangible, interactive applications that leverage the fundamental new possibilities of an emerging Web of Things. After an introduction of the socio-cultural practice of 'Do-it-Yourself' (DiY) as apparent in society, and a discussion on what DiY can mean for the Internet of Things, we introduce a typology of how people can potentially create and customise on top of the Internet of Things. Based on that, we elaborate three concepts forming a basis for new creation paradigms in such smart spaces, potentially leading to new DiY-enabling functions in Internet of Things service creation environments: the *Call-Out Internet of Things*, the *Smart Composables Internet of Things*, and the *Phenomena Internet of Things*. Next to a discussion of applicable state-of-the-art for implementing parts of these concepts, we show first experimental grounding for them, as part of the ongoing exploration process.

3.1 The Meaning of DiY in the Network Society

From the societal practice of DiY a lot of drivers and adoption models can be derived that may be applicable for similar people-driven creation in an Internet of Things world. In this section, we first look broadly at DiY as a cultural practice and discuss some core characteristics. We then make the transposition from the cultural practice to what this may imply for application creation and for context-aware environments, which are the necessary building blocks for reaching the goal of enabling the masses to become creative in smart spaces.

3.1.1 DiY as Socio-Cultural Practice

Although nowadays DiY is commonly associated with youth subcultures, the origin of DiY as an activity can be found in the home improvement and decoration domain. Until the development of dedicated DiY stores in the 1970s, people who wanted to decorate, repair or modify their own home had to venture into the specialised world of the traditional builders merchant (Roush 1999). Companies making and selling tools and materials to amateur rather than professional customers undoubtedly were the promoters of the idea of DiY. In the 1970s lots of DiY shops and magazines were initiated. And this happened with great success. At the basis of the rise of DiY as a cultural practice were different drivers. Firstly, the economic changes brought that more people than only the rich part of the population had money to invest in home interior and decoration. Secondly, there was the fact that work hours became ever more expensive and the related rise of the DiY stores. But there are more than only these economical reasons that made DiY attractive. 'I want to do it myself' is one of the first sentences a young child uses, so it must be a very strong driver in humans in general. DiY confirms people's creative side, and gives them the feeling of 'being their own boss' (Hoftijzer 2009). DiY offers people pleasure by creating personalised artefacts or tune existing applications to their ultimate wishes. Leadbeater and Miller (Leadbeater and Miller 2004) researched another important insight regarding DiY. They claim that participation in gardening, sports and home improvement constitutes a form of everyday resistance to the alienating effects of contemporary society. The contemporary society indeed is characterised by excessive consumerism, globalisation and economic inequalities between persons and groups, alienating us from our environment and ourselves.

We can distinguish two stereotypical types of DiY-ers: the garage-DiY-er and the community-DiY-er. The first is someone who works alone, typically in a personal closed environment like a garage or attic, in a very dedicated way. The second can more often be found in a community of likely interested people. They collaborate in producing their invention of creation, so they are willing to make it public before it is finished and discuss about it with their companions.

We speak of DiY, but out of research we learned that less institutionalised channels, personal networks of family, friends and neighbours are crucial for individual experiences of DiY (Shove et al. 2008). People rely on the help of the so-called 'local warm experts' (Bakardjieva 2005; Steward 2007) or 'lead users' (Von Hippel 2005), terms which are defined in the context of domestication of information and communication technologies (ICT). In a way, DiY always has a *DiT* (Do-it-Together) component in it. Different authors invented names to refer to people active in DiY activities:

- Leadbeater invented the word 'Pro-Am'. A pro-am is an amateur that pursues activities out of the love for it, but at the same time setting a professional standard (Leadbeater and Miller 2004).
- Von Hippel proposed the word 'Lead-User'. A lead-user is at the leading edge of an important market trend, and so is currently experiencing needs that will later be experienced by many users in that market. She/he anticipates relatively high benefits from obtaining a solution to her/his needs, and so may innovate (Von Hippel, 2005).
- Levi-Strauss coined the word 'Bricoleur'. He describes the bricoleur as "someone who uses all the concrete materials he encounters in everyday life and all the earlier experiences of himself and others around him, to find solutions for the problems he is confronted with in everyday life" (Levi-Strauss 1968).
- Bakardjieva and Stewart invented the word 'local warm expert'. A local warm expert is "an Internet/computer technology expert in the professional sense or simply in a relative sense vis-à-vis the less knowledgeable other" (Bakardjieva 2005; Steward 2007).

With the attempt to describe the different roles and activities of a person doing DiY activities it becomes clear that a complex net of practices and social relations are at the basis of a DiY culture. These activities are also related to the kind of DiY activity people are executing; knitting pullovers, making a bench for a dog, designing an operating system, making a YouTube movie, and so on. But, overall, a DiY activity has some common characteristics. It is about *connecting*, about *taking control* and about *diversification.*

3.1.1.1 DiY is About *Connecting*

A core aspect of DiY is the act of 'creating' something. Gauntlett (2010) gives a good insight in the social aspects of creating. He distinguishes three ways on how making is connecting, and, therefore, in essence indicated that DiY is about communication.

1. Making is connecting because you connect things together (materials, ideas or both) to make something new.
2. Making is connecting because arts of creativity usually involve, at some point, a social dimension and connect us with other people.
3. Making is connecting because through making things and sharing them in the world, we increase our engagement and connection with our social and physical environments.

If we look at the changes ICT has brought today to the making is connecting paradigm of DiY, one can see that ICT has the potential of huge impact on DiY. The software culture is very much based on the reuse of code. The recombination of components and mash-up systems are other examples. The *Web 2.0* context made it possible for non-technical end users to create their own weblogs, web pages or *Facebook* profiles. The existence of online communities also is highly important for the DiY communities in the physical world. Not only can DiY-ers rely better on their local network for help, or people with the same interests, via these online communities, their community effectively gets world scale and world level, with *reputation* becoming a stronger factor.

3.1.1.2 DiY is About *Taking Control*

DiY is also about the power of mastering one's work and the tools one needs to succeed in achieving one's goal. Therefore, one needs capabilities as well as tools with particular attributes, *openness* being an important attribute of that. Or, like stated in the *Maker's Bill of Right* (Jalopy 2005): *"If you can't open it, you don't own it"*. In the Bill other important attributes of tools are mentioned that focus on handing over control to the creator:

- *Cases shall be easy to open.*
- *Special tools are allowed only for darn good reasons.*
- *Power from USB is good; power from proprietary power adapters is bad.*
- *Ease of repair shall be a design ideal, not an afterthought.*
- *If it snaps shut, it shall snap open.*

The aim of handing over the control to the creator or end user can be put in the discussions on innovation and technology. Paul Dourish (2006), in his design view, focuses on the fact that users are not to be perceived as passive recipients of predefined technologies, but as actors determined by the circumstances, contexts and consequences of technology use (Dourish 2006). Other trends that confirm the same 'taking control' view are the open innovation process (Chesbrough 2003), the mutual shaping of technology (Williams and Edge, 1996) and co-creation (Hoftijzer 2009).

3.1.1.3 DiY is About *Diversification*

As mentioned earlier, DiY can be perceived as the everyday resistance to the alienating effects of contemporary society. It can be seen as a reaction against excessive consumerism, globalisation and economic inequalities between persons and groups, which alienates us from our environment and from ourselves. While the globalisation makes every shopping street across a whole continent look exactly the same, the need indeed emerges for people to put forward, as an alternative to this 'more-of-the-same', something personal and unique that cannot be bought as a ready-made product in a store. Thus, DiY also is about diversification.

3.1.2 DiY in Software Application Creation

But to what extent is this DiY attitude, practice and culture already a real opportunity in the Internet of Things, as a driver for people to create their own applications in it? Own application creation can currently only be seen as an activity for the happy few. *iPhone* apps are still mostly written by small companies. Applications like *ZohoCreator* and *LongJump* are not aiming at end-users to create applications, but at an audience not much less but professionals. The role of open Application Programming Interfaces (APIs) for application creation cannot be underestimated in this respect, like it is a current topic in the context of *iPhone*, *Blackberry*, *Facebook*, *Twitter*, and other specific ICT environments.

3.1.3 DiY in Smart Spaces

DiY seems an important issue for the topic of context-aware systems and smart spaces. When Claeys and Criel (2009), among others, analysed the future vision on ambient intelligence, or 'smart' applications, two important issues were identified that point to the importance of personal creation of smart behaviour.

First, the vision on context awareness is very much technological driven and often does not take into account the meaning of context for the person that is acting in the particular environment. Since context is not something that describes a setting – *"it's something that people do, the horizon within which the user makes sense of the world"* (Heidegger 1927) – it is not possible to define 'context' for every situation or for different persons at once. This problem lies at the origin of typical context-aware applications today being far from appealing. Because of these intrinsic characteristics, context cannot be defined as a fixed computational structure, and rather is an interesting but hard-to-capture concept.

Second, context-awareness seems to imply loss of control for the person it applies to. Much in contradiction to mostly all other applications, for context awareness there is often no such thing as 'opt in'. While issues as privacy, autonomy and control were in the picture from the start, these issues seem very hard to address. As a result, users often don't have any impact on the feedback loop (Crutzen 2005).

Both these issues are essential arguments for the importance of DiY in smart spaces. The aim is to make people again 'own' their own personal data and let them decide themselves how to use it for context–awareness at any time.

3.2 Research Orientation towards Tangible Creation in Smart Spaces

From the identified trends and drivers concerning the DiY phenomenon and its replications in the networked society of today, as pinpointed in the previous sections, the theme of user-generated – DiY – applications in the Internet of Things is still a very broad research area to tackle. As our multidisciplinary research methodology moreover involves *users* in the validation of mock-ups and proof-of-concepts as well as in the creation process itself, as active participants via e.g. *co-design* and *DiY 'kits'*, there is a need to organise the landscape in more precise creation paradigms for smart spaces, leveraging *tangible user interaction* for that purpose.

Therefore, as a 'landmarks' orientation in this landscape, we identified three architectural concepts that potentially are new enablers towards mass creativity in this area.

The final value assessment will follow from user feedback experimentation in the ongoing work, but already now these concepts can help confining the problem area. In turn, this will make the actual analysis of their potential merits and technological feasibility more practicable. Concrete experimentation around the concepts therefore does not need to resort to one very narrowed-down application domain a priori, but can rather try to apply the concept in multiple concrete domains to enrich it generically, without ending up in an explosively broad DiY scope.

So, in the following sections we discuss the three candidate enabling concepts and add first experimental grounding to them:

- *the Call-Out Internet of Things,*
- *the Smart Composables Internet of Things, and*
- *the Phenomena Internet of Things.*

However, first, as a basis, we introduce a typology of what kinds of DiY creation are imaginable in smart spaces. While acts of DiY show to have an important potential in the Internet of Things, as discussed previously, we should indeed first identify what these DiY creation acts could be in this context. As illustrated by Figure 3.1 below, we can at least distinguish three different, but highly inter-relatable areas for this, as a course typology for DiY creation in the Internet of Things:

- First of all, having a large network of interconnected sensors (and actuators) principally allows for people to incorporate the related data streams in their *DiY applications* that *'use thing data'*. Today, several examples of that exist in the web, as we will discuss further on.
- Secondly, an act of DiY can clearly exist in people *connecting up new sensors* (and actuators) to the Internet of Things, as a form of *DiY installation*. Here

also, several examples exist today, e.g. in sensor network-enabled smart homes, though this is often offered via technologically closed solutions.
- Finally, the ultimate tangible creation experience is deriving from the current trends in DiY electronics, where augmentation and composition as an act of *DiY building smart objects* has become technically feasible. As such, people can be creative in shaping the tangible interaction front-end to the Internet of Things.

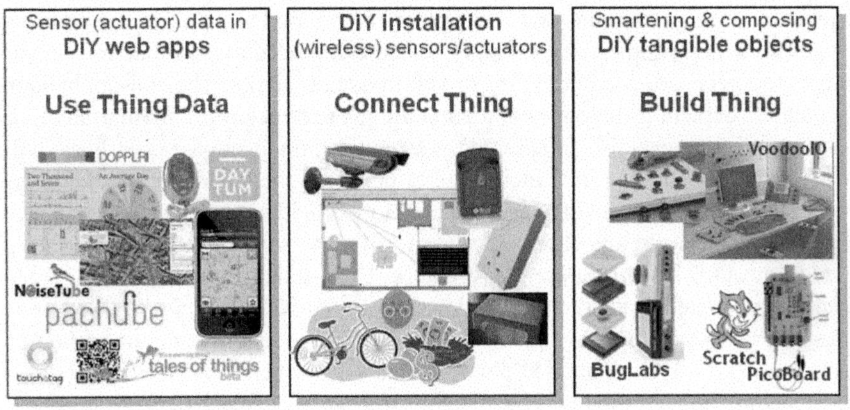

Fig. 3.1 Typology of DiY Creation in the Internet of Things

Inspired by this course typology, in the following sections we formulate three candidate concepts that could enable creativity on top of an Internet of Things.

3.3 Candidate Enabling Concept 1: The Call-out Internet of Things

We define the concept represented by the term *Call-Out Internet of Things* as entailing that the network – or cloud – gets the capability (for people) to expose and exchange call-outs in the user surroundings, as a means to provide individual users and communities with a locative, distributed communication with objects in the environment, and through this, with peer users and communities.

Call-outs, as meant here, may entail the traditional variety of information properties of locations, objects or other aspects of the surroundings, but especially can also be *behavioural descriptions*, describing a local interaction pattern, implying requests for interaction and an opportunity for adding new elements or actors in an open-ended machine or process. A Call-Out Internet of Things would, moreover, support the exchange and reuse of these properties across contexts of place, time or embodiment.

While examples of implementations exist that fall under this definition of the call-out concept, we use it as an instrument to get a deeper understanding of it as a new medium in ambient experiences. Thus, in this section we discuss how this concept is currently applied and what we see as future challenges in mass creativity in the Internet of Things from this perspective.

In fact, call-outs are commonly known in our culture already and are used in a variety of communication applications. People can experience call-outs as a communication medium in their own intimate social space as well as the broader public surroundings. Call-outs have the potential to be used for exclaiming aloud and with surprise, e.g. emotions and feelings, or can be used to post triggers and challenges to other people. For example, commercial electronic billboards in city shopping streets are competing to get their messages across. The call-out balloons in comics, depicting dialogues and supporting the structure of the narrative, are another effective example of attention-grasping communication. As early as in the Middle Ages, Leonardo da Vinci masterly practiced the technique of adding text captions to complex drawings and sketches to explicitly communicate on innovative compositions. In this case, we can see call-outs as a way to expose otherwise hidden meaning and insights into structure. Even today Leonardo's style keeps triggering people's imagination, as illustrated by numerous *Exploded View* drawings or *Cutaway View* drawings available on the Internet. An example is the Leonardo da Vinci styled exploded phone drawing by which artist Kevin Tong captures the imagination of H.G. Wells and the brilliance of Jonathan Ive[20].

In the networked society of today, at least three types of call-outs are practiced. Below, we distinguish roughly three families, using the technological means leveraged for thing/place identification as a categorisation:

- *location-based call-outs*,
- *tag-based call-outs*, and
- *image-based call-outs*.

3.3.1 Location-based Call-outs

Since geo-mapping technologies emerged to be at the disposal of the 'creative' lead user communities, all kind of geographical maps get augmented with layers of personal and community driven annotations, typically as pin-style call-outs. These personally, culturally, and socially driven reflections and annotations augment locative meaning and stimulate interaction in this way. In fact, the Earth's surface is becoming a distributed drawing canvas where people can stick their scribbles on, as a huge, locative mind map.

[20] http://www.isteamphone.com

Google Maps and *Google Earth* applications are used to develop map-based applications which map call-outs in virtual layers depicting locative interactive media. The *Augmented Reality* (AR) browser *Layar*[21] is a good example, where people can browse knowledge layers in overlay to the camera view, position-based. In this way, knowledge attached by people – or commercial organisations – can be experienced in its real geo-spatial context by others, giving the environment new collective meaning, e.g. triggering other people's recollections. With Layar, users can also be routed to locative points of interest by means of 'radar' functions, as a new form of searching. An interesting Layar layer is the application *Tweeps Around*, which queries Twitter for posts labelled with an exact location[22]. With this example, the link to social networks is indeed made, hinting at a trend towards much richer geo-aware variants of the popular communication means. Another currently popular example of that trend is in fact *Foursquare*[23], where people earn community recognition and sometimes rebate vouchers by *checking-in* often, in particular venues such as public places, restaurants and other points of – often commercial – interest, having the community comments at the place as call-outs.

Wikitude World Browser is yet another example of an AR browser, leveraging a location-based style of Wikipedia[24]. A lot of creative development activities are organised and supported, fostering open and collaborative development by the masses extending the Wikipedia-style spirit to a location-based experience. *Wikitude Drive* is the first mobile AR satellite navigation system currently being trialled.

As seen in the examples above, the attachment of crowd-sources data to specific locations in the surroundings by means of rich media overlay of call-outs on geographical maps, by this augmentation making the space 'smart' and communicative in a particularly locative way, is showing to become really valuable since a few years. It is no surprise that open creation platforms, even commercial ones, are now emerging which leverage this 'local value at global fingertips'; an example of such a platform is Google's *AR Wave*[25].

3.3.2 Tag-based Call-outs

Another technique for realising call-outs is to explore the augmented space by means of physical tagging technology, inspired by the human touch paradigm – an act which in itself is again sensationally and emotionally relevant in the user ex-

[21] http://www.layar.com

[22] http://squio.nl/projects/tweeps-around

[23] http://foursquare.com/learn_more

[24] http://www.wikitude.org

[25] http://arwave.org

perience. Historically, this was one of the first expressions of the emergence of an Internet of Things.

Here, the key is that the physical objects are approached at short distance – touch – and that access to augmented media is achieved by reading an attributed object identifier. The reading can be visual, such as one-dimensional barcodes and many variants of so-called *QR codes*, or is done by means of short range radio communication, often *Near Field Communication* (NFC) using *Radio Frequency Identification* (RFID). Today, user support for managing online identifiers and the associated media is offered via various online portal services, such as *Thinglink*[26], *Tales of Things*[27], *ThingD*[28] and the *Touchatag*[29] platform, which pioneered the RFID tagging scene and now offers both business-to-consumer (B2C) as well as business-to-business (B2B) interfaces, e.g. for payment applications.

3.3.3 Image-based Call-outs

One can even go further and use image recognition as the way to identify objects or people in the surroundings, without any further explicit tagging technology. Examples are

- Google's *Goggles* initiative[30] which can visually identify objects[31] as well as recognise text, and
- the Augmented ID technology in the *Recogniser* application of TAT[32] which associated a person's social network and other information with a person's recognised face, in a handy overlay to the image.

3.3.4 The Future of Call-outs

From the examples listed above, we see that call-outs are indeed getting established as a new global, locative interaction medium. It is reasonable to assume that the applications and technologies will evolve to have a conceptual common de-

---

26 http://www.thinglink.com/

27 http://www.talesofthings.com/

28 http://www.thingd.com/

29 http://www.touchatag.com/

30 http://www.google.com/mobile/goggles/

31 In order to avoid privacy issues, Google decided to remove the face recognition feature from Goggles shortly after release.

32 http://www.tat.se/

nominator as an enrichment of our augmented environment. Moreover, with the mixing with social network effects, as seen in several of the mentioned examples, we can expect that locative space will be shaped by DiY creation acts by the masses. Could we say that 'space and place innovations will be democratised' or could we speak about emerging 'Von Hippel' places and spaces[33]?

Even when confining the research space to the identified area of the Call-Out Internet of Things, a number of fundamental research questions remain to be investigated with respect to the enabling concept's meaning and future evolution:

- Will call-out technologies effectively empower people to create any kind of informative knowledge communication *beyond the augmentation by means of classical multimedia*, for example by controlling *haptic feedback* embedded in spatial experiences through, e.g., locative twittering?
- Will people effectively go *beyond the single-point locative augmentation*, towards more composite use of multi-point space, or ad-hoc mind mapping?
- What is the relation between the *tangible object's lifecycle* and its augmented virtual arguments?
- What is the role of call-outs with respect to an *object's intrinsic history*? (See in this respect, e.g., mixed-digital-and-physical-environments[34]).

Finally, can call-outs become a way to pinpoint instructions for acting, or even computing? This gives rise to the concept of the *Smart Composables Internet of Things*, as discussed in the next section.

3.4 Candidate Enabling Concept 2: The Smart Composables Internet of Things

The concept represented by the term *Smart Composables Internet of Things* can be defined as a specialised instance of the Call-Out Internet of Things concept, focusing on knowledge support for a – DiY and industrial – *(de)composition, production or recycling of physical objects*. In a Smart Composables Internet of Things, everyday objects get augmented with crowd- or industry-produced *instructions* and how-to's concerning how they have been or can be produced and composed and how their parts can be reused in other combinations, also in combination with other objects.

Querying can be done by context, possibly in relation to *Phenomena* (see later section on the *Phenomena Internet of Things* concept) and the context of surrounding objects, e.g. with call-outs resulting from Phenomena about frequently used combinations, or nearness of other objects with which known combinations

[33] http://web.mit.edu/evhippel/www/

[34] http://www.slideshare.net/nicolasnova/designing-a-new-ecology-of-mixed-digital-and-physical-environments

exist. A related classification of smart objects based on their awareness, their representation and their interaction can be found in Kortuem et al. (2010).

3.4.1 Object Classification According to Creator and Purpose

Our model in Figure 3.2 arranges smart objects according to their creator and purpose. Sometimes the *creator* is an individual creating an object for personal use, while in other cases the creator is an industrial actor who creates objects for mass consumption. In the figure below, we denote this as *self-made* and *ready-made* smart objects, respectively.

The *purpose* of a smart object may be to play a role in any application – or at least in a broad range of applications – or it may serve as a component in one specific application. We call this *open-ended* versus *specific* smart objects.

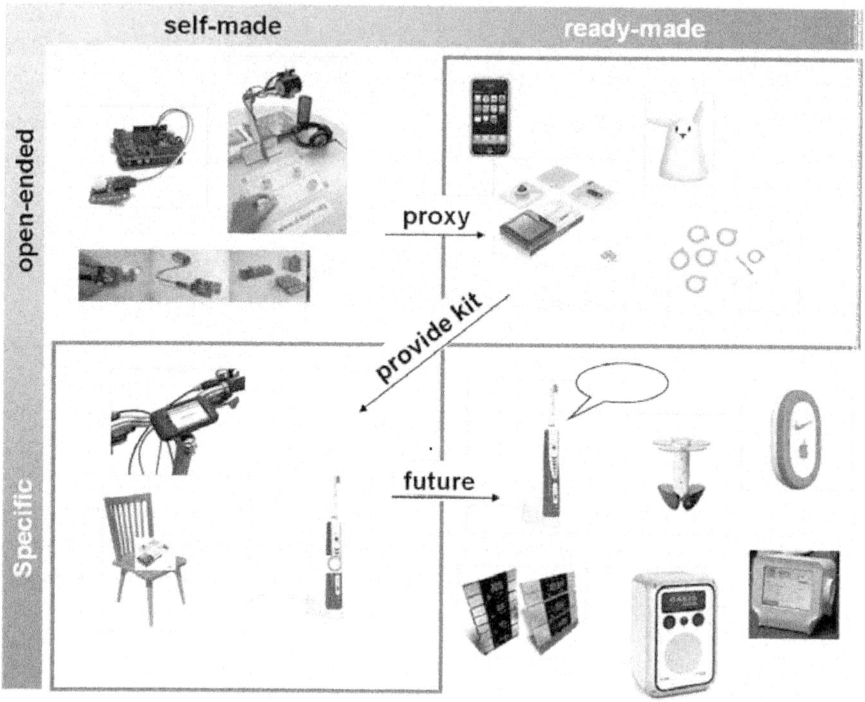

Fig. 3.2 Smart Objects Classification According to Creator and Purpose

Figure 3.2 is filled with today's examples of smart objects. *Littlebits*[35], in the upper left quadrant, is an example of a smart object that can be created by an individual through combination of different electronic components with the purpose of creating any application that the person can think of. In the opposite quadrant, lower right, we see examples of smart objects that are created by industry for a specific domain. *Chumby*[36] is an early example of such a smart object connected to the internet, with applications such as morning wake up calls as well as serving as a window to your favourite social networks.

However, the two most important quadrants for the discussion in the Smart Composables Internet of Things are the upper right and the lower left categories[37]. The *BUG*[38] is an example of a ready-made, open-ended smart object, consisting of a modular hardware kit, out of which individuals can create standalone smart objects by combining kit parts. The BUG can, however, also be used to augment an everyday object for a domain-specific application, thus moving to the lower left quadrant of the diagram. An example of such an augmented object that is self-made for a specific goal, is a chair equipped with the BUG components, for example to detect whether a person is sitting on it or not. The BUG components in such case could, at the same time, be used to, for example, provide the person with an auditory feedback when someone rings the door.

Composables are the smart objects in the upper right quadrant of our diagram. Their strength lies in the creativity that they give to individuals to compose or decompose, and to connect and disconnect with materials, people and society. Although composables are open-ended, they can be components of a *domain-specific kit* that supports the user to create domain-specific smart objects.

If networks of composables exchange data by means of sensors and actuators, according to the Smart Composables Internet of Things concept, their context of use needs to be known in order for anyone to understand what this object interaction means. It is exactly the *sharing* of call-outs that promises to bring support in this respect, attaching meaning to the smart object as well as the data exchanged by it, after *'installation'* as well as during *(de)composition* of the smart object.

Beneath this *user-level* meaningful exchange of data between located, identified objects and object parts, also the *technological* means are needed to do the actual data exchange. Such 'physical mash-up' at the technical data exchange level is typically done via web interfaces. Currently, there is a tendency to consider a REST (Richardson and Ruby 2007) mechanism in this respects, resulting in the

[35] http://littlebits.cc/websiteV1/

[36] http://www.chumby.com/

[37] Note that in this classification we see that object that offer a web-accessible API are sometimes classified either as *open-ended* or as *specific*. This is judged according to their openness for integration in a physical smart object design, as discussed here. So, although classified here as *specific*, for objects such as the Chumby there is nevertheless *another* degree of openness, because of the potential to use – or 'misuse'– them via an API for other applications than originally intended by the manufacturer.

[38] http://www.buglabs.net/

tentative definition of the *Web of Things* (WoT) as a web of bidirectional *RESTful* data exchange between objects (Guinard et al. 2010; Guinard et al. 2009).

3.4.2 Grounding via Experimentation

For the purpose of experimentation grounding of the Smart Composables Internet of Things concept, we organised a workshop with five researchers, starting from four use cases with associated mock-ups of domain specific, self-made composable-augmented objects. All four use cases were based on the use of the same sensor (a greyscale vision sensor) and one actuator (a dispensing actuator). The basic idea behind the experiment is to experience how the act of building such object would proceed, building a 'quick-and-dirty' mock-up of it, and how meaning could be attached to it. The chosen example mock-up cases were inspired by small real-world problems, which we formulated as design goal questions from the user perspective:

1. *The duster case*: how would people create a duster that detects spots on the floor, and automatically cleans them up?
2. *The door case*: how would people augment a door to sprinkle a nice fragrance in the room, whenever the door is opened?
3. *The plant case*: how would people create a flowerpot that automatically provides the contained plant with the right amount of water – in time, and without any manual human intervention?
4. *The garage case*: what can people build to avoid that fresh oil spots in their car garage make their shoes and carpet dirty?

Common to all cases was that we assumed a base object with an initial function (a door is used to close a room, a duster is used to get rid of dust, etc.), to be transformed into a smart object by combining them with composables, adding the additional functionality. The following figures show the artefacts resulting from the mock-up exercise.

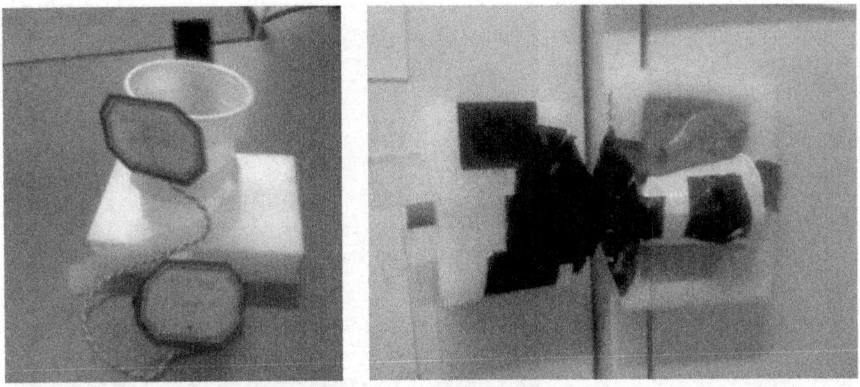

Fig. 3.3 Mock-Ups of a 'Smart Duster' (*left*) and a 'Fragrance Spraying Door' (*right*)

As shown in the left picture of Figure 3.3, one experimenter augmented the duster by attaching the vision sensor to the front of the tool foot, labelling it with a sticker reading 'you are a spot detector', and attaching the dispenser to the duster stick, labelling it with a sticker reading 'you are a spot cleaner', hinting at this approach as a non-technical, DiY way of designing the desired functionality.

The right picture in Figure 3.3 shows composables as attached to a hinging door side by another experimenter, with the vision sensor this time applied to detect the status of the door (i.e. open or closed), and a pink knob representing a fragrance capsule inserted in the dispenser.

The left picture in Figure 3.4 below shows the resulting smart plant pot mock-up, for which the experimenter chose to assume one composable as a platform for the pot to bear a humidity and other sensors as well as a spraying composable hanging over the plant, and to aggregate the sensor data processing, potentially connected to a processing back-end in the network. Again the meant functionality is simply indicated by 'call-out' stickers.

Finally, the smart oil cleaner mock-up is shown in the right picture of Figure 3.4. Here, the experimenter conceptualised an autonomously driving platform for pluggable composables such as the oil stain detector and a sawdust dispenser, the platform having the same sensor data connection and aggregation function as with the smart plant pot.

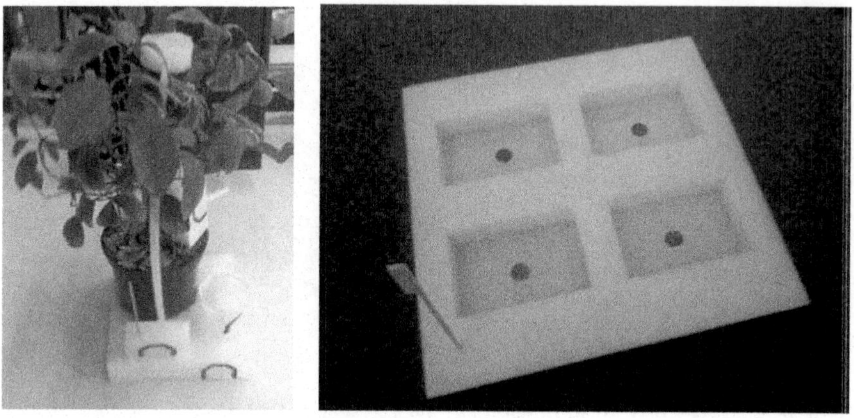

Fig. 3.4 Mock-Ups of a 'Smart Flowerpot' (*left*) and a Smart Oil Cleaner (*right*)

By observing how the experimenters approached the creation process, and the choices they made to get to the specific tangible results they created, the smart object mock-up experiments teach us that there is no single way to give functions to smart objects by means of augmentation with composables. The stickers that some experimenters used for labelling and giving meaning to particular composables or the composed whole, indicate various ways how software or call-out technologies could be used in real smart objects as would be composed by non-technical, kit-supported creators. The notion of a physical platform that makes composable easily pluggable, as introduced in two of the mock-ups, may offer a new approach for the practical implementation of a particular type of *composition* call-out.

3.5 Candidate Enabling Concept 3: The Phenomena Internet of Things

The basic concept as defined by the term *Phenomena Internet of Things* is that the network – or cloud – gets the capability to capture 'phenomena' in the user data, as a means to provide individual users and communities with feedback on patterns in their personal daily life, or in the broader society. Of crucial importance in this respect is which patterns are of real value to users, implying that close user involvement in the iterative identification of these phenomena is essential for maximising the potential of adoption in user-generated or other applications. So, in the *Phenomena Internet of Things*, higher abstractions of user context-awareness, considering long-lived patterns in personal life and society, are aimed to be derived from crowdsourcing across user groups, geography or application domains.

Of course, the crucial question in this perspective is: *Which patterns are of real value to users?* In other words, essential to the identification of these – probably also reusable – patterns is that *users are closely involved in the identification process, iteratively pointing out individual appreciations of proposed patterns*. From that, Phenomena can be identified by the crowd, in line with the continuing *crowdsourcing* trend (Howe 2006), searching such relevant patterns by leveraging massive dimensions of scale, over long time spans, within or across geographical locations and in different application contexts. This Phenomena effect is in particular strengthened by the growing amount of personal data becoming available in the Internet of Things.

3.5.1 Ingredients of the Phenomena Internet of Things

As essential aspects to a Phenomena Internet of Things, we distinguish four 'ingredients' that need to be leveraged in order to obtain valuable Phenomena enablement:

- massive data *collection*,
- user inspection and appreciation *feedback*,
- relevancy improvement from *iteration* on captured feedback, and
- fuelling user-generated *applications* with Phenomena.

In the next subsections we shortly discuss each of the four ingredients.

3.5.1.1 Ingredient 1: Massive Data Collection

The emergence of the Internet of Things and the linking of this swarm of sensors and actuators to the open web, for use in user-generated applications, with examples like *Pachube*[39] and *Noisetube*[40], in combination with the vast range of 2.0-style, crowd-oriented application (and content) creation tools, programming interfaces and application stores, is indeed facilitating a world in which a massive data collection is put to use for individual users as well as society.

In an example such as Noisetube, massive data is collected due to large numbers of people contributing their personal mobile noise measurements.

Clearly missing in this emerging Web-of-Things, is a collective identification of valuable, abstractable patterns, *Phenomena*, which could trigger much richer application possibilities, in the least already because *many such potential patterns are simply not recognised yet as valuable elements for influencing the behaviour of (newly created) applications*, at various expertise levels of the creation process.

---

[39] Pachube, http://www.pachube.com/

[40] Noisetube, http://www.noisetube.net/

Another example is the data collected through detection of user activities in a smart house, with for example hourly, daily or weekly repeated patterns becoming apparent over long time spans.

3.5.1.2 Ingredient 2: User Inspection and Appreciation Feedback

Irrespective of whether users provide data consciously – e.g. by using a sensor explicitly, to measure something in a specific context, or by manually entering such measurements – or unconsciously – e.g. by giving consent to track geographical position, or to automatically detect nearness or touch in an enhanced environment – users should at all times be able to *inspect, control* what data is collected, and *restrict* use, according to varying types of constraints, as identified also as crucially important in Greenfield's *Everyware Theses* (Greenfield 2006). However, as seen in many dedicated applications in the past, the occasional user – in contrast to the 'data organisation fanatic' – needs really simple ways to impose this control (Dey et al. 2006; Claeys and Criel 2008).

Therefore, a sound starting point may be a system defaulting to an assumption of all data being strictly personal and only for personal use, further only requesting user intervention upon specific pattern proposals instead of demanding the configuration of controlling rules for the entire space.

Incentives for the user to get involved in such eased participation can be:

- simply the comfort of visual representation of the data, and standard trend analysis of it, possibly leading to the user's enthusiasm, discovering trends personally perceived as relevant in the analysis of own, community or environmental behaviour over time and place, as well as
- the reuse of such enriched data triggers in personally created, community-built or existing applications.

So, as examples of forms of user feedback one can think of

- users making a personal selection of conventionally analysed data trends or specific data representations in a visualisation application,
- users formulating particular thresholds or other data conditions as trigger for certain application actions,
- users giving simple means to indicate approval or disapproval of application behaviour,
- or still any other form of appreciation feedback.

3.5.1.3 Ingredient 3: Relevancy Improvement from Iteration on Captured Feedback

The key ingredient that should make Phenomena different from unidirectional data mining is indeed the *closing of the loop*, allowing *controlled amplification and re-confirmation by users* on what data and data patterns are increasingly relevant, in general or in particular cases. From this massive re-iteration on user appreciations for captured data as well as preferred filtered/mined visualisation options, *popularity of reoccurring patterns,* as *Phenomena*, can be analysed. The identification of the most relevant patterns as candidate Phenomena even allows for the crystallisation and optimisation of them into *new enabling data brokerage and exposure functions*, that can become new services or products for actors that want to engage in a business based on the Phenomenon.

For example, from the previously listed means for user appreciation feedback, if many of the most popular visualisations as chosen by users are invariant to particular parts of the collected data streams, this may mean that indeed that part of the data is less relevant in general. Or, patterns that support application behaviour that is systematically approved, respectively disapproved by users, may grow, respectively diminish, in importance as candidate Phenomena.

3.5.1.4 Ingredient 4: Fuelling User-generated Applications with Phenomena

In most currently available tag correlation services, the one-to-one relation between tag reading events and web links is the key service value, such as in many of the example web services previously discussed under the section on Call-Out Internet of Things. In contrast to that, a second important incentive for users in leveraging new Phenomena – whether still emerging in a Phenomena network, or already institutionalised in a brokerage service – lies in *applying a Phenomenon in an application.* Either the awareness for the Phenomenon through its use in the application, or the entire application by itself, is then discovered or user-generated. In this way, an additional wave of user value emerges from the identification of the Phenomena.

Many of the existing context-aware applications can be seen as canonical examples of Phenomena-triggered application behaviour.

3.5.2 Links to Current and Historical State-of-the-Art

Although in literature the term ‘Phenomena Network’ was used for event tracking in Wi-Fi network topologies (Bose and Helal 2008), the concept as we have defined above goes beyond what was once applied in particular domains already.

In plain mash-ups[41], the 'Phenomena Network' is restricted to the user profile stored as cookies and preferences in the web browser. In the domain of assisted living, several research activities tried to track user activity over time and monitor health, often assuming a given scenario (a classification task), or analysing specific sequences (a time series analysis task). In the vast amount of literature on computer vision for activity recognition (Moeslund and Granum 2001), motion patterns include variations of neural networks and hidden Markov models. An intensive area of research is the area of smart homes[42,43,44,45], where contextual information is gathered from many different kinds of sensors around the house or office, often also using (layered hidden) Markov models, naive Bayesian networks, or decision trees to identify particular context situations (Desai et al. 2002; Isbell et al. 2004). In the domain of wearable and mobile computing, principal component analysis, Kohonen self-organising maps, k-means clustering, or again first order Markov models are used to detect user status from wearable sensors (Oliver et al. 2002; Krause et al. 2003). Finally, some original approaches use an ontology, e.g. extracted from *WordNet*, to mitigate the problem of activity model incompleteness (Korpip et al. 2003), or use other crowdsourcing techniques to derive an activities vocabulary for unsupervised activity recognition (Munguia-Tapia et al. 2006; Perkowitz et al. 2004).

While closing the loop with users iteratively providing appreciation feedback on identified patterns has already been considered for many applications in the pervasive or ubiquitous computing domain, the more generic approach of leveraging this *Ingredient 2* to identify *Phenomena*, is only starting to be analysed for its potential.

In fact, the personal data analysis aspect in this concept is recently also studied in the new field named *'Personal Informatics'* (Oberkirch 2008; Jones and Coates 2008) with services like *Dopplr*[46], *Fire Eagle*[47] or *Daytum*[48]. These new services, however, do not offer a higher pattern abstraction level, as a commonality for many, and so leave it entirely to the user how to interpret the data in various visualisations, often even exclusively oriented to visualising long-term data trends, without any real-time applications.

---

[41] Many examples of mash-ups are collected at http://www.programmableweb.com/mashups

[42] The Adaptive House. http://www.cs.colorado.edu/7Emozer/house/

[43] The Aware Home. http://awarehome.imtc.gatech.edu/

[44] Easy Living. http://research.microsoft.com/easyliving/

[45] MavHome, Managing an Adaptive Versatile Home. http://mavhome.uta.edu/

[46] http://www.dopplr.com/

[47] http://fireeagle.yahoo.net/

[48] http://daytum.com/

3.5.3 Potential Application Domains

In the following subsections, as further illustrations to the concept, we shortly discuss two example application domains where the Phenomena Internet of Things concept could enable new, improved-behaviour applications.

3.5.3.1 Home Applications Aware of Personal Context

In the domain of personal-context aware smart home applications, we consider the example of 'programming' the home atmosphere by means of the most appropriate selection of background music and lights ambiance. In such an application, the user ideally would like the system to 'understand' which different personal atmospheres– 'My Atmospheres' – should be distinguished, and what music and light actions should be taken upon that. In a classical approach, such an application would require the user to configure, or even design a complex set of context rules, triggered by carefully chosen conditions, and resulting in a well-structured sequence of actions. Crucial for the Phenomena Internet of Things approach thus is that this configuration complexity is hidden from the user dramatically better, by deriving an internal set of rules, conditions and actions, *indirectly rather than explicitly* based on appreciation feedback of the user.

In a first step, the user would assist in monitoring his/her own behaviour at home, e.g. by recording the interactions he/she consciously or unconsciously makes with tagged or otherwise smart objects around the house, possibly by means of DiY augmentations.

From the monitored activity, which may already be tailored by the user to contain especially relevant clues, new patterns are mined, correlating them with (initially) manual lighting and music selections.

When candidate Phenomena are detected, they are presented to the user as a new or clustered candidate 'My Atmosphere', with a matching proposal of a particular music cluster or lighting characteristic.

As a crucial step in the iterative process, the user can start naming the proposed atmospheres, potentially promoting them to effectively become a (possibly temporary) 'My Atmosphere'. The user not only becomes aware of the underlying patterns the system discovers, but also implicitly appreciates their relevance from the personal user perspective.

While iterating in this way, the system gets to know what underlying Phenomena best trigger or determine the personal home atmospheres, and gets to know which atmospheres the user indeed confirms to be representing a concept in the user's mind.

Also by detecting the user's activity to explicitly *deviate* from the regular activity patterns in presumably active atmospheres, for example, the system matures to become more and more reliable, without, at any time, taking away user control, and, at all times, allowing him/her to use this control to eventually steer the desired light and music settings as of a selected atmosphere.

Beyond that stage, users have at their disposal a well-trained system that detects and possibly even predicts personally defined atmospheres, which they now can start leveraging in other smart home applications, like presence-based communication applications or home energy management services. From use of the data in these additional applications, new types of user feedback, yet refining the personal atmosphere model, can follow, making the system still more accurate.

3.5.3.2 Massive City Data for 'Optimal' Traffic Behaviour

A totally different application domain, yet allowing also for applying the Phenomena Internet of Things concept, is the domain of multimodal (public transport, cars, pedestrians, bicycles, etc.) traffic optimisation in urban realms, leveraging large amounts of Internet of Things flow data.

Here, the process starts with the classical analysis of crowd traffic patterns and visualisation of them in an attractive way on geographical maps for citizens. Phenomena, in this context, could be public phenomena concerning e.g. the behaviour of the crowd transport activity, in the occurrence of particular local events (accidents, road blocks, etc.), changing weather conditions or still other conditions. Based on these, a routing application could derive route change advice to users, and get feedback by people actually following the advice or not, or giving other appreciation when an advised route is eventually followed.

In this way, a model is grown about the collective awareness and behaviour upon changing city traffic conditions. From the user feedback, the system learns what traffic phenomena are relevant to people, making them change route plans, or it learns about obstructions it did not explicitly detect in the first place (very local peak hour traffic congestion effects, unregistered road works or damage, etc.).

Here again, other applications can start using the identified Phenomena as advanced context triggers for smart adaptation.

3.5.4 Grounding via Experimentation

In the context of the Phenomena Internet of Things concept, we started an experiment supported by the City SensPod sensors[49] – as also used in Fing's Villes 2.0[50] project – with the ultimate goal to demonstrate an example where citizens co-produce overall city environmental data, consequently finding themselves em-

---

[49] Sensaris, http://www.sensaris.com/

[50] La Montre Verte, http://www.lamontreverte.org

powered to change their own ecological behaviour. In this way, the citizens also influence decisions and actions of local government and other stakeholders. In fact, in line with the notion of *Phenomena*, the community's *Situational Awareness* (Endsley 1995) is increased. Figure 3.5 shows the City SensPod – packed with an extendible set of sensors for noise, metal oxide, humidity, COx, NOx and GPS location – and illustrates that raw data, which is obtained via Bluetooth.

With the experiment we target large scale experimentation in the city for

- obtaining citizens' feedback, observing the changes in their behaviour, as well as possible promotion of social actions upon getting a city view from the data; and
- evaluating the platform for (anonymously) collecting data and deriving patterns from it, to be identified ultimately as *Phenomena*, and to be applied in user-generated applications, e.g. for 'green route' navigation.

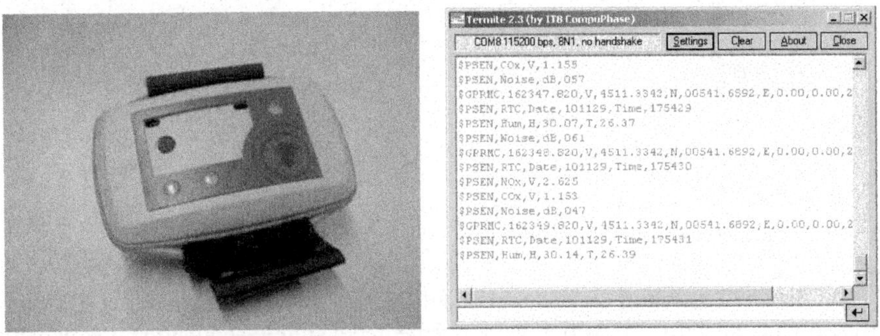

Fig. 3.5 City SensPod and an Impression on Raw Data Collection

The Prototype System

From the target goals stipulated above, we built a first working prototype in an initial technical step towards the evaluation of the Phenomena Internet of Things concept in a user community.

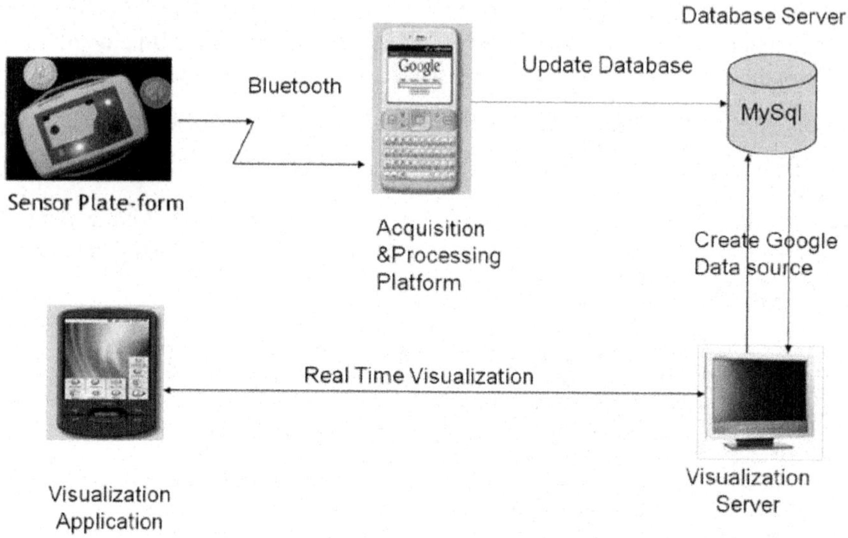

Fig. 3.6 Phenomena Prototype Architecture

As depicted in Figure 3.6, we mounted the City Senspod device to a PC via a Bluetooth connection, on which we implemented a simple acquisition and parsing service, allowing a developer to quickly parse any event coming from the device and transform it into a *tuple* of type *{Measurement Timestamp, Measurement Type, Measurement Value}*. The stream of *tuples* is saved at a remotely located *MySql* database, from which real-time visualisation or any – possibly third party – application is made possible using a Google visualisation API, taking the *MySql* server as an external data source.

Currently, it is still up to the developer to design and make his/her own visualisations. For instance, as shown in Figure 3.7, we implemented a real-time, online visualisation of noise data. The visualisation on the right in the figure shows a more sophisticated instrumentation, allowing users to selects parameters according to their preferences.

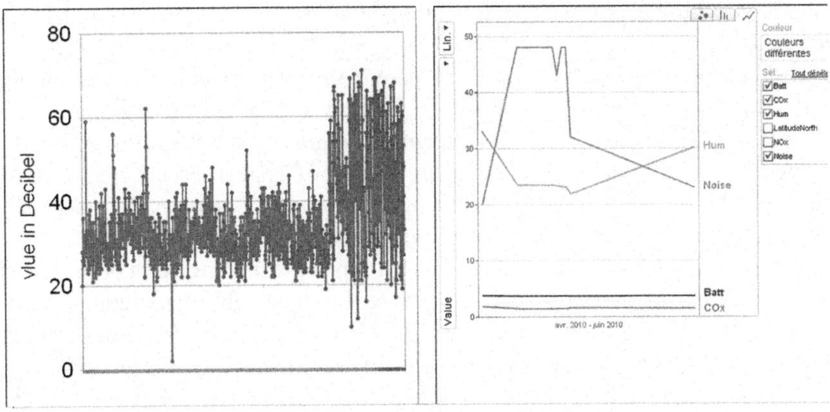

Fig. 3.7 First Simple Visualisations of Potential Phenomena

As a first step to evaluate the Phenomena Internet of Things concept, the current implementation still needs to incorporate the key element of collective identification of valuable patterns and the presentation thereof to the user for feedback, or for use in user-generated application eventually. For this purpose, we plan to build a data broker agent, also cooperating with a Bell Labs research department specialising in the data mining aspects and related relevant techniques. In line with Dinoff et al. (2007) and Kim et al. (2009), we plan to model the subject's (be it a human, a software agent or a smart object) learnt habits and intentions, for the first order identification of candidate Phenomena. We expect to get stable and consistent results from this for each subject, in line with the MIT *Reality Mining* project's[51] finding that people have predictable *Eigenbehaviours*.

The versatility and intuitivity of the visual representation is another aspect we plan enhancing, in order to meet the Ingredient 2 discussed before. We aim to provide the users with an easy control tool for monitoring the own personal and environmental data, especially from sensors and devices as specified by the user him/herself, as we believe this to provide a much more motivating experience, which can naturally entail user appreciation feedback. In this way, we envision to start exploring this most important new dimension of the Phenomena Internet of Things concept.

3.6 Conclusion

Motivated by the socio-cultural practice of *'Do-it-Yourself'* (DiY) as apparent in society, we presented a first analysis towards the enablement of mass creativity in the Internet of Things in this chapter, leveraging the DiY movement. With the new possibilities emerging with a *Web of Things* approach, DiY and the Internet of

[51] http://reality.media.mit.edu

Thing promise to become a powerful combination that has the potential to boost to mass scale.

From a typology of how people can potentially create and customise applications, services and objects on top of the Internet of Things, we elaborated three concepts forming a basis for new creation paradigms in smart spaces, potentially leading to new DiY-enabling functions in Internet of Things service creation environments: the *Call-Out Internet of Things*, the *Smart Composables Internet of Things*, and the *Phenomena Internet of Things*.

Considering the applicable state-of-the-art for implementing parts of these concepts and with first experimental grounding for them, the exploration process around these enabling concepts is ongoing, and several challenges clearly remain. Most notably, while the close involvement of the user in any Internet of Things service deployment and even real user participation in the creation process as analysed in this chapter may be key in tackling the expected *privacy issues*, these issues can be expected to grow to be the biggest challenge in the ever smarter world sensing and automating everything around us. On a more technical level, the underlying architectures also will need to be able to handle the massive sharing of personal or public sensor data, as users will assume performance and ubiquity as a condition for value. Evolution beyond the postulated enabling concepts discussed can as well be expected in the future, giving rise to yet more challenges to come at a more conceptual user acceptance level.

Acknowlegements

This Alcatel-Lucent Bell Labs work is supported by the ITEA2 Eureka cluster, with funding support by the Flemish authority IWT, as part of the European ITEA2 project 08005, DiY Smart Experiences (DiYSE), conducted by 40 partners from 7 European countries. More information about the project can be found in the dedicated book chapter on the project and at the project's public website http://www.dyse.org.

References

Bakardjieva M (2005) Internet Society: The Internet in Everyday Life. London, Thousand Oaks, New Delhi, Sage

Bose R, Helal A (2008) Distributed Localized Detection and Tracking of Phenomena Clouds using Wireless Sensor Networks. Proceeding of the 6th ACM Conference on Embedded Networked Sensor Systems

Chesbrough H (2003) Open Innovation: The New Imperative for Creating and Profiting from Technology. Boston, Harvard Business School Press

Claeys L, Criel J (2008) Context aware computing: Future living as a social application. In: Withworth B, Demoor A (eds) Handbook of research on socio-technical design

Claeys L, Criel J (2009) Future Living in a Participatory Way. In: Withworth B, De Moor, A (eds) Handbook of Research on Socio-Technical Design and Social Networking Systems. Information Science Reference, Hershey, New York

Crutzen C (2005) Intelligente Ambiance, tussen hemel en hel: een verlossing? Paper presented at Ambient Intelligence: een ego-harnas?, Open University Nederland, Faculty Informatics

Desai N Kaowthumrong K, Lebsack.J, Shah N Han R (2002) Automated Selection of Remote Control User Interfaces in Pervasive Smart Spaces. Proceedings of HCIC Workshop on Pervasive Smart Spaces, 2002

Dey AK, Sohn T, Streng S, Kodama J (2006) iCAP: interactive prototyping of context-aware applications. Proc. of Pervasive 2006

Dinoff R, Ho TK, Hull R, Kumar B, Lieuwen D, Santos P (2007) Intuitive network applications: Learning for personalized converged services involving social networks. J Comput 2:72-84

Dourish P (2006) Implications for Design. Paper presented at the CHI 2006, Montreal, Quebec, Canada

Endsley MR (1995) Toward a theory of situation awareness in dynamic systems. Hum Factors 37:32–64. doi:10.1518/001872095779049543

Gauntlett D, (2010) Making is Connecting [to be published]. Abstract from http://www.makingisconnecting.org/gauntlett2010-extract2.pdf. Accessed 1 June 2010

Greenfield A (2006) Everyware: The Dawning Age of Ubiquitous Computing. New Riders Publishing, Berkeley

Guinard D, Trifa V, Pham T, Liechti O (2009) Towards Physical Mashups in the Web of Things. Proceedings of INSS 2009 (IEEE Sixth International Conference on Networked Sensing Systems). Pittsburgh, USA, June 2009 http://www.vs.inf.ethz.ch/publ/papers/guinardSensorMashups09.pdf. Accessed 20 September 2010

Guinard D, Trifa V, Wilde E (2010) Architecting a Mashable Open World Wide Web of Things. Technical Report No. 663. Department of Computer Science, ETH Zurich, February 2010. http://www.vs.inf.ethz.ch/publ/papers/WoT.pdf. Accessed 20 September 2010

Heidegger M (1927) Sein und Zeit, Tübingen

Hoftijzer JW (2009) The Implications of doing it yourself. A changing structure in business and consumption. Proceedings of the First International conference on Integration of Design, Engineering and management for Innovation IDEMI09. Porto, Portugal

Howe J (2006) The Rise of Crowdsourcing, Wired, Issue 14.06, June 2006, http://www.wired.com/wired/archive/14.06/crowds.html. Accessed 20 September 2010

Isbell CL Jr, Omojukon O, Pierce JS (2004) From Devices to Tasks: Automatic Task Prediction for Personalized Appliance Control. Pers Ubiquitous Comput 8:146-153

Jalopy M (2005) Owner's manifesto: the maker's Bill of Rights. http://makezine.com/04/ownyourown/. Accessed 20 September 2010

Jones M , Coates T (2008) Polite, Pertinent, and... Pretty: Designing for the New-wave of Personal Informatics, http://www.slideshare.net/blackbeltjones/polite-pertinent-and-pretty-designing-for-the-newwave-of-personal-informatics-493301. Accessed 20 September 2010

Kim E, Helal A, Cook D (2009) Human Activity Recognition and Pattern Discovery. IEEE Pervasive Comput Mag. doi:10.1109/MPRV.2010.7

Korpip P, Koskinen M, Peltola J, Mäkelä SM, Seppänen T (2003) Bayesian approach to sensor-based context awareness. Pers Ubiquitous Comput 7:113-124

Kortuem G, Kawsar F, Sundramoorthy V, Fitton D (2010) Smart Objects as Building Blocks for the Internet of Things. IEEE Internet Comput 14:44-51.doi:10.1109/MIC.2009.143

Krause A, Siewiorek DP, Smailgaic A,Farringdon J (2003) Unsupervised Dynamic Identification of Physiological and Activity Context in Wearable Computing.Proceedings of the Seventh IEEE International Symposium on Wearable Computers (ISWC'03), Oct 2003

Leadbeater C, Miller P (2004) The Pro-Am Revolution. How enthusiasts are changing the way our economy and society work. Demos. http://www.demos.co.uk/files/proamrevolutionfinal.pdf. Accessed 20 September 2010

Levi-Strauss C (1968) Het wilde denken. Amsterdam: Meulenhoff

Moeslund TB, Granum E (2001) A Survey of Computer Vision Based Human Motion Capture. Comput Vis Image Underst 81:231-268

Munguia-Tapia E, Choudhury T, Philipose M (2006) Building Reliable Activity Models using Hierarchical Shrinkage and Mined Ontology. Proceedings of Pervasive. May 2006, Dublin, Ireland.

Oberkirch B (2008) Under Sousveillance: Personal Informatics & Techniques of the Self. Defrag 2008.http://www.slideshare.net/brianoberkirch/under-sousveillance-personal-informatics-techniques-of-the-self-presentation. Accessed 20 September 2010

Oliver N, Garg A, Horvitz E (2002) Layered Representations for Recognizing Office activity. Proceedings of the Fourth IEEE International Conference on Multimodal Interaction (ICMI), October 2002

Perkowitz M, Philipose M, Patterson DJ, Fishkin KP (2004) Mining Models of Human Activities from the Web. Proc. of the Thirteenth International World Wide Web Conference (WWW 2004), May 2004.

Richardson L, Ruby S (2007) RESTful Web Services. O'Reilly

Roush C (1999) Inside Home depot: How One Company Revolutionized an Industry through the Relentless Pursuit of Growth. New York: McGraw Hill

Shove E, Watson M, Hand M, Ingram J (2008) The design of everyday life. Oxford: Berg

Steward J (2007) Local Experts in the Domestication of Information and Communication Technologies. Inf Commun Soc 10:547-569

Von Hippel E (2005) Democratizing Innovation. Massachusetts: MIT Press

Williams R, Edge D (1996) The social shaping of technology. Res Policy 25:865-899. http://www.rcss.ed.ac.uk/technology/SSTRPfull.doc. Accessed 20 September 2010

4 The Toolkit Approach for End-user Participation in the Internet of Things

Irena Pletikosa Cvijikj, Florian Michahelles

Information Management, Department of Management, Technology and Economics, ETH Zürich, Switzerland

Abstract Today, there are many end-user programming tools available, but in the Internet of Things domain, this concept is relatively new. Some pioneer examples include solutions, such as d.tools and Pachube, but also Web2.0, Mash-ups, Twitter and Facebook are suitable backplanes for this kind of applications. Another level of development support is various hardware concepts and solutions, such as RFIDs, Arduino, Violet, NFC, barcodes and many more. Appropriate user programmability could transform a system, multiplying the effectiveness of programmers and users. This article discusses how end-users can be empowered with new building blocks and tools, analogous to those that were emerging during the early phases of Internet growth. Accelerators, frameworks and toolkits are introduced, which would allow everybody to participate in the Internet of Things in the same manner as in the Internet through Wikis, Blogs etc.

4.1 From Internet to Internet of Things

Back in the 1960s, a group of visionaries saw great potential in allowing computers to share information, which contributed to the creation of what is today known as Internet. The first network, ARPANET, was developed for a very special purpose: the connection of just four major computers at universities in southwestern US (Salus 1995). This early form of Internet was founded by the government and was used only by computer experts, scientists and librarians for research, education, military and government purposes. There were no personal computers and anyone who used it, had to learn how to use a very complex system. Commercial uses were prohibited unless they directly served the goals of research and education. This policy continued until the late 80's, when the major breakthrough came with the introduction of Berners-Lee's World Wide Web. The new protocol he proposed gave simple information access to the general public by embedding hypertext into the Internet. In the following years, the lowering of barriers has made it simple enough, not just to use the Internet, but also to shape it and to add and generate new services. The second generation of the Internet, called Web 2.0 or *social web*, with its key supporting technologies, Ajax, Wiki, Blog, RSS, and Atom became ubiquitous, faster, and increasingly accessible to non-technical communities, thus introducing the rapid content generation and distribution, which lead to the richness of information of today's Internet.

It is evident that from its beginning the Internet was changing very fast and nowadays it is still evolving. However, instead of just connecting computers and/or wearable devices, it grew from a network linking digital information to a network relating digital information to real world physical items. This new network is called *Internet of Things* and provides embedding of physical reality into the Internet and information into the physical reality.

The concept of the Internet of Things became popular in 1999, when the Auto-ID Center at Massachusetts Institute of Technology (MIT) designed the RFID technology. Kevin Ashton, co-founder and director of MIT, in an article published in Forbes Magazine, in 2002, said,

> "...we need an internet for things, a standardised way for computers to understand the real world..."

The name of this article contained the first documented usage of the term Internet of Things in literature. In the following years this idea became more popular (Friedemann and Flörkemeier 2009) and in 2009 it was recognised as a general transformation of the Internet by the EU Commission (European Commission 2009).

Smart objects play the central role in the Internet of Things idea. Equipped with information and communication technology, everyday items provide a new quality by featuring tiny computers. These objects can store their context, they are networked together, they are able to access Internet services and they interact among themselves and with human beings. In order to connect everyday objects and devices to large databases and networks, a simple, unobtrusive and cost-effective system of item identification is required. Radio Frequency Identification (RFID) offers this functionality.

Based on this concept, many traditional industries, such as logistics, manufacturing and retail, have increased their effectiveness of the production cycle by implementing smart devices through RFID and barcode technologies. The cost of rudimentary RFID tags promises to drop to roughly $0.05/unit in the next several years (Sarma 2001), while tags as small as 0.4mm × 0.4mm, and thin enough to be embedded in paper are already commercially available (Takaragi et al. 2001). Such improvements in cost and size will lead to a second wave of applications including vertical market applications, ubiquitous positioning and physical world web, i.e. the real Internet of Things. However, at the moment the Internet of Things is still mostly governed by business applications.

On the other hand, new generations of smart phones, sensor networks, open-source Application Programming Interfaces (APIs) and toolkits are becoming more and more pervasive. Still, these devices are mostly not customised to meet specific user expectations. Emerging trends of user programming (Scaffidi et al. 2005) give the opportunity to non-professional end-users of making additions to products, according to their specific needs. However, to let individuals play a main role in the Internet of Things, there are still a number of problems and challenges to overcome.

4.2 Problems and Challenges

The development of tools and techniques is a key focus for the concept of *participatory design* (PD). End-user tools and techniques should promote a practice where researchers and design professionals will be able to learn about users' work, and where users will be able to take an active part in technology design. To achieve this, these tools should avoid traditional design techniques with abstract representations, and instead allow users to more easily experiment with various design possibilities in a cost effective way (Kensing and Blomberg 1998).

Gronbaek et al. (Gronbaek et al. 1997) suggest the use of cooperative prototyping, where users and designers collectively explore the functionality and form of applications as well as their relations to the work in question. This type of cooperation requires access to adequate prototyping tools, which would act as *catalysts* and *triggers* in discussions about the relations between work and technology (Mogensen 1992), (Trigg et al. 1991). Tools and techniques used in PD projects should all have the common aim of providing designers and users with a way of connecting current and future work practices with envisioned new technologies.

Despite all the challenges, the need for innovation has been recognised and supported. Since Internet of Things systems will be designed, managed and used by multiple stakeholders, driven by different business models and various interests, these systems should (European Commission 2009):

- allow new applications to be built on top of existing systems,
- allow new systems to be deployed in parallel with existing systems, and
- allow an adequate level of interoperability, so that innovative and competitive cross-domain systems and applications can be developed.

Pioneer projects in this field, some of which are presented later in this chapter, have already been developed. Still, real-case user-driven scenarios do not exist yet, leading to a situation where the uptake of the technology itself is slowed down. To overcome this situation, end-users must be empowered with new building blocks and tools that were analogously emerging during the Internet growth.

4.3 Towards a Participatory Approach

The process of designing and developing new solutions presents a big challenge for scientists and manufacturers, even when it is about simple, everyday objects. When designing a new product, solutions are usually based on observations and usability conclusions. However, once the product has left the laboratory, the situation is completely different. The problem lies in the fact that designers are often drawn into the trap of trying to find uses for the tools, and deploying the coolest new features, forgetting their primary focus should be on providing value to the end user. However, the large variety of end users, usage conditions and scenarios usually leads to confusion and dissatisfaction regarding the usability of the product in the real world environment.

The concept of personalisation offers the solution for the described situation. Still, a designer working on a task of personalisation on an existing application, or building a new personalised application, is poised to make the classic error of putting technology before the needs of the end users (Kramer et al. 2000). Based on these observations, a new idea has grown, focusing on involving end users into the development process.

4.3.1 User-centered Design

User-centered design (UCD) is a broad term, used to describe a design philosophy and a variety of methods in which the needs, wants, and limitations of end users are placed at the center of attention at each stage of the design process. UCD differs from other approaches in trying to optimise the solutions based on how people can, want or need to use them, rather than forcing the users to change their working habits in order to comply with the offered approach. UCD is based on involving the users in different stages of the design process, from gathering ideas for functional requirements and usability testing, to direct involvement into the development process itself.

The term UCD was initially used by Donald Norman from the University of California San Diego (UCSD) in the 1980s and became widely used after the publication of his co-authored book "User-Centered System Design: New Perspectives on Human-Computer Interaction" (Norman and Draper 1986). Norman's work explained that involving actual users in the development process, preferably in the environment in which the product would be used, was a natural evolution in the field of UCD. Based on these statements, users became a central part of the development process resulting in more effective, efficient and safer products (Abras et al. 2004).

4.3.1.1 User-centered Principles and Activities

Today, there is an international standard, *ISO 13407: Human-centered design process* (ISO 13407 1999), that defines a general process for including human-centered activities throughout a development life-cycle. This standard describes four principles of UCD:

- active involvement of users,
- appropriate allocation of function to system and to user,
- iteration of design solutions, and
- multi-disciplinary design;

and four UCD activities:

- requirements gathering, understanding and specifying the context of use;
- requirements specification, specifying the user and organisational requirements;
- design, producing designs and prototypes; and
- evaluation, carrying out user-based assessment of the site.

The process involves iterating through these activities until the objectives are satisfied.

As a form of UDC performed during the design activity, PD, focusing on the participation of users in the development process, has gained strong acceptance, particularly in Scandinavian countries.

4.3.1.2 Participatory Design

PD applications are diverse in their perspectives, backgrounds, and areas of concern, leading to a lack of a single definition. However, in its essence, PD represents an approach towards assessment, design and development of various systems in which people, destined to actually use these systems, play the major role in designing and in the decision-making process. In other words, users are the co-designers of the systems.

The PD approach emerged in the mid 1970s in Scandinavia, under the name *cooperative design*. It was born out of the labor union's push for workers to have more democratic control in their work environment (Ehn 1989). However, when presented to the US community, due to the strong separation between managers and workers, *participatory* was the more appropriate description of the process, where managers and workers did not sit and work together, but rather work separately on the same problems, thus participating in the solution finding process. Pioneer projects included:

- Norwegian Iron and Metal Workers Union (NJMF) project, that took a first move from traditional research to working with people (Ehn and Kyng 1987),

- Utopia project (Bodker et al. 1987), (Ehn 1988), whose major achievements were the experience-based design methods, developed through the focus on hands-on experiences, emphasising the need for technical and organisational alternatives, and
- Florence project (Bjerknes and Bratteteig 1995) which has started a long line of Scandinavian research projects in the health sector, giving voice to nurses during the development of work and IT in hospitals.

Since then, PD projects have varied with respect to how and why workers have participated, from being limited to providing designers with access to workers' skills and experiences, where workers have little or no control over the design process or its outcome, to fully participating in the process. In all cases, worker participation is considered central to the value and, therefore, the success of the project (Kensing and Blomberg 1998).

Caused by the differences that can arise between users and designers, sometimes the users are unable to understand the language of the designers. Therefore, the development of innovative tools and techniques is a key focus for PD projects and their extension into the specific context of particular projects has become part of PD researchers' repertoire for action. PD techniques involve informal presentation of relations between technology and work, including visualisations (Brun-Cottan and Wall 1995), toolkits, prototypes and mockups. These tools and techniques promote a practice where both, the technology and the work organisation are in focus, and where users are able to take an active part in technology design.

At the dawn of the 21st century, technology is an inseparable part of everyday life, used at work, at home, in school, and on the move. This poses a new challenge to PD to embrace the fact that much technology development no longer happens as a design of isolated systems in well-defined working environments (Beck 2002), but instead community based development is becoming an emerging trend. This produces new techniques in development processes, such as open-source development, end-user programming, crowdsourcing and others, discussed below.

4.3.2 Open-source Development

Open source (OS) is a development method for software that harnesses the power of distributed peer review and transparency of process. The promise of OS is software or products of better quality, higher reliability, more flexibility, lower cost, and an end to predatory vendor lock-in (Open Source Initiative).

The concept of free software is an old one and can be traced back to the 1950s. First computers served only as research tools at universities, software was exchanged freely and programmers were paid for the programming as a process, and not for the software itself. When computers reached the business world, software became commercialised and developers started charging for each installation copy. In 1984, Richard Stallmann, a researcher at MIT, founded the Free Software Foundation (FSF) and the GNU project (GNU manifesto), thus providing the foundations for today's OS movement. Where proprietary commercial software vendors saw an economic opportunity that must be protected, Stallman saw scientific knowledge that must be shared and distributed (Dibona et al. 1999). The OS software movement has received enormous attention in the last several years. It is often characterised as a fundamentally new way to develop software (Raymond 1999) that poses a serious challenge (Dibona et al. 1999) to the commercial software businesses, which dominate most software markets today (Mockus et al. 2002). According to SourceForge.net, as of August, 2010, more than 240,000 software projects have been registered to use their services by more than 2.6 million registered users.

The OS development model fundamentally differs from the approaches and economics of traditional software development. First, the usual goal of an OS project is to create a system that is useful or interesting to those who are working on it (Godfrey and Tu 2000). Many successful OS software products have been and are being developed, distributed, and supported by an internet-based community of programmers, i.e. users themselves (Lakhani and von Hippel 2003). Developers are often unpaid volunteers who contribute towards the project as a hobby and there is no direct compensation for their work (Hars and Ou 2002). The question that this poses is the motivation for OS development? Eric Raymond reports (Raymond 1999) that there are at least three basic motives for writing or contributing to the OS projects: user's direct need for the software and software improvements, enjoyment of the work itself and the enhanced reputation.

The most fascinating development in the OS movement today is not necessarily the success of companies like Red Hat or Sendmail Inc. Instead, seeing major corporations like IBM and Oracle, turning their attention to OS as a business opportunity is intriguing. There is only one explanation for this behavior: innovation (Dibona et al. 1999). The new concept based on this, also known as *open innovation,* is using the OS as the most natural network for innovations (von Hippel 2002).

4.3.3 End-user Programming

A further type of community-based development is end-user programming (EUP). One way to define *programming* is as the process of transforming a mental plan of desired actions for a computer into a representation that can be understood by the computer (Hoc and Nguyen-Xuan 1990). Back in the 1940s, at the beginning of the era of computers, programming was done only by a small number of scientists, i.e. professional software developers. Since that time, software industry has been growing rapidly, and computer programming has become a technical skill of millions. In parallel, a second, powerful trend has begun to take shape, the so-called *end-user programming* (Myers and Ko 2009).

To understand the concept of end-user programming, it is important to explain the difference between professional and end-user programmers. While professional programmers develop software as a part of their jobs, end-user programmers are people who also write programs, but not as their primary job function. They are not formally trained in programming, yet need to program in order to accomplish their daily tasks. Spreadsheets are considered the major success story in end-user programming (Erwig 2009); however, many end-user programmers also use other special-purpose languages, or are even faced with the requirement of learning professional programming languages to achieve their goals. Despite the differences in priorities between professionals and end-user programmers, they both face the same software engineering challenges.

End-user programming has become so ubiquitous, that today there are more end-user programmers than there are professional programmers. According to the expert estimations (Scaffidi et al. 2005), in 2012, 90 million Americans will use computers on workplaces, significantly exceeding the 3 million anticipated professional programmers. Over 13 million workers will "do programming" in a self-reporting sense; and more than 55 million people will use spreadsheets and databases.

Due to the variety of end-user programmer profiles and backgrounds, there is not a single method for end-user programming. Instead, several techniques are being used, including programming by demonstration, visual and natural programming and many domain-specific languages and formalisms (End-User Programming).

What are the benefits of this approach? The obvious and most important one is: users know their problems best. Therefore, software products could become simpler, and at the same time more reliable. Only the general features will be supported by the issuing company, while details will be developed by end-user programmers, thus providing a richness of features that would otherwise be difficult to explain to a programmer. Allowing users to add their programs would give them freedom and responsibility at the same time. Therefore, it is beneficial for both users and product developers, to use these techniques and provide end-users with the possibility to shape products according to their needs.

4.3.4 Crowdsourcing

> "There is an incredible story to be told about human ingenuity! The first step to its unfolding is to reject the binary notion of client/designer. ... The old-fashioned notion of an individual with a dream of perfection is being replaced by distributed problem solving and team-based multi-disciplinary practice. The reality for advanced design today is

> dominated by three ideas: distributed, plural, collaborative. … Problems are taken up everywhere, solutions are developed and tested and contributed to the global commons, and those ideas are tested against other solutions. The effect of this is to imagine a future for design that is both more modest and more ambitious." (Mau 2004)

The term *crowdsourcing* was initially used by Jeff Howe and Mark Robinson in 2006 for describing a new web-based, distributed problem-solving and production model that has emerged in recent years. Howe offers the following definition:

> "Crowdsourcing is the act of taking a job traditionally performed by a designated agent (usually an employee) and outsourcing it to an undefined, generally large group of people in the form of an open call." (Crowdsourcing)

In other words, crowdsourcing is a process engaging the following major steps, (1) company posts a problem online, (2) a vast number of individuals offer solutions, (3) winning ideas are selected and awarded (Wikipedia). The difference between crowdsourcing and ordinary outsourcing is that a task or problem is outsourced to an undefined public rather than a specific other body. On the other hand, crowdsourcing differs from the OS practice. Problems solved and products designed by the crowd become the property of companies, who turn large profits off from this crowd labor. Technological advances are breaking down the cost barriers that once separated amateurs from professionals. New markets for the efforts of non-professionals open, as smart companies discover ways to tap the latent talent of the crowd. The labor isn't always free, but it costs a lot less than paying traditional employees (Howe 2006).

Crowdsourcing platforms appear to be a good place for young professionals or amateur affirmation. Learning, resource exchanges, comparison, recognition by peers or by companies, capitalisation of proposals and successes remain the main stimulation for creation, even if participants have a low chance of earning financial rewards (Trompette et al. 2008). Based on the principle of opening the design/development process to the crowd, initiated either by the company or the community, crowdsourcing provides the way to access external knowledge for the purposes of innovation, thus complying with the principles of open innovations.

4.3.5 Living Labs

Although the majority considers that the concept of Living Lab originates from William Mitchell, professor at MIT, Boston, it seems that the term Living Laboratory was initially used in literature by Abowd and his colleagues at Georgia Institute of Technology to refer to real-world contexts, in which users were given the opportunity to European state-of-the art technology (Abowd 1999). The concept was based on the idea of smart/future homes, serving as "real home" settings where real people were observed in their usage of emerging technologies. Markopoulos and Rauterberg gave broader definition by envisioning the living lab as

"... a planned research infrastructure that is pivotal for user-system interaction research in the next decade" (Markopoulos and Rauterberg 2000).

They have described the Living Labs as platforms for collaborative research that will serve as a basis for development and testing of novel technologies.

Today, the literature distinguishes between two different interpretations of the Living Labs concept (Folstad 2008):

1. Contextualised co-creation: Living Labs supporting context research and co-creation with users, and
2. Testbed associations: Living Labs as extensions to testbeds, where testbed applications are accessed in contexts familiar to the users.

The tendency to view Living Labs as environments for user-driven innovation is extremely interesting, since it meets both the industry needs for involvement of end-users in the early phases of ICT innovation and the end-users' needs for obtaining personalised solutions complying with their specific interests.

According to Folstad (2008), the main goals of a great majority of Living Labs identified in the literature are:

- Evaluate or validate new ICT solutions with users;
- Gain insight in unexpected ICT uses and new service opportunities;
- Experience and experiment with ICT solutions in contexts familiar to the users;
- Medium- or long-term studies with users.

It is interesting to note that all of these goals support the idea presented in this chapter: end-user participation in the Internet of Things domain of ITC development.

According to Eriksson et al. (Eriksson et al.2005) Living Labs differ from the other known user participation approaches. While the previously described approaches mostly involve the user in the development of the end-products, the Living Labs concept refers to an R&D methodology where innovations are created and validated in collaborative multi-contextual real-world environments with the individual in focus.

The concept of Living Labs is still emerging. The continuous growth of the number of organisations and initiatives whose intentions are to promote the concept and to provide support for collaboration between existing Living Labs is evidence to this. Examples of these organisations include Living Labs Global[52], European Network of Living Labs (EnoLL)[53] and Community-Based Living Labs to Enhance SMEs Innovation in Europe (CO-LLABS)[54]. Certain activities have also been undertaken by the European Commission[55].

52 http://www.livinglabs-europe.com/

53 http://www.openlivinglabs.eu/

54 http://www.ami-communities.eu/wiki/CO-LLABS

55 http://ec.europa.eu/information_society/activities/livinglabs/index_en.htm

Until now, Living Labs have been used in R&D environments for variety of purposes, e.g., ubiquitous computing, mobile ICT, retail innovations, online communities and collaborative work-support systems. However, we also see that the Living Lab approach seem to be suitable to meet the evolving needs for user participation in the field of Internet of Things.

4.4 Innovations to Users via Toolkits

A general trend toward heterogeneous consumer needs makes product development increasingly difficult (von Hippel 2001). Product developers face the problem that users hold an essential, but rather "sticky" portion of information required for product development. Von Hippel (Von Hippel 1994) defines *sticky information* as:

> "…the information that is costly to acquire, transfer, and use in a new location".

The degree of stickiness is defined as the incremental expenditure required for transferring a certain unit of information to a specified locus in a form that is useable to the information seeker. When this cost is low, information stickiness is low; when it is high, stickiness is high. The traditional development approach typically engages companies in costly market research, because they assume homogeneity of needs within a market segment and thus can amortise over many consumers. This results in creation of somewhat different products for each segment, each intended to address the average customer needs in the selected segment (Jeppesen 2005). However, research on the issue concludes that a large share (about 50%) of the total variation in consumer needs will typically remain unaddressed in within-segment variation (von Hippel and Katz 2002).

In the manufacturing domain, user toolkits for innovation were recently proposed as a means to eliminate the (costly) exchange of need-related information between users and manufactures in the product development process. Two lines of argumentation have been brought forth to explain the potential benefits of toolkits for innovation and design: (1) the heterogeneity of customer preferences; and (2) the problems associated with shifting preference information from the customer to the manufacturer (Franke and Piller 2004). This approach allows manufacturers to serve the "markets of one", and to handle large and small customers in the same way, at the same time promising opportunity for entrepreneurs.

User toolkits for innovation emerged in a primitive form in the high-tech field of custom integrated circuit (IC) design in the 1980s as a result of enormous growth of price as custom IC products grew larger and more complex. Today, they range from food industry to software toolkits, allowing consumers to design key features by themselves. Depending on the type of toolkit, the outcome might be a product (Park et al. 2000) or an innovation (Thomke and von Hippel 2002). However, regardless of the underlying research area, the supporting rationale is the same: the toolkit allows the customer to take an active part in product development.

The user toolkits method is built around the idea of relocating a segment of problem-solving tasks with sticky need-related information to the consumer setting. The intention is to eliminate the need for information transfer and iterations throughout the development process by outsourcing tasks of product development to consumers. Toolkits should divide tasks, so that consumers primarily carry out tasks related to those areas of development that involve their sticky information. Letting consumers carry out essential *design-by-trial-and-error processes* avoids costly iteration and speeds up the process.

Von Hippel (Von Hippel 2001) argues that an effective toolkit for user innovation should enable five objectives. First, toolkits should enable users to carry out complete cycles of trial-and-error learning. Second, they will offer a well defined "solution space" that encompasses the specific designs. Third requirement is that well-designed toolkits must be "user friendly" in the sense that users do not need to engage in much additional training to use them competently. Fourth, they will contain libraries of commonly used modules, thus allowing the user to focus his efforts on the truly unique elements of that design. Fifth and finally, properly-designed toolkits will ensure that custom products and services designed by users can be produced on manufacturer production equipment without requiring revisions by manufacturer-based engineers.

Although an increase in opportunities for consumer involvement seems to raise the need for supporting consumers, there is a promising solution to this problem, namely, the establishment of consumer-consumer support interaction (Jeppesen 2005). A case study of Westwood Studios, an outlier in terms of firm support to consumers, shows that consumers who use toolkits may be willing to support each other. This complies with previous experiences on the field of OS development.

The new wave of ubiquitous computing, where everyday objects are augmented with computing capabilities, poses new challenges for designing interactive technologies. In terms of Internet of Things, collaborative programming in all of its forms is a growing research field, aiming to involve end-users, i.e. individuals already in possession of the relevant need-related information, and allowing them to actively participate in the development of the next generation of Internet of Things. The challenge is to create toolkits and frameworks that would provide reusable building blocks for recurring sub-tasks pertinent to users' problems. To achieve the high level of participation and excellent quality of resulting products, an environment has to be provided where barriers to contribute are low.

4.5 Existing Toolkits

This chapter is about accelerators, frameworks and toolkits that would allow everybody to participate in the Internet of Things in the same manner as it can already be done on Internet through Wikis, Blogs, etc. Typical software with end-user programmability features should at least have an editor, an interpreter or compiler, error checking and debugging tools, documentation and version management tools, as the bare minimum requirements. Today there are many EUP tools available (End-User Programming), but in the Internet of Things domain, this concept is relatively new. Some pioneer examples include solutions, such as d.tools and Pachube, but also Web2.0, Mash-ups, Twitter and Facebook are suitable backplanes for this kind of applications. Another aspect is the emerging usage of wireless devices and sensor technologies. Today, they can be found everywhere, including the wearable devices and smart phones. This served as inspiration for various hardware concepts and solutions, such as Arduino, Violet, NFC, barcodes, RFIDs, and many more.

4.5.1 I/O Boards and HW Based Systems

4.5.1.1 Wiring

> "Wiring is an open source programming environment and electronics I/O board for exploring the electronic arts, tangible media, teaching and learning computer programming and prototyping with electronics. It illustrates the concept of programming with electronics and the physical realm of hardware control which are necessary to explore physical interaction design and tangible media aspects." (Wiring)

Wiring was designed as a part of a master thesis by Hernando Barragán, at the Interaction Design Institute Ivrea, in 2004 (Barragán 2004). Wiring is based on OS principles. Its small I/O board represents a cheap standalone computer with many connection capabilities. The board can be used to control all kinds of sensors and actuators: sensors allow the board to acquire information from the surrounding environment, while actuators allow the board to create changes in the physical world (lights, motors, heating devices, etc). Wiring can also interact with other devices, such as PC/Mac, GPS, barcode readers, etc. It can be programmed using the Processing (Processing.org) language and numerous available libraries.

Processing is an OS programming language and environment for people who want to program images, animation, and sound. It is used by students, artists, designers, architects, researchers, and hobbyists for learning, prototyping, and production. It is created to teach fundamentals of computer programming within a visual context and to serve as a software sketchbook and professional production tool. Processing is developed by artists and designers as an alternative to commercial software tools in the same domain.

Wiring has been used for numerous solutions (Wiring - Exhibition Archives). It has also been used as a basis for another prototyping platform: Arduino (Arduino).

4.5.1.2 Arduino

> "Arduino is a tool for making computers that can sense and control more of the physical world than your desktop computer. It's an open-source physical computing platform based on a simple microcontroller board, and a development environment for writing software for the board. It's intended for artists, designers, hobbyists, and anyone interested in creating interactive objects or environments". (Arduino)

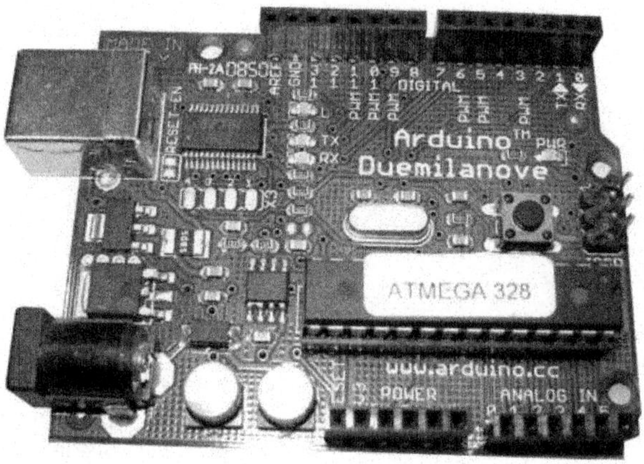

Fig. 4.1 Arduino Duemilanove Board Based on the ATmega168/ATmega328 Microcontroller

The Arduino microcontroller was originally created as an educational platform for a class project at the Interaction Design Institute Ivrea in 2005 in order to engage artistic and design-oriented minds. It was based on the previous work of the Wiring microcontroller, focusing on simplicity, a goal in pursuit of designing for a non-technical audience (Gibb 2010).

Arduino is a combined software/hardware platform. It can receive input from a variety of sensors and can affect its surroundings by controlling lights, motors, and other actuators. The microcontroller on the hardware board can be programmed using the Arduino programming language and the Arduino Integrated Development Environment (IDE) (written in Java and based on Processing (Processing.org). Arduino supports two working modes, stand-alone or connected to a computer via USB cable.

Arduino offers everything that is needed for ubiquitous computing. That is the reason why its usage has overgrown the initial art and design solutions space. One very popular toolkit based on Arduino is LilyPad, described in detail in next section.

4.5.1.3 LilyPad

> "The LilyPad Arduino is a system for experimenting with embedded computation that allows users to build their own soft wearables by sewing fabric-mounted microcontroller, sensor and actuator modules together with conductive thread...The kit was designed to engage kids (and adults) in computing and electronics and teach them fundamental skills in these areas by allowing them to creatively experiment with e-textiles in the same way that the Mindstorms kit allows people to experiment with robotics." (Buechley et al. 2008)

LilyPad was initially designed and developed by Leah Buechley and SparkFun Electronics in 2003, providing large connecting pads, to create an interface between small electronic components and textiles, to be sewn into clothing. Various input, output, power, and sensor LilyPads are available.

A LilyPad board is based on the ATmega168V or the Atmega328V. Programming can be done using the Arduino IDE software. There are also several libraries that allow users to easily control an assortment of sensors and output devices. To program the LilyPad, a user clips it to a USB device that supplies the patch with power and facilitates computer patch communication. In order to create wearable electronic fashion items, at least the following modules are required: mainboard, power supply and a USB connection to download the software from a computer to the LilyPad mainboard.

LilyPad was used in several user studies (Buechley and Eisenberg 2008) and various end-user projects available on the Internet. Some of them are: a sonar garment for providing navigation assistance to visually impaired people by navigating the built environment, a blinking bike safety patch, an interactive passion sensing scarf, and many more.

4.5.1.4 MAKE Controller Kit

The MAKE Controller Kit (Making Things) represents a new generation of OS hardware platforms, successor to the Teleo (Making Things – Teleo) modular kits. This system was developed in collaboration with MAKE magazine and specifically embraces the "Do It Yourself" (DIY) subculture.

Fig. 4.2 The Make Controller Kit v2.0 Assembled with Controller and Application Boards[56]

The MAKE Controller Kit is targeted at enthusiasts and hobbyists. It is built out of two boards; a general controller board plugs into a specific application board and offers extensive features and interfaces. The controller board makes almost all signals from the chip available while the application board has application specific hardware (motor control, networking (Ethernet/USB/CAN/Serial/SPI) and a circuit protection.

This toolkit provides the possibility of designing a specific application board, including only the required features. Otherwise, the MAKE application board allows users to interface directly to the relevant devices (sensors, high current outputs for motors, etc.) and not worry that they're going to break the delicate controller board.

The MAKE controller kit comprises the software development environment. It can work as an interface to the PC when connected by Ethernet or a USB connection, or can be programmed to run standalone programs. A simplified API is available to program the board in C (freeRTOS operating system (FreeRTOS-A FreeRTOS)), so that the most difficult aspects of microcontroller programming are taken care of for less experienced users. Otherwise, full access to the chip is available for experienced coders.

The future steps in MAKE controller kit development consider providing further simplified programming environment similar to Wiring/Arduino.

[56] http://www.makingthings.com/store/make-controller/make-controller-kit.html

4.5.1.5 Phidgets

"Physical widgets, or phidgets, comprise devices and software that are almost direct analogs of graphical user interface widgets. Like widgets, phidgets abstract and package input and output devices: they hide implementation and construction details while exposing functionality through a well-defined API." (Greenberg and Fitchett 2001)

Fig. 4.3 The Phidget Single Board Computer (SBC) with an Integrated PhidgetInterfaceKit 8/8/8, 4 Full-speed USB Ports and a Network Connection[57]

In other words, Phidgets are a set of plug and play building blocks for interfacing the physical and the virtual worlds via low cost USB sensing and control from your PC. The system arose out of a research project at the University of Calgary in Canada and has later been commercialised (Phidgets, Inc.).

Phidgets includes USB-based hardware boards for input (e.g., temperature, movement, light intensity, RFID tags, switches, etc.) and output actuators (e.g., servo motors, LED indicators, LCD text displays). Its architecture and API let programmers discover, observe and control all phidgets connected to a single computer.

Phidgets are connected to a computer via USB, and are identified by the computer as a USB device. Each device knows and can transmit its phidget type, as well as an identification number that is unique for a phidget instance of that type. On the software side, all the required components are packed as an ActiveX COM Component.

Unlike widgets, each phidget component requires a corresponding visual component, providing a visual on-screen interface for interactive end-user control. Additionally, a connection manager tracks how devices appear on-line and there is a way to link a software phidget with its physical counterpart. And finally, there is a possibility for running in simulation mode, in order to allow the programmer to

[57] http://www.phidgets.com/products.php?category=0&product_id=1070

develop, debug and test a physical interface even when no physical device is present (Fitchett and Greenberg 2001).

The system has an extensive library of APIs and can be used with a large number of applications, even with other toolkits in some cases (Marquardt and Greenberg 2007). Using Phidgets enables programmers to rapidly develop physical interfaces without the need for extent knowledge in electronics design issues.

4.5.1.6 I-CubeX

> "I-CubeX proprietary system is based on the MIDI communication protocol and offers modular components covering a large field of applications. With more than 10 years of experience in real-time sensor data gathering, I-CubeX is renowned for its ease of use, its variety of sensors and its robustness. It is widely used as a tool for prototyping, experimentation, research and teaching". (I-CubeX Online Store - Resources)

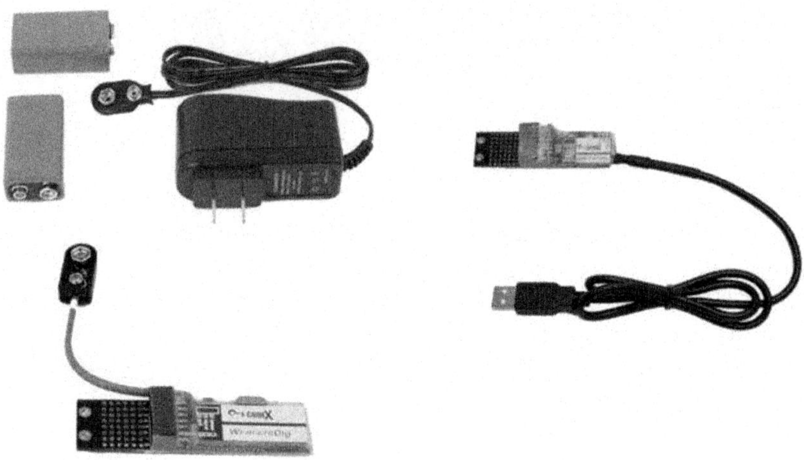

Fig. 4.4 Left: I-CubeX Wi-micro System, Including a Wi-microDig Analog to Digital Encoder with Wireless Bluetooth Transmitter, Cable, 9V Batteries and a BatteryPack-800; right: USB-micro System, Including a USB-microDig Analog Sensor Interface[58]

I-CubeX arose out of a research project in 1995 (Mulder 1995) directed by Axel Mulder at the Department of Kinesiology, Simon Fraser University, to address the need for better tools for artists to create interactive art and for musicians to more easily create or modify musical instruments. While I-CubeX helped to open up access to technology for artists interested in sensor technology; it in itself inspired others to create new technology.

[58] http://www.partly-cloudy.com/misc/

I-CubeX comprises a system of sensors, actuators and interfaces that are configured by a personal computer. Using MIDI, Bluetooth or USB as the basis for all communication, the complexity is managed by a variety of software tools, including an end-user configuration editor, Max (software) plug-ins, and a C++ API, which allows applications to be developed in Mac OS X, Linux and Windows operating systems.

Today, this system is intended for production or serious integration (I-CubeX Online Store – Demos) and the possibilities offered by this system are quite large. It can be used in science to obtain human performance data, and study animal and human responses to environments that change depending on the sensor data. In engineering, it can be used to develop prototypes of interactive products and to create innovative control surfaces enabling new ways to interact with multimedia. And in the arts, its environment of origin, it can create a responsive environment, build an alternate musical controller and develop novel interactive media pieces.

4.5.1.7 Comparison

Table 4.1 gives a comparison of the previously presented HW based toolkits.

Name	Wiring	Arduino	LilyPad	Make Controller[59]	Phidgets [60]	I-CubeX
Components	IO board, SW platform	IO board, SW platform	IO board, SW platform	2 IO boards, SW platform	IO board, sensors, motors, SW platform	digitisers, sensors, SW platform
OpenSource	yes	yes	yes	yes	no	no
Microcontrollers	ATmega128 ATmega1281 ATmega2561	ATmega8 ATmga168 ATmga328 ATmega1280	ATmega168V ATmega328	Atmel SAM7X processor, ARM7	PhidgetSBC	N/A
Memory	128K	16/32 KB (ATmega168/ATmega328)	16/32 KB (ATmega168/ATmega328)	256K	SDRAM 64MB	N/A
Programming Lan-	Wiring	Arduino	Arduino	C/C++	C/C++, Java	N/A

[59] Data corresponds to the Interface/Application Board.

[60] Data corresponds to Phidget SBC with integrated InterfaceKit 8/8/8.

guage						
Board OS	-	-	-	FreeRTOS	Linux	N/A
Tools Support	Wiring IDE	Arduino IDE	Arduino IDE	mcbuilder	Phidget IDE	Mac OS / Windows editor
PWM (analog) Outputs	6	6	6	4	N/A	N/A
Analog Inputs	8	6	6	8	8	N/A
Digital I/O pins	54	14	14	35 / 8	8I + 8O	N/A
MIDI Ports	-	-	-	-	-	IN/OUT
USB Ports	1	1	1	1	4	via MIDI-USB Adapter
Power	7-12V/USB	7-12V/USB	2.7-5.5V/USB	6-12V	6-15V	7.5V
External Interrupts	8	2	2	N/A	N/A	N/A
Hardware Serial Port	2	1	1	2.5-2/1	via Serial Adapter	N/A

Table 4.1 Comparison of HW Centered Prototyping Systems

Phidgets and I-CubeX distinguish from other toolkits since they represent complete systems with "ready-to-use" sensor and actuator/motor components, thus representing low-end solutions. Therefore, these two systems can easily be used by non-technical persons, without the need to go into detail regarding complex schematics and electronic issues. Additionally, I-CubeX differs from other systems in two more aspects. First, it provides only MIDI interfaces; other communication possibilities (USB, Bluetooth) are available via appropriate adapters. Second, I-CubeX has support only for Windows and Mac OSX, while all the other systems also have Linux support.

While Phidgets and I-CubeX are low-end solutions, high-end solutions, i.e. Wiring, Arduino, LilyPad and MAKE Controller Kit, provide more freedom for their users in the process of developing customised solutions.

4.5.2 SW Based Solutions

4.5.2.1 d.tools

d.tools is a hardware and software system that can be described as

> "…a design tool that embodies an iterative-design-centered approach to prototyping physical UIs." (Hartmann et al. 2006)

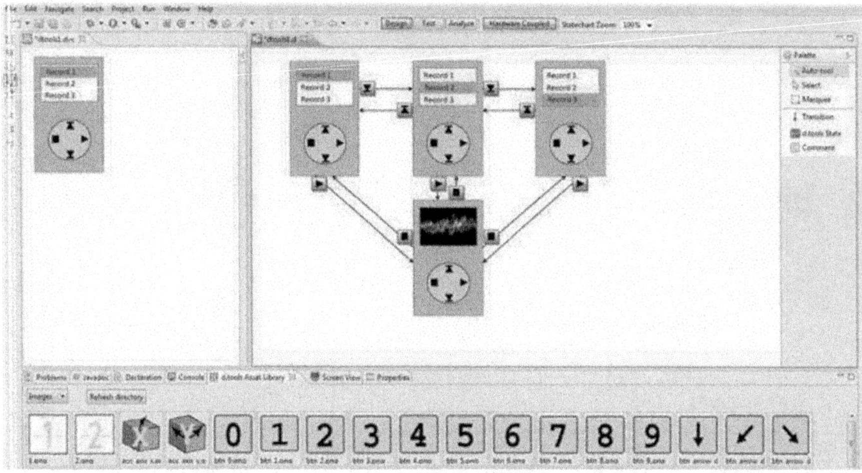

Fig. 4.5 The d.tools Visual Authoring Environment Showing a State-chart for an iPod Shuffle Prototype[61]

d.tools was built to support design thinking rather than implementation thinking. With d.tools, designers place physical controllers (e.g., buttons, sliders), sensors (e.g., accelerometers), and output devices (e.g., LEDs, LCD screens) directly into the prototypes, and program their behavior visually in the provided software workbench. The d.tools visual authoring environment is implemented in Java J2SE 5.0 as an Eclipse IDE plug-in using its Graphical Editing Framework (GEF). The d.tools interface comprises a device designer, a state-chart designer, and associated views for specifying properties.

d.tools employs a PC as a proxy for embedded processors, so designers can focus on user experience-related tasks rather than implementation-related details. The d.tools library includes an extensible set of smart components that cover a wide range of input and output technologies. It provides plug-and-play prototyping of hardware components by adding microcontrollers to each component and networking the components on a common bus.

[61] http://hci.stanford.edu/research/dtools/gallery.html

d.tools can now be connected to other commercially available hardware platforms, such as Wiring boards, Arduino boards and Phidgets Interface Kits (d.tools).

4.5.2.2 iStuff

> "iStuff is a toolkit for physical devices that extends the ideas of supporting wireless devices, a loose coupling between input and application logic, and the ability to develop physical interactions that function across an entire ubiquitous computing environment." (Ballagas et al. 2003)

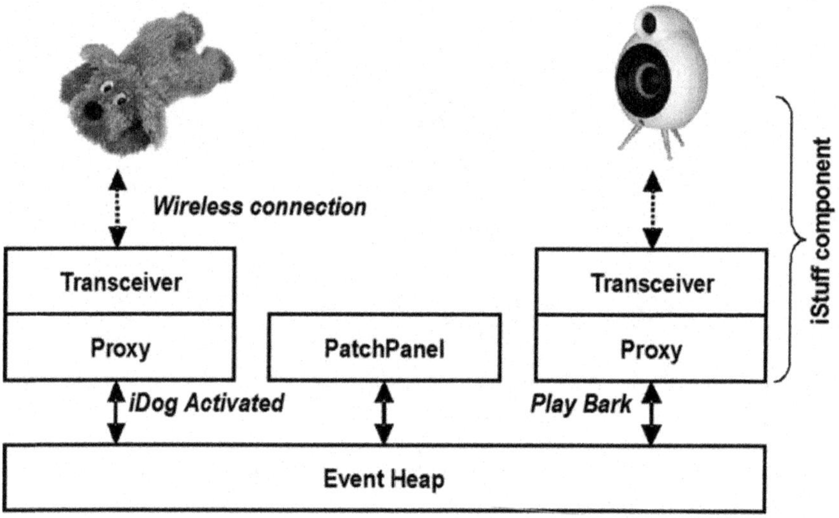

Fig. 4.6 The iStuff Components Architecture[62]

The iStuff toolkit was developed in a research conducted on a Computer Science Department at Stanford University in 2003. It was designed on top of iROS, a TCP and Java based middleware that allows multiple machines and applications to exchange information. Recently, its functionality has been extended with toolkits for mobile phone interactions (Ballagas et al. 2007).

iStuff leverages an existing interactive workspace infrastructure, making it lightweight and platform independent. The supporting software framework includes a dynamically configurable intermediary to simplify the mapping of input and output devices, such as buttons, sliders, wands, speakers, buzzers, microphones, etc., with their respective software proxies in order to create iStuff com-

[62] http://hci.rwth-aachen.de/istuff/tutorial.php

ponents. iStuff, in conjunction with the Patch Panel (Ballagas et al. 2004), enables standard UIs to be controlled by novel inputs.

4.5.2.3 Lego Mindstorms

Lego Mindstorms (LEGO.com MINDSTORMS) is a line of programmable robotics/construction toys, manufactured by the Lego Group. The hardware and software roots of the Mindstorms Robotics Invention System Kit go back to the programmable brick created at the MIT Media Lab.

Fig. 4.7 Lego MINDSTORMS NXT in a Mobile Robot Configuration

The first Lego's visual programming environment, called LEGOsheets, was created by the University of Colorado in 1994 (Gindling et al. 1995). The initial Mindstorms Robotics Invention System Kit consisted of two motors, two touch sensors, and one light sensor. Today's NXT version has three servo motors and one sensor, each for touch, light, sound, and distance.

The programmable LEGO brick is the RCX, which transforms models into robots and controls their actions. LEGO provides two tools for programming the RCX. The first one is a development environment for programming the RCX with an interface that models programming as a process of dragging the puzzle pieces together to produce a chain (complete program). This GUI environment supports the basic programming constructs, such as loops, subroutines (though not true procedure calls), and concurrency. LEGO's second programming tool is a library for generating Visual Basic programs to control the RCX.

Lego Mindstorms may be used to build a model of an embedded system with computer-controlled electromechanical parts. Many kinds of real-life embedded systems, from elevator controllers to industrial robots, may be modeled using Mindstorms.

4.5.2.4 Pachube

> "Pachube is a web service that enables storing, sharing & discovering real-time sensor, energy and environment data from objects, devices & buildings around the world. It represents a convenient, secure & scalable platform that helps in building the Internet of Things." (Pachube)

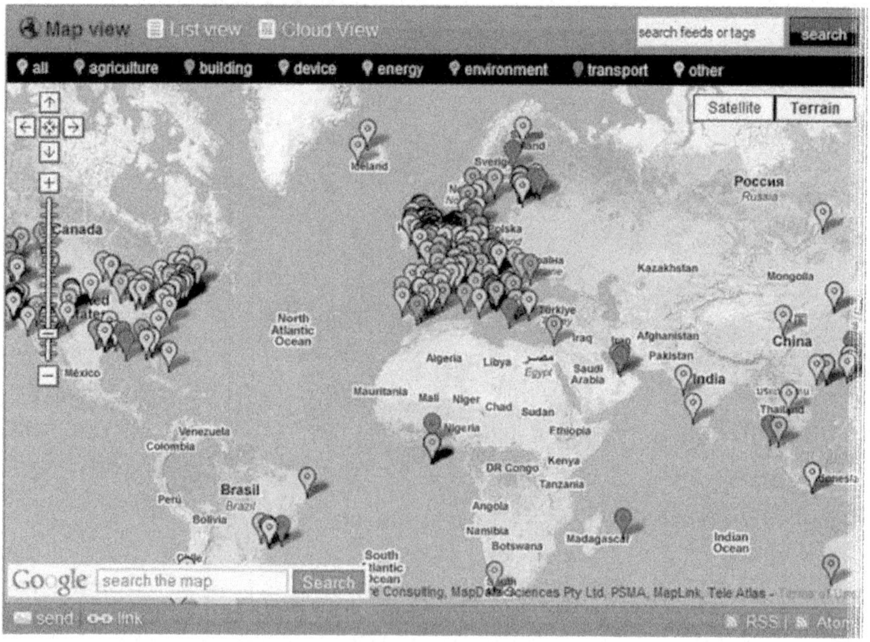

Fig. 4.8 Location of Pachube Sensors All over the World

The key idea behind the concept is to facilitate interaction between remote environments, both physical and virtual. Pachube enables a direct connection between these two environments, but it can also be used to facilitate many-to-many connections enabling any participating project to plug-in to any other project participating in real time, so that they could "talk" to each other and exchange data.

Apart from being used in physical environments, it also enables people to embed this data in web-pages and thus in effect to the blog sensor data.

Pachube uses Extended Environments Markup Language (EEML), which extends the construction industry protocol IFC. An extensive RESTful API makes it possible to both serve and request the data in all formats. Integration with other tools, such as Arduino, is also possible.

4.5.2.5 Comparison

The key issue to be discussed when comparing the SW prototyping platform is the support for the hardware platforms that they offer. Table 4.2 shows a comparison of all the relevant aspects of the previously presented SW based solutions that enable fast prototyping.

Name	d.tools	iStuff	Lego Mindstorms	Pachube
Components	Eclipse IDE plug-in	iStuff IDE	LabVIEW Graphical Programming Microsoft Robotics Studio IAR Embedded Workbench	Internet of Things platform, RESTful based API
SW Platforms	Java	Java	NBC[63]	Java, Ruby, PHP, .NET, Processing, Xquery, …
HW Platforms	Wiring, Arduino, Phidgets	Phidgets, MAKE Controller Kit, Smart-its, Powerbook Tilt Sensor, …	N/A	Zigbee, Phidgets, Arduino (+ WiShield / Danger shield), Sun SPOT, Home Automation Hub, …
OpenSource	yes	yes	no	no
Dedicated HW	yes	no	yes	no

[63] Next Byte Codes

Community	no	no	no	yes

Table 4.2 Comparison of SW Centered Prototyping Platforms.

Based on this comparison, we can conclude that there is greater variety in approaches when it comes to SW centered toolkits than HW centered ones. Lego Mindstorms is a closed platform, providing the possibility for usage only with the available LEGO HW components (sensors and servo motors).

Pachube, on the other hand, is not a suitable platform for developing customised solutions, but rather a platform for sharing the data on the Web inside the provided Internet of Things community. However, it provides integration of customised solutions based on the previously described HW toolkits.

D.tools and iStuff are the closest ones to the idea of toolkit based prototyping, providing direct support for creating and programming systems on the basis of the lower level HW based toolkits described in the previous section.

4.6 Discussion

Building on the potential benefits that the Internet of Things offers, poses a number of challenges, not only due to the nature of the underlying technologies, but also to the sheer scale of their deployment. Although it is essential for the mass deployment and diffusion, technological standardisation is still in its infancy, or remains fragmented. Successful standardisation in RFID was initiated by the Auto-ID Center and is now under the governance of EPC Global. The ZigBee Alliance, among others, contributed in the standardisation of wireless sensor networks, but in the case of nanotechnology and robotics, the situation is still very fragmented (International Telecommunication Union 2005).

On the other hand, research on toolkits for user design and innovation has been going on for the last 30 years. The conceptualisation of this phenomenon and the exploration of possibilities, limitations, and the underlying theoretical patterns of these new instruments only constitutes an initial step in this area. Although there are many research articles on technical aspects of toolkits and the production environment, many questions concerning the design side of individualisation and innovation still remain unanswered.

Greatest challenges and objectives to achieve when it comes to the toolkits usage for end-user participation are:

- *Diversifying and expanding the number of toolkits.* The greater the diversity and number of existing toolkits, the greater the diversity and number of toolkit users will be, leading to mass adoption. Modes of exploiting the toolkit approach depend directly on what the toolkit allows the user to do. The comparison of some of the existing toolkits presented in the previous chapter gave an indication that there is a great variety between the existing toolkits, not just from a perspective whether they are software or hardware based solutions, but also from the perspective of the level and form of user involvement in the innovation and prototyping process. Studies (Prügl and Schreier 2006) have already shown that "one toolkit may not serve all users effectively". End-users often try to surpass the limits of the design freedom provided in firm-constructed toolkits by employing tools from related fields and by expanding the scope of existing tools or even creating their own toolkits. Thus, different types of users employ different types of tools that in turn lead to different types of innovation activities, which support the previous discussion.

- *Focusing on low-end toolkits.* Highly technically skilled individuals represent a small percentage of the users' pool. Therefore, low-end toolkits could significantly increase the number of participating end-users. High-end toolkits offer a theoretically unlimited solution space within the production capabilities of the manufacturer, thus resulting in new functions or even new products. Innovations derived from this type of toolkit could be marketed successfully to a broader target group. However, not all users have the required skills to use such toolkits. Therefore, low-end toolkits might be more valuable when it comes to proposing new, individualised solutions. Additionally, not all users need radically new solutions. In the previous chapter we gave a brief overview of some of the most popular existing toolkits in the Internet of Things domain. However, the majority belong to the category of high-end toolkits, requiring a certain level of effort to learn how to use and configure them in order to create personalised solutions. Further development, with the emphasis on low-end toolkits, is required in order to achieve their mass popularisation and usage.
- *Exploitation of existing markets through individualisation.* Another possible way to exploit toolkits in the Internet of Things domain is to allow incremental innovations, product adaptations and individualisation. Individualisation platforms can be seen as valuable contribution due to the simplicity of usage. This approach has successfully been used in some industries, e.g., designing your personal watch, sneakers, etc., and has shown to be popular. Hence, in order to target a greater number of participants, the major field of toolkit applications should be the (further) exploitation of existing markets through individualisation.
- *Living Labs as innovation platform.* Living Labs can provide the missing ideas for real-life applications in the Internet of Things domain. Since the Internet of Things is still a relatively new and growing field, there is a requirement for the generation of ideas for possible applications in the Internet of Things domain. These ideas could significantly benefit the creation of new toolkits for participation. A popular approach for generating ideas, related to the open innovation process is Living Labs approach, presented previously in this chapter. Living Labs present an appealing opportunity for both generating ideas and at the same time creating and testing new solutions, based on the observation of the everyday activities and end-user requirements.

- *Revealing the source code.* In parallel to supplying a diversified toolkit, revealing the source code of the toolkit itself to users would enable them to push a toolkit's design limits. OS development was proven to be an efficient approach in supporting the open-innovation concept. Applied to the Internet of Things, it could speed up the process of integrating the Internet of Things in everyday life.

The common challenges related with the concept of involving end-users in the process of building solutions for new technologies are not the only ones that the Internet of Things is confronting. The complexity of the Internet of Things vision itself posses great challenges upon practitioners of the Internet of Things related EUP. The Internet of Things should not be seen as an extension of today's Internet, but rather as a number of new independent systems that operate with their own infrastructures (and partly rely on existing Internet infrastructures). The Internet of Things differs from the traditional Internet in many aspects that would influence the development of prototyping tools, from the characteristics and dimensions of the used hardware, growing size and diversity of "end nodes" of the smart objects' networks, standardisation issues for object identification and services it covers. The prototyping approach could give solution to at least some of these issues. Development of plug-and-play software and hardware prototyping platforms based on high level object abstraction could solve the challenges related to heterogeneity and complexity of Internet of Things network nodes, as well as the diversity of modes of communication.

From the business point of view, applying the described approach would lead to decreasing the product development costs and achieving greater customer satisfaction. Appropriate user programmability could transform a system, multiplying the effectiveness of programmers and users. Concepts like volunteering and OS have been used successfully for years. These observations could indicate that toolkit end-user participation is the right direction to go towards the future of the Internet of Things.

4.7 Conclusion

In this chapter we tried to answer the question: how to proceed in order to achieve the vision known as Internet of Things and identified toolkits as a major component in realising the vision. We have described how the Internet grew to its current form and then into the Internet of Things and we argued about the major problems and challenges faced upon its future growth. Then we gave a brief overview of the theoretical background regarding EUP and PD methodologies. In section 4.5 we illustrated early application examples of the previously described approach and finally we discussed specific issues that should be reconsidered when trying to apply PD in the Internet of Things domain.

Based on the given discussion, our general conclusion is that creating prototyping tools and techniques may be the right approach for achieving faster growth of the Internet of Things network and corresponding applications. When realising the described PD techniques, the general ideas of individualisation, in terms of providing users with different types of toolkits, and the widely adopted open software/hardware concepts should be considered. Standardisation of the Internet of Things network itself, with its high diversity and size, should be the initial step in this direction.

References

Abowd GD (1999) Classroom 2000: An experiment with the instrumentation of a living educational environment. IBM Sys J 38:508-530

Abras C, Maloney-Krichmar D, Preece J (2004 (in press)) User-Centered Design. In: Bainbridge W (ed) Encyclopedia of Human-Computer Interaction. Thousand Oaks: Sage Publications

Arduino. http://www.arduino.cc/. Accessed 14 June 2010

Ballagas R, Ringel M, Stone M, Borchers J (2003) iStuff: a physical user interface toolkit for ubiquitous computing environments. CHI: ACM Conference on Human Factors in Computing Systems, CHI Letters 5

Ballagas R, Szybalski A, Fox A (2004) Patch Panel: Enabling control-flow interoperability in ubicomp environments. Proceedings of PerCom Second IEEE International Conference on Pervasive Computing and Communications

Ballagas R, Memon F, Reiners R, Borchers J (2007) iStuff Mobile: Rapidly prototyping new mobile phone interfaces for ubiquitous computing. SIGCHI Conference on Human Factors in Computing Systems

Barragán H (2004) Wiring: Prototyping physical interaction design. Thesis, Interaction Design Institute, Ivrea, Italy

Beck E (2002) P for Political - Participation is not enough. SJIS 14

Bjerknes G, Bratteteig T (1995) User participation and democracy: A discussion of Scandinavian research on system development. Scand J Inf Syst 7:73–98

Bodker S, Ehn P, Kammersgaard J, Kyng M, Sundblad Y (1987) A Utopian experience. In: Bjerknes G, Ehn P,Kyng M (eds) Computers and democracy: A Scandinavian challenge. Aldershot, UK: Avebury

Brun-Cottan F, Wall P (1995) Using video to re-present the user. Commun ACM 38:61–71

Buechley L, Eisenberg M (2008) The LilyPad Arduino: Toward wearable engineering for everyone. IEEE Pervasive Comput 7:12-15

Buechley L, Eisenberg M, Catchen J, Crockett A (2008) The LilyPad Arduino: Using computational textiles to investigate engagement, aesthetics, and diversity in computer science education. Proceedings of the SIGCHI conference on Human factors in computing systems

Crowdsourcing. http://crowdsourcing.typepad.com/. Accessed 14 June 2010

Dibona C, Ockman S, Stone M (1999) Open Sources: Voices from the open source revolution. O'Reilly, Sebastopol, California

d.tools: Enabling rapid prototyping for physical interaction design. http://hci.stanford.edu/research/dtools/. Accessed 14 June 2010

Ehn P (1988) Work-oriented design of computer artifacts. Falköping: Arbetslivscentrum/Almqvist & Wiksell International, Hillsdale, NJ: Lawrence Erlbaum Associates

Ehn P (1989) Work-oriented design of computer artifacts, 2nd edn. Erlbaum

Ehn P, Kyng M (1987) The collective resource approach to systems design. In: Bjerknes G, Ehn P, Kyng M (eds), Computers and Democracy - A Scandinavian Challenge. Aldershot, UK: Avebury

End-User Programming. http://www.cs.uml.edu/~hgoodell/EndUser. Accessed 14 June 2010

Eriksson M, Niitamo VP, Kulkki S (2005) State-of-the-art in utilizing Living Labs approach to user-centric ICT innovation - a European approach, Centre of Distance Spanning Technology at Luleå University of Technology, Sweden, Nokia Oy, Centre for Knowledge and Innovation Research at Helsinki School of Economics, Finland

Erwig M (2009) Software engineering for spreadsheets. IEEE Softw Arch 26:25-30

European commission (2009) Internet of Things - An action plan for Europe. http://ec.europa.eu/information_society/policy/rfid/documents/commiot2009.pdf. Accessed 14 June 2010

Fitchett C, Greenberg S (2001) The Phidget architecture: Rapid development of physical user interfaces. UbiTools'01, Workshop on Application Models and Programming Tools for Ubiquitous Computing. UBICOMP

Folstad A (2008) Living Labs for Innovation and Development of Information and Communication Technology: A Literature Review. Electron J Virtual Organ Netw 10:99-131

Franke N, Piller F (2004) Value creation by toolkits for user innovation and design: The case of the watch market. J Prod Innov Manag 21:401-415

FreeRTOS-A Free RTOS for ARM7, ARM9, Cortex-M3, MSP430, MicroBlaze, AVR, x86, PIC32, PIC24, dsPIC, H8S, HCS12 and 8051. http://www.freertos.org/. Accessed 14 June 2010

Friedemann M, Flörkemeier C (2009) Vom Internet der Computer zum Internet der Dinge. Inform-Spektrum 33:107-121

Gibb AM (2010) New media art, design, and the Arduino microcontroller: A malleable tool. Thesis, School of Art and Design, Pratt Institute

Gindling J, Ioannidou A, Loh J, Lokkebo O, Repenning A (1995) LEGOsheets: A rule-based programming, simulation and manipulation environment for the LEGO programming brick. Proceedings of IEEE Symposium on Visual Languages

GNU manifesto. http://www.gnu.org/gnu/manifesto.html. Accessed 14 June 2010

Godfrey MW, Tu Q (2000) Evolution in open source software: A case study. Proceedings of the International Conference on Software Maintenance, ICSM 2000

Greenberg S, Fitchett C (2001) Phidgets: Incorporating physical devices into the interface. Proc. UIST 2001

Gronbaek K, Kyng M, Mogensen P (1997) Toward a cooperative experimental system development approach. In: Kyng M, Mathiassen L (eds) Computers and design in context. Cambridge, MA: MIT Press

Hars A, Ou S (2002) Working for free? Motivations for participating in Open-Source projects. Int J Electron Commer 6:25–39

Hartmann B, Klemmer SR, Bernstein M, Abdulla L, Burr B, Robinson-Mosher A, Gee J (2006) Reflective physical prototyping through integrated design, test, and analysis. Proc. UIST 2006

Hoc JM, Nguyen-Xuan A (1990) Language semantics, mental models and analogy. In: Hoc JM, Green TRG, Samurçay R, Gilmore DJ (eds) Psychology of Programming Psychology of Programming. Academic Press, London

Howe J (2006) The Rise of Crowdsourcing, Wired, Issue 14.06. http://www.wired.com/wired/archive/14.06/crowds.html. Accessed 14 June 2010

I-CubeX Online Store - Resources: About I-CubeX. http://infusionsystems.com/catalog/info_pages.php?pages_id=117. Accessed 14 June 2010

I-CubeX Online Store - Demos. http://infusionsystems.com/catalog/info_pages.php/pages_id/137. Accessed 14 June 2010

International Telecommunication Union (2005) ITU Internet Reports 2005: The Internet of Things. http://www.itu.int/osg/spu/publications/internetofthings/. Accessed 27 September 2010

ISO (1999) ISO 13407: Human centered design processes for interactive systems. http://www.iso.org/iso/catalogue_detail.htm?csnumber=21197. Accessed 27 September 2010

Jeppesen LB (2005) User toolkits for innovation: Consumers support each other. J Prod Innov Manag 22:347-362

Kensing F, Blomberg J (1998) Participatory design: issues and concerns. Comp Support Coop Work 7:167-185

Kramer J, Noronha S, Vergo J (2000) A user-centered design approach to personalization. ACM Computing Surveys, 43: 44-48

Lakhani K, von Hippel E (2003) How open source software works: Free user-to-user assistance. Res Policy 32:923-943

LEGO.com MINDSTORMS. http://mindstorms.lego.com/. Accessed 14 June 2010

Making Things. http://www.makingthings.com/. Accessed 14 June 2010

Making Things - Teleo. http://www.makingthings.com/teleo/. Accessed 14 June 2010

Markopoulos P, Rauterberg GWM (2000) LivingLab: A white paper, IPO Annual Progress Report 35

Marquardt N, Greenberg S (2007) Distributed physical interfaces with shared phidgets. Proc. of TEI'2007

Mau B (2004) Massive Change. Phaidon Press Ltd., London

Mockus A, Fielding R, Herbsleb J (2002) Two case studies of open source software development: Apache and Mozilla. ACM Trans Softw Eng Methodol 11:1–38

Mogensen P (1992) Towards a prototyping approach in systems development. Scand J Inf Syst 4:31–53

Mulder A (1995) The I-Cube system: Moving towards sensor technology for artists. Proc, the 6th International Symposium on Electronic Art

Myers BA, Ko AJ (2009) The past, present and future of programming in HCI. Human-Computer Interaction Consortium

Norman DA, Draper SW (1986) User-centered system design: New perspectives on human-computer interaction. Lawrence Earlbaum Associates, Hillsdale, NJ, Editors

Pachube: connecting environments, patching the planet. http://www.pachube.com/. Accessed 14 June 2010

Park CW, Jun SY, MacInnis DJ (2000) Choosing what I want versus rejecting what I do not want: An application of decision framing to product option choice decisions. J Mark Res 37:187–202

Phidgets, Inc. http://www.phidgets.com/. Accessed 14 June 2010

Processing.org. http://www.processing.org/. Accessed 14 June 2010

Prügl R, Schreier M (2006) Learning from leading-edge customers at The Sims: opening up the innovation process using toolkits. R&D Manag 36:237-250

Raymond ES (1999) The cathedral and the bazaar. http://www.tuxedo.org/~esr/writings/cathedralbazaar/. Accessed 14 June 2010

Salus PH (1995) Casting the Net: from ARPANET to Internet and beyond. Addison-Wesley

Sarma SE (2001) Towards the five-cent tag. Technical Report MIT-AUTOID-WH-006, Auto-ID Labs, 2001

Scaffidi C, Shaw M, Myers B (2005) Estimating the numbers of end users and end user programmers. IEEE Symposium on Visual Languages and Human-Centric Computing

SourceForge.net. http://sourceforge.net/. Accessed 27 September 2010

Takaragi K, Usami M, Imura R, Itsuki R, Satoh T (2001) An ultra small individual recognition security chip. IEEE Micro 21:43–49

Thomke S, von Hippel E (2002). Customers as innovators: A new way to create value. Harv Bus Rev 80:74–81

Trigg RH, Bodker S, Gronbaek K (1991) Open-ended interaction in cooperative prototyping: A video-based analysis. Scand J Inf Syst 3:63– 86
Trompette P, Chanal V, Pelissier C (2008) Crowdsourcing as a way to access external knowledge for innovation. 24th EGOS Colloquium, Amsterdam: France
von Hippel E (1994) Sticky information and the locus of problem solving: Implications for Innovation. Manag Sci 40:429–439
von Hippel E (2001) Perspective: User toolkits for innovation. Prod Innov Manag 18:247-257
von Hippel E (2002) Open source projects as user innovation networks. MIT Sloan School of Management Working Paper 4366-02
von Hippel E, Katz R (2002) Shifting innovation to users via toolkits. Manag Sci 48:821-833
Wiring. http://wiring.org.co/. Accessed 14 June 2010
Wiring - Exhibition Archives. http://wiring.org.co/exhibition/. Accessed 14 June 2010

5 From the Internet of Things to the Web of Things: Resource-oriented Architecture and Best Practices

Dominique Guinard[1,2], Vlad Trifa[1,2], Friedemann Mattern[1], Erik Wilde[3]

[1]Institute for Pervasive Computing, ETH Zurich

[2]SAP Research, Zurich

[3]School of Information, UC Berkeley

Abstract Creating networks of “smart things” found in the physical world (e.g., with RFID, wireless sensor and actuator networks, embedded devices) on a large scale has become the goal of a variety of recent research activities. Rather than exposing real-world data and functionality through vertical system designs, we propose to make them an integral part of the Web. As a result, smart things become easier to build upon. In such an architecture, popular Web technologies (e.g., HTML, JavaScript, Ajax, PHP, Ruby) can be used to build applications involving smart things, and users can leverage well-known Web mechanisms (e.g., browsing, searching, bookmarking, caching, linking) to interact with and share these devices. In this chapter, we describe the *Web of Things (WoT)* architecture and best practices based on the RESTful principles that have already contributed to the popular success, scalability, and evolvability of the Web. We discuss several prototypes using these principles, which connect environmental sensor nodes, energy monitoring systems, and RFID-tagged objects to the Web. We also show how Web-enabled smart things can be used in lightweight ad-hoc applications, called “physical Mashups”, and discuss some of the remaining challenges towards the global World Wide Web of Things.

5.1 From the Internet of Things to the Web of Things

As more and more devices are getting connected to the Internet, the next logical step is to use the World Wide Web and its associated technologies as a platform for smart things (i.e., sensor and actuator networks, embedded devices, electronic appliances and digitally enhanced everyday objects). Several years ago, in the Cool Town project, Kindberg et al. (Kindberg et al. 2002) proposed to link physical objects with Web pages containing information and associated services. Using infrared interfaces or bar codes on objects, users could retrieve the URI of the associated page simply by interacting with the object. Another way to use the Web for real-world objects is to incorporate smart things into a standardised Web service architecture (using standards, such as SOAP, WSDL, UDDI) (Guinard et al. 2010d). In practice, this would often be too heavy and complex for simple objects.

Instead of these heavyweight Web services (SOAP/WSDL, etc.), often referred to as WS-* technologies, recent "Web of Things" projects (Wilde 2007; Guinard et al. 2010c; Luckenbach et al. 2005; Stirbu 2008) have explored simple embedded Hypertext Transfer Protocol (HTTP) servers and Web 2.0 technology. In fact, recent embedded Web servers with advanced features (such as concurrent connections or server push for event notifications), can be implemented with only 8 KB of memory and no operating system support, thanks to efficient cross-layer TCP/HTTP optimisations, and can therefore run on tiny embedded systems, such as smart cards (Duquennoy et al. 2009). Since embedded Web servers in an Internet of Things generally have fewer resources than Web clients, such as browsers or mobile phones, Asynchronous JavaScript and XML (Ajax) has proven to be a good way of transferring some of the server workload to the client.

So far, projects and initiatives, subsumed here under the umbrella term "Internet of Things", have focused mainly on establishing connectivity in a variety of challenging and constrained networking environments. A promising next step is to build scalable interaction models on top of this basic network connectivity and thus focus on the application layer. In the Web of Things concept, smart things and their services are fully integrated in the Web by reusing and adapting technologies and patterns commonly used for traditional Web content. More precisely, tiny Web servers are embedded into smart things and the REST architectural style (Richardson and Ruby 2007; Fielding 2000) is applied to resources in the physical world (Guinard et al. 2010c; Luckenbach et al. 2005; Duquennoy et al. 2009; Hui and Culler 2008). The essence of REST is to focus on creating loosely coupled services on the Web, so that they can be easily reused. REST is the architectural style of the Web (implemented by URIs, HTTP, and standardised media types, such as HTML and Extensible Markup Language (XML) and uses URIs for identifying resources on the Web. It abstracts services in a uniform interface (HTTP's methods) from their application-specific semantics and provides mechanisms for clients to select the best possible representations for interactions. This makes it an ideal candidate to build a "universal" architecture and Application Programming Interface (API) for smart things. As we will explain in this chapter, the services that smart things expose on the Web usually take the form of a structured XML document or a JavaScript Object Notation (JSON) object, which are directly machine-readable. These formats can be understood not only by machines, but are also reasonably accessible to people; provided meaningful markup elements and variable names are used and documentation is made available. They can also be supplemented with semantic information using microformats, so that smart things can not only communicate on the Web, but also provide a user-friendly representation of themselves. This makes it possible to interact with them via Web browsers and thus explore the world of smart things with its many relationships (via links to other related things). Dynamically generated real-world data on smart objects can be displayed on such "representative" Web pages, and then processed with Web 2.0 tools. For example, things can be indexed like Web pages via their representations, users can "google" for them, and their URI can be emailed to friends or it can be bookmarked. The physical objects themselves can become active and publish blogs or inform each other using services, such as Twitter.[64] The general idea is that the Web is being used as a decentralised information system for easily exposing new services and applications, made possible, directly or indirectly, by smart things.

The Web-enablement of smart things delivers more flexibility and customisation possibilities for end-users. As an example, tech-savvy end-users, at ease with new technologies, can easily build small applications on top of their appliances. Following the trend of Web 2.0 participatory services, in particular Web Mashups (Zang et al. 2008), users can create applications mixing real-world devices, such as home appliances, with virtual services on the Web. This type of applications is often referred to as physical Mashup (Wilde 2007, Guinard et al. 2010c). As an example, a music system could be connected to Facebook or Twitter in order to post the songs one mostly listens to. On the Web, this type of small, ad-hoc application is usually created through a Mashup editor (e.g., Yahoo Pipes[65]), which is a Web platform that enables tech-saavy users (i.e., proficient users of technology) to visually create simple rules to compose Web sites and data sources. We describe how these principles and tools can also be applied to empower the user to create physical Mashups on top of their things.

In Section 2 and 3 we provide a "cookbook" describing the design steps towards embedding smart things into the Web. We also discuss a number of patterns and illustrate them via real prototypes that we have developed over the past few years. In Section 4, we use three concrete prototypes to exemplify how developers, domain-experts, and tech-savvy users can all benefit from a composable Web of Things. Finally, in Section 5 and 6 we discuss the remaining challenges towards implementing a World Wide Web of Things.

[64] http://www.twitter.com

[65] http://pipes.yahoo.com/pipes/

5.2 Designing RESTful Smart Things

The "Web of Things" can be realised by applying principles of Web architecture, so that real-world objects and embedded devices can blend seamlessly into the Web. Instead of using the Web as a transport infrastructure – as done when using WS-* Web services – we aim at making devices an integral part of the Web and its infrastructure and tools by using HTTP as an application layer protocol.

The main contribution of the "Web of Things" approach is to offer a foundation for the next step beyond basic network connectivity. We hope that the Web of Things can do for real-world resources what the Web did for information resources: basic connectivity was a necessary, but not a sufficient condition for the Internet to grow as spectacularly as it is still growing today; it was the architecture of the Web that allowed data and services to be shared in a way that was unheard of before, and that spurred the decentralised growth of what was made available on the Web.

In this section, we describe the use of REST (Fielding 2000) as a universal interaction architecture, so that interactions with smart things can be built around universally supported methods (Pautasso and Wilde 2009).

In the following, we provide a set of guidelines to Web-enable smart things and illustrate them with concrete examples of implemented prototypes. As case study, we describe how we Web-enabled a wireless sensor network (Sun SPOT[66]). These guidelines are based on the concepts of *Resource Oriented Architecture (ROA)*, described by Richardson and Ruby (Richardson and Ruby 2007). Our main goal is to focus on how these concepts can be applied and adapted in order to apply to smart things.

5.2.1 Modeling Functionality as Linked Resources

The central idea of REST revolves around the notion of a resource as *any component of an application that is worth being uniquely identified and linked to*. On the Web, the identification of resources relies on Uniform Resource Identifiers (URIs), and representations retrieved through resource interactions contain links to other resources, so that applications can follow links through an interconnected web of resources. Clients of RESTful services are supposed to follow these links, just like one browses Web pages, in order to find resources to interact with. This allows clients to "explore" a service simply by browsing it, and in many cases, services will use a variety of link types to establish different relationships between resources.

---

[66] http://www.sunspotworld.com

In the case of the Sun SPOT, each node has a few sensors (light, temperature, accelerometer, etc.), actuators (digital outputs, LEDs, etc.), and a number of internal components (radio, battery). Each of these components is modeled as a resource and assigned a URI. For instance, typing a URI such as

```
http://.../sunspots/spot1/sensors/light
```

in a browser requests a representation of the resource *light* of the resource *sensors* of *spot1*. Resources are primarily structured hierarchically and each resource also provides links back to its parent and forward to its children. As an example, the resource

```
http://.../sunspots/spot1/sensors/
```

provides a list of links to all the sensors offered by *spot1*. This interlinking of resources that is established through both, resource links and hierarchical URI, is not strictly necessary, but well-designed URIs make it easier for developers to "understand" resource relationship and even allow non-link based "ad-hoc interactions", such as "hacking" a URI by removing some structure and still expecting for it to work somehow.[67]

In a nutshell, the first step when Web-enabling a smart thing is to design its *resource network*. Identification of resources and their relationships are the two important aspects of this step.

5.2.2 Representing Resources

Resources are abstract entities and are not bound to any particular representation. Thus, several formats can be used to represent a single resource. However, agreed-upon resource representation formats make it much easier for a decentralised system of clients and servers to interact without the need for individual negotiations. On the Web, media type support in HTTP and the Hypertext Markup Language (HTML) allow peers to cooperate without individual agreements. It further allows clients to navigate amongst the resources using hyperlinks.

For machine-to-machine communication, other media types, such as the XML and the JSON have gained widespread support across services and client plat-

[67] In some browsers this "URI hacking" is even part of the UI, where a "go up" function in the browser simply removes anything behind the last slash character in the current URI and expects that the Web site will serve a useful representation at that guessed URI.

forms. JSON is a lightweight alternative to XML that is widely used in Web 2.0 applications.[68]

In the case of smart things, we suggest support for at least an HTML representation to ensure browsability by humans. Note that since HTML is a rather verbose format, it might not be directly served by the things themselves, but by intermediate proxies, as described in Section 0. For machine-to-machine communications, we suggest using JSON. Since JSON is a more lightweight format compared to XML, we believe that it is better adapted to devices with limited capabilities such as smart things. Furthermore, it can directly be parsed to JavaScript objects. This makes it an ideal candidate for integration into Web Mashups.

In the Sun SPOT example, each resource provides both, an HTML and a JSON representation. As an example, the listing in Figure 5.1a shows the JSON representation of the *temperature* resource of a Sun SPOT and Figure 5.1b shows the same resource represented as an HTML page with links to parents, subresources, and related resources.

```
{"resource":
{"methods":["GET"],
"name":"Temperature",
"children":[],
"content":
[{"description":"Current Temperature",
"name":"Current Ambient Temperature",
"value":"27.75"}]}}
```

Fig. 5.1a JSON Representation of the Temperature Resource of a Sun SPOT

68 http://www.json.org

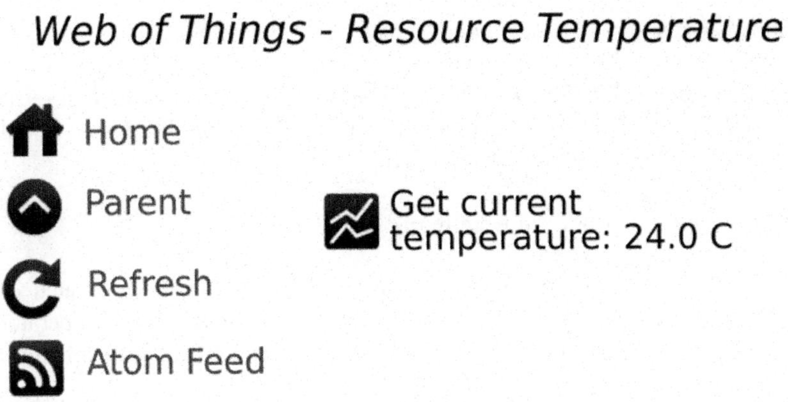

Fig. 5.1b HTML Representation (Rendered by a Browser) of the Temperature Resource of a Sun SPOT Containing Links to Parent and Related Resources

5.2.3 Servicing Through a Uniform Interface

In REST, interacting with resources and retrieving their representations all happens through a uniform interface which specifies a service contract between the clients and servers. The uniform interface is based on the identification (and thus interaction) of resources, and in case of the Web, this interface is defined by the HTTP. We concentrate on three particular parts of this interface: operations, content-negotiation, and status codes.

5.2.3.1 Operations

HTTP provides four main methods to interact with resources, often also referred to as "verbs": *GET*, *PUT*, *POST*, and *DELETE*. *GET* is used to retrieve the representation of a resource. *PUT* is used to update the state of an existing resource or to create a resource by providing its identifier. *POST* creates a new resource without specifying any identifier. *DELETE* is used to remove (or "unbind") a resource.

In the Web of Things, these operations map rather naturally, since smart things usually offer quite simple and atomic operations. As an example, a *GET* on

```
http://.../spot1/sensors/temperature
```

returns the temperature observed by *spot1*, i.e., it retrieves the current representation of the temperature resource. A *PUT* on

http://.../sunspots/spot1/actuators/leds/1

with the updated JSON representation *{"status":"on"}* (which was first retrieved with a GET on */leds/1*) switches on the first LED of the Sun SPOT, i.e., it updates the state of the LED resource. A *POST* on

http://.../spot1/sensors/temperature/rules

with a JSON representation of the rule as *{"threshold":35}* encapsulated in the HTTP body, creates a rule that will notify the caller whenever the temperature is higher than 35 degrees, i.e., it creates a new rule resource without explicitly providing an identifier. Finally, a *DELETE* on

http://.../spot

is used to shutdown the node, or a *DELETE* on

http://.../spot1/sensors/temperature/rules/1

is used to remove rule number 1.

Additionally, another less-known verb is specified in HTTP and implemented by most Web servers: *OPTIONS* can be used to retrieve the operations that are allowed on a resource. In a programmable Web of Things, this feature is quite useful, since it allows applications to find out at runtime what operations are allowed for any URI. As an example, an *OPTIONS* request on

http://.../sunspots/spot1/sensors/tilt

returns *GET, OPTIONS.*

5.2.3.2 Content Negotiation

HTTP also specifies a mechanism for clients and servers to communicate about the requested and provided representations for any given resource; this mechanism is called *content negotiation.* Since content negotiation is built into the uniform interface of HTTP, clients and servers have agreed-upon ways in which they can exchange information about requested and available resource representations, and the negotiation allows clients and servers to choose the best representation for a given scenario.

A typical content-negotiation for the Sun SPOTs looks as follows. The client begins with a *GET* request on

```
http://.../spot1/sensors/temperature/rules
```

It also sets the *Accept* header of the HTTP request to a weighted list of media types it understands, for example to: *application/json;q=1, application/xml;q=0.5*. The server then tries to serve the best possible format it knows about and specifies it in the *Content-Type* of the HTTP response. In our case, the Sun SPOT cannot offer XML and would thus return a JSON representation and set the HTTP header *Content-Type: application/json.*

5.2.3.3 Status Codes

Finally, the status of a response is represented by standardised status codes sent back as part of the header in the HTTP message. There exist several dozens of codes which each have well-known meaning for HTTP clients. In a Web of Things, this is very valuable since it gives us a lightweight but yet powerful way of notifying abnormal requests execution.

As an example, a *POST* request on

```
http://.../sunspots/spot1/sensors/acceleration
```

returns a 405 status code that the client has to interpret as the notification that "the method specified in the request is not allowed for the resource identified by the request URI."

5.2.4 Syndicating Things

Many applications for smart things require syndicating information about objects or collections of objects. With Atom, the Web has a standardised and RESTful model for interacting with collections, and the Atom Publishing Protocol (AtomPub) extends Atom's read-only interactions with methods for write access to collections. Because Atom is RESTful, interactions with Atom feeds can be based on simple GET operations which can then be cached. Atom enables decoupled scenarios by allowing clients to monitor smart things by subscribing to feeds and polling a feed on a remote server, instead of directly polling data from each device.

We implemented this model for the Sun SPOTs, since it fits the interaction model of sensor networks. Thus, the nodes can be controlled (e.g., turning LEDs on, enabling the digital outputs, etc.) using synchronous HTTP calls (client pull) as explained before, but can also be monitored by subscribing to feeds (node push). For example, a subscription to a feed can be done by creating a new "rule" on a sensor resource and *POST*ing a threshold (e.g., > 100).

```
http://.../sunspots/spot1/sensors/light/rules
```

In response, the Sun SPOT returns a URI to an Atom feed. Every time the threshold is reached, the node pushes a JSON message to the Atom server using AtomPub. This allows for thousands of clients to monitor a single sensor by *outsourcing* the processing onto an intermediate, more powerful server.

5.2.5 Things Calling Back: Web Hooks

While Atom allows asynchronous communication between clients and smart things, clients still need to pull the feed server on a regular basis to get data. In addition to being inefficient in terms of communications, this might be problematic for scenarios where the focus is on monitoring. This is often the case with applications communicating with wireless sensor networks.

For those applications, we suggest supporting HTTP callbacks, sometimes called *Web hooks.*[69] *Web hooks* are a mechanism for clients and applications that want to receive notifications from other Web sites using user-defined callbacks over HTTP. Users can specify a callback URI where the application will POST data to once an event occurs. This mechanism has been used by the PayPal service which allows you to specify a URI to be triggered by the service once payment has been accepted.

As an example, let us consider again the case of creating a new rule on a Sun SPOT:

```
http://.../sunspots/spot1/sensors/light/rules
```

Now, alongside with the rule, the client *POST*s a URI on which it will listen for incoming messages. Every time the threshold is reached, the node (or an intermediate) will push a JSON message to the given URI(s).

Using Web hooks is a first step towards bi-directional, real-time interaction with smart things. However, this model has a number of limitations as it requires from clients to have a public URI where data can be posted to, which is rarely the case when clients are behind a firewall. We will discuss further solutions in Section 0.

---

[69] http://www.webhooks.org

5.3 Web-enabling Constrained Devices

Although Web servers are likely to be embedded into more and more devices, we cannot assume that every smart device will directly offer a RESTful interface. In some cases, it makes sense to hide the platform-dependent protocol to access the resources of a particular device, and to expose them as RESTful service provided by a gateway. The actual interactions behind that RESTful service are invisible and often will include specialised protocols for the specific implementation scenario. REST defines the notion of *intermediaries* as a core part of the architectural style, and therefore such a design can easily be achieved by implementing the RESTful service on intermediaries. By using either *proxies* or *reverse proxies*, it is furthermore possible to establish such an intermediary from the client or from the server side, effectively introducing a robust pattern for wrapping non-RESTful services in RESTful abstractions.

In practice, two solutions are possible: Web connectivity directly on the smart things, or indirectly through a proxy. Previous work has shown that serving content using Web servers on resource-constrained devices is feasible (Duquennoy et al. 2009). Also, in the foreseeable future, most embedded platforms will have native support for TCP/IP connectivity (in particular with 6LowPAN (Hui and Culler 2008), therefore, a Web server on most devices is a reasonable assumption. This approach is sometimes desirable, as there is no need to translate HTTP requests from Web clients into the appropriate protocol for the different devices, and thus devices can be directly integrated and make their RESTful APIs directly accessible on the Web, as shown in the right part of Figure 5.2.

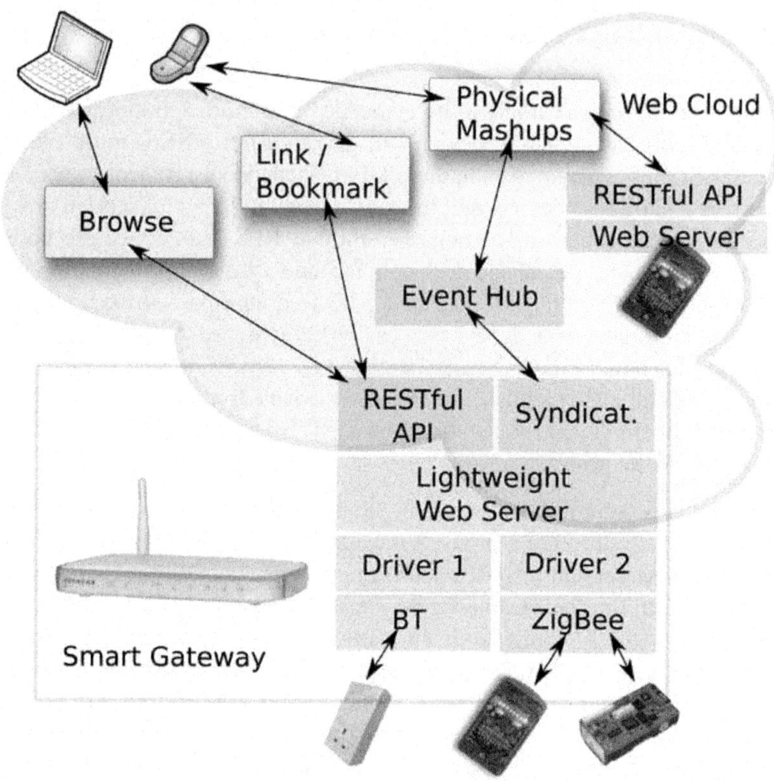

Fig. 5.2 Web and Internet Integration with Smart Gateways and Direct Integration

However, when an on-board HTTP server is not possible or not desirable, Web integration takes place using a reverse proxy that bridges devices that are not directly accessible as Web resources. We call such as proxy a *Smart Gateway* (Trifa et al. 2009) to account for the fact that it is a network component that does more than only data forwarding. A Smart Gateway is a Web server that hides the actual communication between networked devices (e.g., Bluetooth or Zigbee) and the clients through the use of dedicated drivers behind a RESTful service. From the Web clients' perspective, the actual Web-enabling process is fully transparent, as interactions are HTTP in both cases.

As an example, consider a request to a sensor node coming from the Web through the RESTful service. The gateway maps this request to a request into the proprietary API of the node and transmits it using the communication protocol understood by the sensor node. A Smart Gateway can support several types of devices through a driver architecture, as shown in Figure 5.3, where the gateway supports three types of devices and their corresponding communication protocols. Ideally, gateways should have a small memory footprint to be integrated into embedded computers already present in network infrastructures, such as wireless routers, set-top boxes, or Network Attached Storage (NAS) devices.

Aside from connecting limited devices to the Web, a Smart Gateway can also provide more complex functions to devices such as orchestration and composition of several low-level services, offered by various devices into higher-level services available through the RESTful service. For example, if an embedded device measures the energy consumption of appliances, the Smart Gateway could provide a service that returns the total energy consumption as a sum of the data collected by all the devices connected to the gateway. Additionally, a gateway could take care of notifying all the URI call-backs (or Web hooks) whenever a given condition is met.

Example: A Smart Gateway for Smart Meters
A prototype for a smart meter infrastructure illustrates the application of the WoT architecture and the concept of Smart Gateways for monitoring and controlling the energy consumption of households. We used intelligent power sockets, called *Plogg*[70], which can measure the electricity consumption of the appliance plugged into them. Each Plogg is also a wireless sensor node that communicates over Bluetooth or Zigbee. However, the integration interface offered by the Ploggs is proprietary, which makes the development of applications using Ploggs rather tedious, and does not allow for easy Web integration.

[70] http://www.plogginternational.com

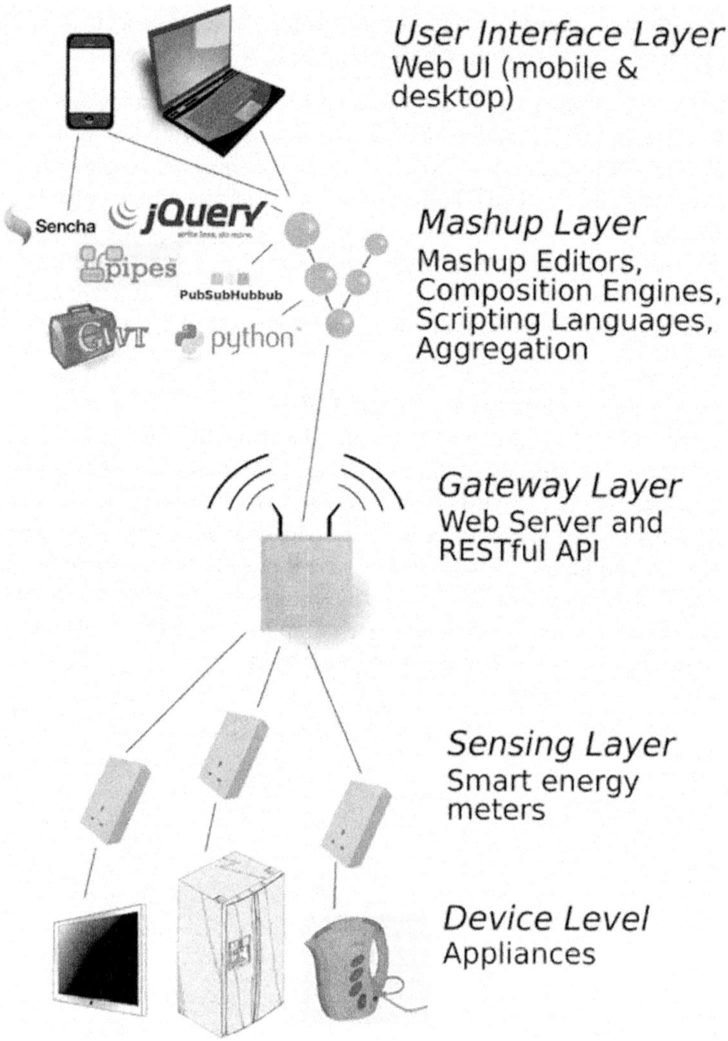

Fig. 5.3 Appliances Attached to Ploggs Power Outlets Which Communicate with a Smart Gateway Offering the Ploggs' Functionalities as RESTful Web Services

The Web-oriented architecture we have implemented using the Ploggs is based on five main layers as shown in Figure 5.3. The Device Layer is composed of appliances we want to monitor and control through the system. In the Sensing Layer, each of these appliances is then plugged into a Plogg sensor node. In the Gateway Layer, the Ploggs are discovered and managed by a Smart Gateway as described before. In the Mashup layer the Ploggs' services are composed together to create an energy monitoring and control application, using Web scripting languages or composition tools. Finally, this application is made available through a Web User

Interface in a Web browser (e.g., on a mobile phone, a desktop computer, a tablet PC, etc.)

The Smart Gateway in this example is a C++ application running on an embedded machine, whose role is to automatically find all the Ploggs in the environment and make them available as Web resources. The gateway first periodically looks for the Ploggs in the area by scanning the environment for Bluetooth devices. The next step is to expose them as RESTful resources. A small footprint Web server (Mongoose[71]) is used to enable access to the Ploggs' functionalities over the Web, simply by mapping URIs to the various requests of the native Plogg Bluetooth API.

In addition to discovering the Ploggs and mapping their functionalities to URIs, the Smart Gateway has two other important features. First, it offers *local* aggregates of device-level services. For example, the gateway offers a service that returns the combined electricity consumption of all the Ploggs found at any given time. The second feature is that the gateway can represent resources in various formats. By default an HTML page with links to the resources, is returned, this ensures browsability. Using this representation the user can literally "browse" with any Web client the structure of smart meters to identify the one he or she wants to use and directly test the Ploggs by clicking on links (e.g., for the HTTP *GET* method) or filling forms (e.g., for the *POST* method). Alternatively, the Smart Gateway can also represent results of resources like JSON, to ease the integration with other Web applications.

To illustrate the concept from a client point of view, let us briefly describe an example of interaction between a client application (e.g., written in Ajax) and the Ploggs' RESTful Smart Gateway. First, the client contacts the root URI of the application

```
http://.../EnergieVisible/SmartMeters/
```

with the *GET* method. The server responds with the list of all the smart meters connected to the gateway.

Afterwards, the client selects from that list the device it wants to interact with identified by a URI

```
http://.../EnergieVisible/SmartMeters/RoomLamp
```

alongside with the format it wants to get back (using HTTP content negotiation, see Section 5.2.3). By issuing a *GET* request on this resource with the *Accept* header set to *application/json;q=1,* it gets back a JSON representation as shown in Figure 5.4 below. In the response message of this listing, the client finds energy

[71] http://code.google.com/p/mongoose

consumption data (e.g., current consumption, global consumption, etc.) as well as hyperlinks to related resources. Using these links, the client can discover other related “services”.

```
GET /EnergieVisible/SmartMeters/RoomLamp
[...] HTTP/1.x 200 OK
Content-Type: application/json
{
  "deviceName":  "RoomLamp",
  "currentWatts":  60.52,
  "KWh":  40.3,
  "maxWattage":  80.56
  "links":
  [{"aggregate":  "../all"},
  {"load": "../load"},
  {"status":  "/status"}]
}, {...}]
```

Fig. 5.4 JSON Representation of a Plogg connected to a Lamp

As an example, by contacting

http://.../RoomLamp/status

with the standard OPTIONS method, the client gets back a list of the methods allowed on the status resource (e.g., *Allow: GET, HEAD, POST, PUT*). By sending the *PUT* method to this URI alongside with the representation (e.g., JSON) {“*statu”:"off”*}, the appliance plugged into the Plogg is turned off.

The Web-enabling of the Ploggs through a Smart Gateway allows building fully Web-based energy monitoring applications. It also enables simple interactions, such as bookmarking connected appliances, and control or monitor them from any device (e.g., a mobile phone, an embedded computer, a wireless sensor node, etc.), offering a standard Web browser or understanding the HTTP protocol.

5.4 Physical Mashups: Recomposing the Physical World

In this section, we illustrate how the Web of Things concepts and architecture facilitates the creation of Mashups in the physical world. A Web Mashup is an application that takes several Web resources and uses them to create a new application. Unlike traditional forms of integration, Mashups focus mainly on opportunistic integration occurring on the Web for an end-user’s personal use and generally for non-critical applications (Yu et al. 2008). They are usually created ad-hoc, using lightweight and well-known Web technologies, such as JavaScript and HTML, and contribute to serving short terms needs. As an example, a Mashup can be created to display, on Google Maps, the location of all the pictures posted to Flickr.[72]

By extending the Mashup concept to physical objects and applying RESTful patterns to smart things, we allow their seamless integration into the Web, thus enabling a new range of applications based on this unified view of a Web of information resources and physical objects. We call this concept "physical Mashup", because it is directly inspired from Web 2.0 Mashups.

In this section, we present three Mashups representing three different use cases. In the first prototype, we create an energy monitoring and control system based on the Ploggs Smart Gateway. In the second, we show how domain experts (e.g., product managers, marketing executives, etc.) can leverage such tools to build a business intelligence platform suited to their business needs. In the last example, we show how end-users could use a visual physical Mashup editor to dynamically "configure" their home appliances.

5.4.1 *Energy Aware Mashup: "Energie Visible"*

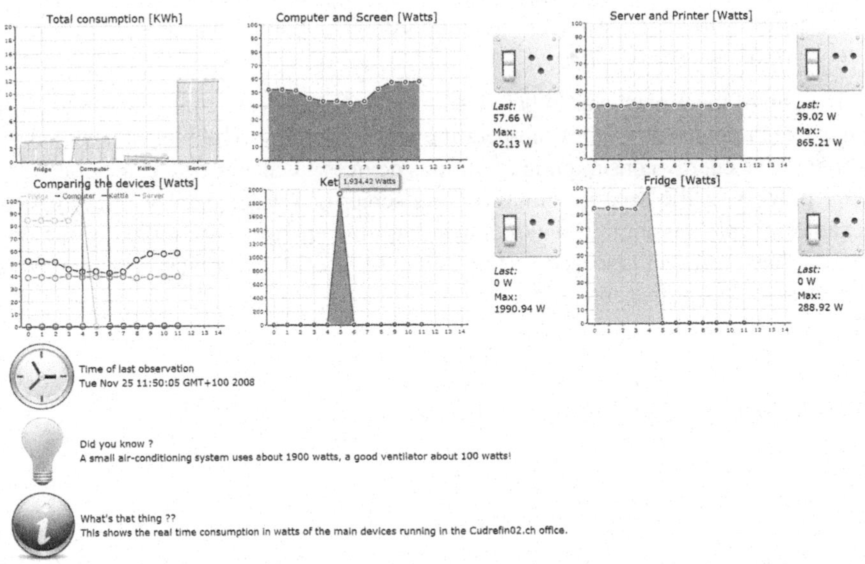

Fig. 5.5 The Web-Based User Interface for Monitoring and Controlling the Ploggs

---

[72] http://www.flickr.com

In this first example, we create a Mashup to help households to understand their energy consumption and to be able to remotely monitor and control it.

The idea of the "Energie Visible"[73] project is to offer a Web dashboard that enables people to visualise and control the energy consumption of their household appliances. The dashboard is shown in Figure 5.5 and provides six real-time interactive graphs. The four graphs on the right side provide detailed information about the current electricity consumption of all the detected Ploggs.

Thanks to the Ploggs Smart Gateway described before, the dashboard can be implemented using any Web scripting language or tool (PHP, Ruby, Python, JavaScript, etc.). The Energie visible application was built using Google Web Toolkit (GWT)[74], which is a platform for developing JavaScript Web applications in Java, and provides a large number of easily customisable widgets. To display the current energy consumption in real time, the application simply sends HTTP *GET* requests to the gateway

```
http://.../EnergieVisible/SmartMeters/all.json
```

on a regular basis or subscribes to this resource using Web hooks. The resulting feed entry is then dispatched to the corresponding graphs widgets, which can directly parse JSON, and extract the relevant data in it to be displayed.

The "Energie Visible" prototype was deployed at the headquarters of a private foundation working on sustainability (cudrefin02[75]) and has now been running reliably since November 2008.

The aim of the project was to help visitors and members to better understand how much each device consumes in operation and in standby. The Ploggs are used to monitor the energy consumption of various devices, such as a fridge, a kettle, several printers, a file server, computers and screens. A large display in the office enables people passing by to experiment with the energy consumption of the devices. The staff can also access the user interface of any Plogg with the Web browser of their office computer.

5.4.2 Business Intelligence Mashup: RESTful EPCIS

[73] The project is available on http://www.webofthings.com/energievisible

[74] http://code.google.com/webtoolkit/

[75] http://cudrefin02.ch

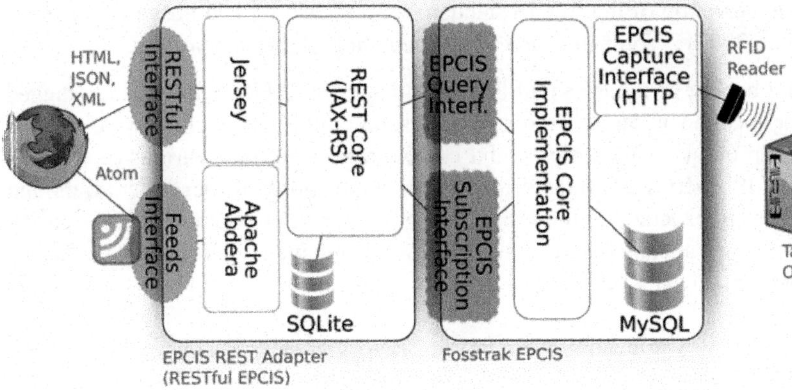

Fig. 5.6 Architecture of the RESTful EPCIS Based on the Jersey RESTful Framework and Deployed on Top of the Fosstrak EPCIS

The Electronic Product Code (EPC) Network (Floerkemeier et al. 2007) is a set of standards established by industrial key players towards a uniform platform for tracking and discovering RFID-tagged objects and goods in supply chains. This network offers a standardised server-side EPC Information Service (EPCIS) for managing and offering access to track and trace RFID events. Implementations of EPCIS provide a standard query and capture API through WS-* Web Services.

In order to integrate not only embedded devices, but also RFID-tagged everyday items into the Web of Things, we use the concepts presented to turn the EPCIS into a "Smart Gateway". This helps to better grasp the benefits of a seamless Web integration based on REST, as opposed to using HTTP as a transport protocol only (as WS-* Web Services use it).

The EPCIS offers three core features. First, it offers an interface to query for RFID events. The WS-* interface, however, does not allow to directly query for RFID events using Web languages, such as JavaScript or HTML. More importantly, it does not allow to explore the EPCIS using a Web browser, or to search for tagged objects or exchange links pointing to traces of tagged objects. To remedy the problem, we implemented a RESTful translation of the EPCIS WS-* interface.

As shown in Figure 5.6, the RESTful EPCIS (Guinard et al. 2010b) is a software module based on Jersey[76], a software framework for building RESTful applications. Clients of the RESTful EPCIS, such as browsers or Web applications, can query for tagged objects directly using REST and its uniform HTTP interface. Requests are then translated by the framework into WS-* calls on the standard EPCIS interface. This allows for the RESTful EPCIS to serve data provided by

[76] http://jersey.dev.java.net

any implementation of the EPCIS standard. In our case we use Fosstrak (Floerkemeier et al. 2007)[77], an open source implementation of the standard.

The first benefit of the RESTful EPCIS is that every RFID event, reader, tagged object or location is turned into a Web resource and gets a globally resolvable URI, which uniquely identifies it and can be used to retrieve various representations. EPCIS queries are transformed into compositions of these identifiers and can be directly executed in the browser, sent by email, or bookmarked. As an example, a factory manager who wants to know what tagged objects enter his factory can bookmark a URI, such as:

```
http://.../epcis/rest/location/urn:company:factory1/reader/urn:company:entrance:1
```

Furthermore, these URIs are linked together through their representations in order to reflect the relationships of the physical world. This makes the RESTful EPCIS directly browsable. Indeed, in addition to the XML representation of tagged objects offered by the standard, it also provides HTML, JSON and Atom representations. With the HTML representation, end-users can literally browse tagged things and their traces simply by following hyperlinks in the very same way as they browse the Web of documents. For example, a location offers links to co-located RFID readers.

With the Atom representation, end-users can formulate queries by browsing the hyperlinked EPCIS and obtain the updated results represented as Atom feeds, which browsers can understand and directly subscribe, too. As an example, a product manager can create a feed in order to be automatically updated in his browser whenever one of his products is ready to be shipped. He can then use the URI of the feed to send it to his most important customers so that they could track the goods' progress as well. This is a simple but very useful use case, which would require a dedicated client to be developed and installed by each customer in the case of the WS-* based EPCIS.

5.4.3 A Mashup Editor for the Smart Home

Tech-savvy users can create Web Mashups using "Mashup editors", such as Yahoo Pipes. These editors usually provide visual components representing Web sites and operations (add, filter, etc.) that users only needs to connect (or *pipe*) together to create new applications. We wanted to apply the same principles to allow users to create physical Mashups without requiring any programming skills.

[77] http://www.fosstrak.org

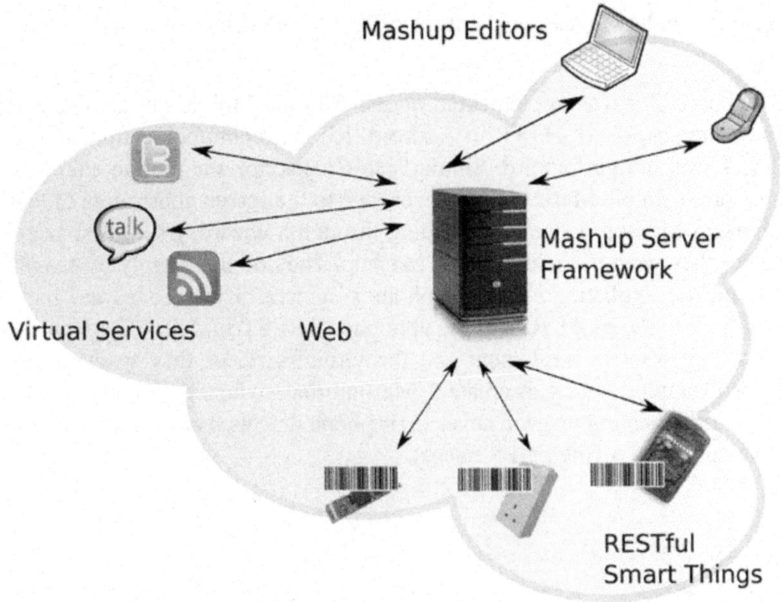

Fig. 5.7 The Physical Mashup Framework

We briefly introduce our physical Mashup architecture and two Mashup editors built on top of it. As shown in Figure 5.7, the system is composed of four main parts. We first have RESTful, Web-enabled, smart things and appliances. In our prototype, we tag them with small 2D barcodes in order to ease their identification with mobile phones. We then have "virtual" services on the Web, such as Twitter, Google Visualisation API, Google Talk, etc. In the middle, the Mashup server framework allows to compose services of different smart appliances as well as virtual services on the Web. It is in charge of executing the workflows created by end-users in their Mashup applications. It discovers, listens, and interacts with the devices over their RESTful API. The last components are the Mashup editors themselves, which allow users to create Mashup applications very easily.

We implemented two Mashup editors using this architecture. The first one is based on the Clickscript project.[78] A Firefox plugin written on top of an Ajax library allows people to visually create Web Mashups by connecting building blocks of resources (Web sites) and operations (greater than, if/then, loops, etc.). Since it is written in JavaScript, Clickscript cannot use resources based on proprietary service protocols. However, it can easily access RESTful services, such as those provided by Web-enabled smart appliances. This makes it straightforward to create Clickscript building blocks that represent smart appliances. The Mashup shown in Figure 5.8 gets the room temperature by *GET*ting the temperature re-

[78] http://www.clickscript.ch

source. If it is below 36 degrees, it turns off the Web-enabled air-conditioning system.

The second editor was implemented on the Android Mobile Phone. Once again, thanks to the support of HTTP in Android, RESTful communication with smart appliances was straightforward. Similarly to Clickscript, the mobile editor allows the creation of simple Mashups. However, due to the screen constraints of the mobile phone, a Mashup is created by going through a wizard. Users first select the appliances they want to include in the Mashup. They do this simply by scanning a barcode on the appliance using the phone's camera. These codes are basically pointing back to the root URLs of the appliance's RESTful APIs. They then set up the rules they want to implement and the virtual services they want to interact with. For example, users can create a Mashup that switches on their appliances, e.g, turning the heating up, whenever their phone detects that they are moving towards home (based on their GPS traces).

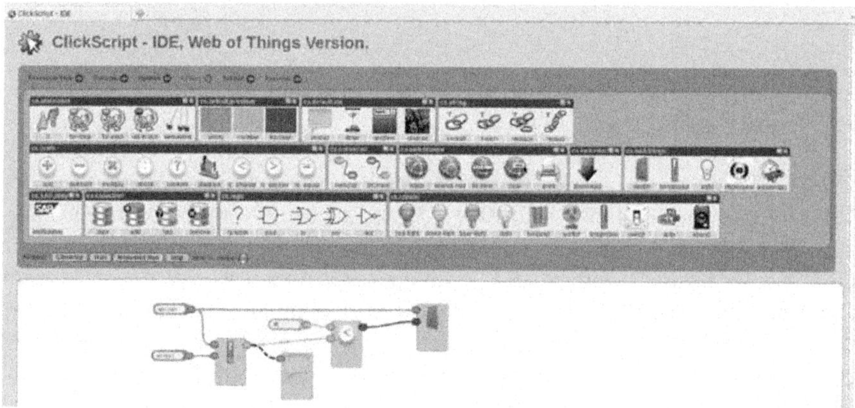

Fig. 5.8 Using the Clickscript Mashup Editor to Create a Physical Mashup by Connecting Building Blocks Directly to a Browser

5.5 Advanced Concepts: The Future Web of Things

So far, we have shown how Web standards and design principles can be leveraged for smart things. While this seems to be a rather adequate architecture for the Web of Things, many open challenges remain. In this section, we explore three such challenges, and sketch potential solutions for each. We begin by discussing the needs for *real-time* data of many smart things applications. Then, we address the challenges of *finding* and *understanding* services available in a global Web of Things. We finally look at mechanisms for *sharing* smart things.

5.5.1 Real-time Web of Things

HTTP is a stateless client/server protocol where interactions are always initiated by the client, and there is no protocol context bigger than a request/response exchange. This interaction model is well-suited for control-oriented applications where clients read/write data from/to embedded devices. However, this client-initiated interaction models seem inappropriate for bi-directional event-based and streaming systems, where data must be sent asynchronously to the clients as soon as it is produced.

For example, many pervasive scenarios must deal with real-time information to combine stored or streaming data from various sources to detect spatial or temporal patterns, as is the case in many environmental monitoring applications. As such applications are often event-based and embedded devices usually have a low-duty cycle (i.e., sleep most of the time), smart things should also be able to push data to clients (rather than being continuously polled). To support the complex, data-centric queries required for such scenarios, more flexible data models are required to expose sensor data streams over the Web. In this section, we explore the recent developments in the real-time Web to build such a data model that is more suited to the data-centric, stream-based nature of sensor-driven applications.

As mentioned before, using syndication protocols, such as Atom, improves the model when monitoring, since devices can publish data asynchronously using AtomPub on an intermediate server or Smart Gateway. Nevertheless, clients still have to pull data from Atom servers. Web streaming media protocols (RTP/RTSP) have enabled transmission of potentially infinite data objects, such as Internet radio stations. Sensor streams are similar to streaming media in this respect. However, streaming media mainly support *play* and *pause* commands, which are insufficient for sensor streams where more elaborate control commands are needed. The Extensible Messaging and Presence Protocol (XMPP)[79] is an open standard for real-time communication based on exchanges of XML messages, and powers a wide range of applications including instant messaging (Google Talk is based on XMPP). Although widely used and successful, XMPP is a fairly complex standard, which is often too heavy for the limited resources of embedded devices used in sensor networks.

An alternative type of Web applications that attempt to eliminate the limitations of the traditional HTTP polling has become increasingly popular. This model, called *Comet*[80] (also called HTTP streaming or server push), enables a Web server to push data back to the browser without the client requesting it explicitly. Since browsers are not designed with server-sent events in mind, Web application de-

[79] http://www.xmpp.org

[80] http://www.tinyurl.com/tc95h

velopers have tried to work around several specification loopholes to implement Comet-like behavior, each with different benefits and drawbacks. One general idea is that a Web server does not terminate the TCP connection after response data has been served to a client, but leaves the connection open to send further events.

Based on this brief overview, one can observe that the tradeoff between scalability and query expressiveness is also present in the Web world. However, as the recent developments in Web techniques have allowed to build efficient and scalable publish/subscribe systems, we suggest that a Web-based pub/sub model could be used to connect sensor networks with applications. PubSubHubbub (PuSH)[81] is a simple, open pub/sub protocol as an extension to Atom and RSS. Parties (servers) speaking the PuSH protocol can get near-instant notifications (via callbacks) when a feed they are interested in is updated. PuSH can also be used as a general-purpose messaging protocol for devices (Trifa et al. 2010).

The following model can be used to enable Web-based stream processing applications where users can post queries using an HTTP request to one or more sensors. The HTTP request shown in Figure 5.9, collects the light and temperature sensor readings twice per second (the `ds.freq=2` Hz parameter) only if the light sensor value is not over "200" and the temperature reading is less than "19":

```
POST /datastreams/ HTTP/1.1
Content-Type: application/x-www-form-urlencoded

ds.device=purpleSensor
&ds.data=temperature,light
&ds.freq=2
&ds.filter=light <= 200 && temperature < 19
```

Fig. 5.9 HTTP Request Collecting Light and Temperature Sensor Readings

As a result, a specific pub/sub feed will be created on a pub/sub broker as a stream (sequence of messages) in which all the data matching the request will be pushed by the stream processing engine. This allows decoupling the application from the stream processing engine, which can be easily replaced, as long as it supports the same interface to process Web requests and also can push the matching data into the pub/sub broker.

All the data samples corresponding to these queries are then pushed into a feed on the message broker, where users can subscribe using the PuSH protocol. They will then receive the data from the stream pushed from the broker via callbacks.

---

[81] http://code.google.com/p/pubsubhubbub

Although HTTP was not designed for real-time stream delivery, exploratory research in the Web of Things area shows promising results when using Web standards to interact with distributed sensors and actuators (Trifa et al. 2010). The loss in raw performance and latency, due to verbose HTTP requests, is compensated by allowing sensor networks to be exposed in an easily accessible and universal way. Additionally, thanks to the many advantages offered by Web standards, such as transparent proxies, declarative Web-based queries can be mapped to the specialised processing features of sensor networks, therefore, one can still take advantage of the optimisations and advanced processing implemented within sensor networks and other stream processing systems.

While it is clear that a Web of Things needs more developments and standards in the areas that we have described, the developments of recent years and the foreseeable future of HTML5 and its Web Sockets and Server-Sent Events is a sign of developments moving in the right direction for the WoT. However, it is an important task for Internet of Things researchers to identify the shortcomings of the current Web architecture and propose solutions that work well for monitoring the real world and still integrate well with the Web.

5.5.2 Finding and Describing Smart Things

Another major challenge for a global Web of Things is searching and finding relevant devices among billions of smart things that will be connected to the Web. Finding them by browsing HTML pages with hyperlinks is literally impossible in this case, hence the idea of searching for smart things. Searching for things is significantly more complicated than searching for documents, as things are tightly bound to contextual information, such as location, are often moving from one context to another, and have no obvious easily indexable properties, such as human-readable text in the case of documents.

Beyond location, smart things need a mechanism to describe themselves and their services to be (automatically) discovered and used. But what is the best way to describe a thing on the Web so that both, humans and machines, can understand what services it provides? This problem is not inherent to smart things, but more generally a complex problem of describing services, which has always been an important challenge to be tackled in the Web research community, usually in the area of the *Semantic Web.*[82]

To overcome the rather limited descriptive power of resources on the Web, several languages have been proposed, such as RDF[83] or Microformats[84]. Designed

---

[82] http://www.w3.org/standards/semanticweb/

[83] http://www.w3.org/RDF/

for both, human and machines, Microformats provide a simple way to add semantics to Web resources. There is not one single Microformat, but rather a number of them, each one for a particular domain; a "geo" and "adr" microformat for describing places or an "hProduct" and "hReview" microformat for describing products and what people think about them. Each Microformat undergoes a "standardisation" process that ensures its content to be widely understood and used, if accepted.

Microformats are especially interesting in a Web of Things for two reasons; first they are directly embedded into Web pages and thus can be used to semantically annotate the HTML representation of a thing's RESTful API. Secondly, Microformats (as well as RDFa) are increasingly supported by search engines, such as Google and Yahoo, where it is used to enhance the search results. For example, the "Geo" Microformat could be used to localise search results close to you or, in our context, to localise smart things in your direct vicinity.

More concretely, we use a compound of several microformats to describe our smart things. This helps the things to be searched by humans using traditional or dedicated search engines, but it also helps them being "discovered" and understood by software applications in order to automatically use them. As an example, in Figure 5.10 we use 5 microformats to describe a Sun SPOT and embed this semantic information directly in the HTML representation of the SPOT resources.

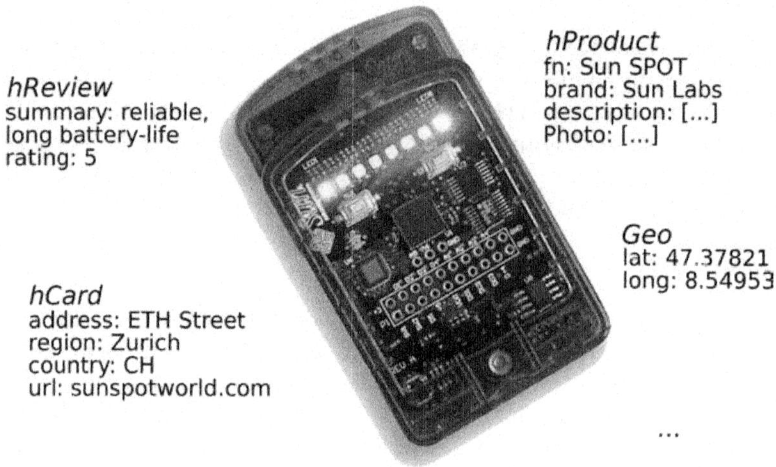

Fig. 5.10 Compound Microformats for Describing a Sun SPOT Using the Geo, hCard, hProduct and hReview Microformats

[84] http://www.microformats.org

The listing shown in Figure 5.11 shows how to define the formal name (fn) of the Sun SPOT as well as an authoritative URL, where more information about the device can be found. We provide this semantic markup in the HTML representation of a Sun SPOT:

```
<span class=”fn”>Sun SPOT</span>
<span class=”URL>
    <a href=”http://sunspotworld.com</a>
</span>
```

Fig. 5.11 Snippet of the HTML Representation of a Sun SPOT Including the hProduct Microformats

While there is still much research to be undertaken to be able to search for and discover smart things, the recent developments of the Web standards are going in the right direction for globally supporting such semantic descriptions. Indeed, a derivative form of the already well supported Microformats, called Microdata,[85] might be part of the HTML 5 standard and might be widely adopted and understood by most next generation Web browsers and other Web clients.

5.5.3 Sharing Smart Things

The success of Web 2.0 Mashups depends on the trend for Web 2.0 service providers (e.g., Google, Twitter, Wordpress, etc.) to provide access to some of their services through relatively simple, often RESTful, open APIs on the Web. Mashup developers often share their Mashups on the Web and expose them through open APIs as well, making the service ecosystem grow with each application and Mashup. Figure 5.12 shows the simplified component architecture of a Social Access Controller (SAC), which serves as authentication proxy between clients and smart things.

To ensure the success of physical Mashups, they need to replicate the same level of openness. However, enabling such an open model for a Web of Things requires a sharing mechanism for physical things supporting access control to the RESTful services provided by devices. For example, one could share the energy consumption sensors in one’s house with the community. However, this is a potentially risky process, given that these devices are part of our everyday life and their public sharing might result in serious privacy implications (if almost no energy has been used recently, the home owners may be on vacation and burglars might look for these kinds of patterns). HTTP already provides authentication mechanisms

[85] http://dev.w3.org/html5/md/

(e.g., HTTP Authentication[86]) based on credentials and server-managed user groups. While this solution is already available for free on most (embedded) Web servers, it still presents a number of drawbacks in the WoT context. First, for a large number of smart things it becomes quite unmanageable to share credentials for each of them. Then, as the shared resources are not advertised anywhere, sharing also requires the use of secondary channels, such as sending emails containing credentials to people. Several platforms, such as SenseWeb (Luo et al. 2008)or Pachube[87] propose to overcome these limitations by providing a central platform for people to share their sensor data. However, these approaches are based on a centralised data repository and are not designed to support decentralisation and direct interaction with smart things.

A promising solution is to leverage existing social structures of social networks (e.g., Facebook, Linkedin, Twitter, etc.) and their (open) APIs to share things. Using social networks enables users to share things with people they know and trust (e.g., relatives, friends, colleagues, fellow researchers, etc.), without the need to recreate yet another social network or user database from scratch on a new online service. Additionally, this enables advertising and sharing through a unique channel: you can use various well-known social networks to inform your friends about the sensors you shared with them by automatically posting messages to their profile or newsfeed.

The SAC platform (Guinard et al. 2010a) is an implementation of this idea. SAC is an authentication proxy between clients (e.g., Web browsers) and smart things. Rather than maintaining its own database or list of trusted connections and credentials – as it would be done with simple HTTP authentication – SAC connects to a number of social networks (e.g., Twitter, Facebook, LinkedIn, etc.) to extract all potential users and groups one could share with.

---

[86] http://www.ietf.org/rfc/rfc2617.txt

[87] http://www.pachube.com

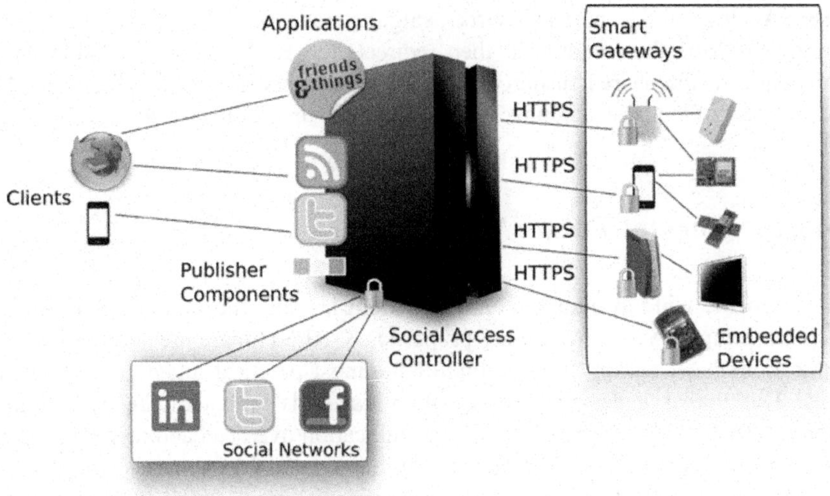

Fig. 5.12 Simplified Component Architecture of the SAC

This is possible as most social networks offer a Web API (e.g., Facebook Connect[88]). Providing an open Web API is one of the success factors of social networks themselves. Indeed, these APIs allow third-party Web applications to be built using partial data extracted from the social networks and thus to enhance the functionality of the social networks.

The sharing process occurs in three phases. First, the smart things owner accesses SAC by logging in, using at least one of his social networks credentials. SAC then uses delegated authentication with the social network to identify the owner. Afterwards, the smart thing to be shared has to be crawled in order to identify the resources and capabilities of its RESTful services, i.e., which functionalities can be shared for that thing. Finally, the user generates the access control list of the smart thing by selecting which friends can interact with what resource.

When an owner shares resources with a trusted connection, the latter is informed about it directly on their social network. In case of Facebook, it publishes a message to the news feed of the friend. In case of Twitter it simply tweets a message to the trusted connection (e.g., "Rachel shared her Ploggs Energy sensors with you"). The posted message also contains a *link* that redirects to the shared resource. The link does not point to the smart thing directly but to an instance of SAC that acts as the authentication proxy, as shown in Figure 5.12 When a trusted connection uses the provided link, SAC will verify its identity. If the friend is logged in successfully with one of their social networks, SAC will internally check whether this person also has access to the requested resource. If it is the

[88] http://developers.facebook.com/connect.php

case, SAC logs on the shared resource using the credentials provided by the owner when registering the resource. It then redirects the HTTP request of the trusted connection to the shared resource. Finally, it redirects the result directly to the HTTP client of the trusted connection, for example to a Web browser.

5.6 Discussing the Future Web of Things

Thanks to the wide availability of HTTP libraries and clients, and to the loose coupling, simplicity, and scalability properties of RESTful architectures, RESTful applications have rapidly become one of the most practical integration architectures. This makes it desirable to use Web standards for interacting with smart things. Although HTTP introduces a communication overhead and increases average response latency, it is still sufficient for many pervasive scenarios where longer delays do not affect user experience (Drytkiewicz et al. 2004; Priyantha et al. 2008). Previous work (Trifa et al. 2009; Yazar and Dunkels 2009) has shown that the performance of using HTTP as a data exchange protocol is largely sufficient for common pervasive scenarios, especially when only a few concurrent users are accessing the same resource simultaneously (200 ms mean response time with 100 concurrent users on a 1.1 GHz server running a Smart Gateway). We have also shown that caching techniques can significantly improve the performance of concurrent sensor data reading by using tools used for massively scalable Web sites (Trifa et al. 2009). These techniques can be directly applied to Web devices, given that devices have on-board HTTP support.

Web 2.0 Mashups have significantly lowered the entry barrier for the development of Web applications, which is now accessible to non-programmers. It should be noted that a resource-oriented approach should not be universally considered as the miracle solution for every problem. In particular, scenarios with very specific requirements, such as high performance real-time communications, might benefit from tightly coupled systems based on different system architectures. However, for less constrained applications, where massive scalability, ad-hoc interaction, and serendipitous re-use are necessary, Web standards allow any device to speak the same language as other services on the Web. This makes the integration of the real-world with any other Web content much easier, so that physical things can be bookmarked, browsed, searched for, and used just like any other Web resource.

Based on our personal experience, the drawbacks of Web architectures are easily offset by a notable simplification of the application design, integration, and deployment processes (Guinard et al. 2009), in particular when comparing RESTful devices with other systems for embedded devices, such as WS-* Web services. As an example, the Plogg RESTful Gateway and the Sun SPOTs have been used by external development teams who read about our project on our Web site. In the first case, the idea was to build a mobile energy monitoring application based on the iPhone that communicates with the Ploggs. In the second case, the goal was to demonstrate the use of a browser-based JavaScript Mashup editor with real-world services. According to interviews we conducted with these developers, their experience confirmed ours. They enjoyed using the RESTful smart things, in particular the ease of use of a RESTful Web API versus a different kind of API. For the iPhone application, a native API to Bluetooth did not exist at that time. However, like for almost any platform an HTTP (and JSON) library was available. One of the developers mentioned a learning curve for REST but emphasised the fact that it was still rather simple and that once it was learnt, the same principles could be used to interact with a large number of services. They finally noted the direct integration to HTML and Web browsers as one of the most prevalent benefits.

5.7 Conclusion

In this chapter, we suggested that Web technologies are – contrary to popular belief – a suitable protocol for building applications on top of services offered by smart things. After summarising the core design principles of Web architecture, we proposed an architecture for the Web of Things based on the concepts of REST, syndication for smart things, Web Hooks, and Smart Gateways. We demonstrate the idea with several prototypes.

Thanks to the loose-coupling, simplicity and scalability of RESTful architectures, and the wide availability of HTTP libraries and clients, RESTful architectures are becoming one of the most ubiquitous and lightweight integration platforms. Because of this, using Web standards to interact with smart things seems to be increasingly adequate. Although HTTP introduces a communication overhead and increases average latency, it is sufficient for many pervasive scenarios where such longer delays do not affect user experience.

Introducing support for Web standards at the device-level is beneficial for developing a new generation of networked devices that are much simpler to deploy, program, and reuse. Applying the same design principles that supported the success of the Web, in particular openness, connectedness, and simplicity, can significantly leverage the ubiquity and versatility of the Web as a common ground for supporting interactions between devices and applications. Furthermore, as most mobile devices have already Web connectivity and Web browsers, and most programming environments support HTTP, we tap into the very large Web developer community as potential application developers for the Web of Things.

References

Drytkiewicz W, Radusch I, Arbanowski S, Popescu-Zeletin R (2004) pREST: a REST-based protocol for pervasive systems. Proceedings of the IEEE International Conference on Mobile Ad-hoc and Sensor Systems

Duquennoy S, Grimaud G, Vandewalle J (2009) The Web of Things: interconnecting devices with high usability and performance. Proceedings of the 6th IEEE International Conference on Embedded Software and Systems (ICESS'09). HangZhou, Zhejiang, China

Fielding RT (2000), Architectural styles and the design of network-based software architectures. Ph.D. Thesis, University of California. Irvine, USA

Floerkemeier C, Lampe M, Roduner C (2007) Facilitating RFID Development with the Accada Prototyping Platform. Proceedings of the Fifth IEEE International Conference on Pervasive Computing and Communications Workshops. IEEE Computer Society

Guinard D, Trifa V, Pham T, Liechti O (2009) Towards Physical Mashups in the Web of Things. Proc. of the 6th International Conference on Networked Sensing Systems (INSS). Pittsburgh, USA

Guinard D, Fischer M, Trifa V (2010a) Sharing Using Social Networks in a Composable Web of Things. Proceedings of the 1st IEEE International Workshop on the Web of Things (WoT 2010) at IEEE PerCom, Mannheim, Germany

Guinard D, Mueller M, Pasquier J (2010b) Giving RFID a REST: Building a Web-Enabled EPCIS. Proceedings of the IEEE International Conference on the Internet of Things (IOT 2010). Tokyo, Japan

Guinard D, Trifa V, Wilde E (2010c) A Resource Oriented Architecture for the Web of Things. Proceedings of IoT 2010, IEEE International Conference on the Internet of Things. Tokyo, Japan

Guinard D, Trifa M, Karnouskos S, Spiess P, Savio D (2010d) Interacting with the SOA-Based Internet of Things: Discovery, Query, Selection, and On-Demand Provisioning of Web Services., IEEE Transactions on Services Computing. 3, 223–235

Hui J, Culler D (2008) Extending IP to low-power, wireless personal area networks. IEEE Internet Comput 12:37-45

Hui J, Culler D (2008) IP is dead, long live IP for wireless sensor networks. Proceedings of the 6th ACM conference on embedded network sensor systems. ACM, Raleigh, NC, USA

Kindberg T, Barton J, Morgan J, Becker G, Caswell D, Debaty P, Gopal G, Frid M, Krishnan V, Morris H, Schettino J, Serra B, Spasojevic M (2002) People, places, things: web presence for the real world. Mob Netw Appl 7:365-376

Luckenbach T, Gober P, Arbanowski S, Kotsopoulos A, Kim K (2005) TinyREST - A protocol for integrating sensor networks into the internet. Proceedings of the Workshop on Real-World Wireless Sensor Network: SICS. Stockholm, Sweden

Luo L, Kansal A, Nath S, Zhao F (2008) Sharing and exploring sensor streams over geocentric interfaces. Proceedings of the 16th ACM SIGSPATIAL international conference on advances in geographic information systems. ACM, Irvine, California

Pautasso C, Wilde E (2009) Why is the Web Loosely Coupled? A Multi-Faceted Metric for Service Design. Proceedings of the 18th International World Wide Web Conference (WWW2009). Madrid, Spain

Priyantha NB, Kansal A, Goraczko M, Zhao F (2008) Tiny web services: design and implementation of interoperable and evolvable sensor networks. Proceedings of the 6th ACM conference on embedded network sensor systems. ACM, Raleigh, NC, USA

Richardson L, Ruby S (2007) RESTful Web Services. O'Reilly Media, Inc

Stirbu V (2008) Towards a RESTful Plug and Play Experience in the Web of Things. Proceedings of the IEEE International Conference on Semantic Computing

Trifa V, Wieland S, Guinard D, Bohnert TM (2009) Design and Implementation of a Gateway for Web-based Interaction and Management of Embedded Devices. Proceedings of the 2nd

International Workshop on Sensor Network Engineering (IWSNE 09). Marina del Rey, CA, USA
Trifa V, Guinard D, Davidovski V, Kamilaris A, Delchev I (2010) Web-based Messaging Mechanisms for Open and Scalable Distributed Sensing Applications. Proceedings of the 10th International Conference on Web Engineering (ICWE 2010). Vienna, Austria
Wilde E (2007) Putting Things to REST. School of Information. UC Berkeley
Yazar D, Dunkels A (2009) Efficient Application Integration in IP-based Sensor Networks. Proceedings of ACM BuildSys, the First ACM Workshop On Embedded Sensing Systems For Energy-Efficiency In Buildings, BuildSys. Berkeley, USA
Yu J, Benatallah B, Casati F, Daniel F (2008) Understanding Mashup Development. IEEE Internet Comput 12:44-52
Zang N, Rosson MB, Nasser V (2008) Mashups: who? what? why?. Proceedings of CHI '08 extended abstracts on Human factors in computing systems. ACM, Florence, Italy

6 A Service-oriented, Semantic Approach to Data Integration for an Internet of Things Supporting Autonomous Cooperating Logistics Processes

Karl A. Hribernik, Carl Hans, Christoph Kramer, Klaus-Dieter Thoben

BIBA – Bremer Institut für Produktion und Logistik GmbH, Germany

Abstract The core vision put forward by the Internet of Things, of networked, intelligent objects capable of taking autonomous decisions based on decentral information processing, resonates strongly with research in the field of autonomous cooperating logistics processes. The characteristics of the IT landscape underlying autonomous cooperating logistics processes pose a number of challenges towards data integration. The heterogeneity of the data sources, their highly distributed nature, along with their availability, make the application of traditional approaches problematic. The field of semantic data integration offers potential solutions to address these issues. This contribution aims to examine in what way an adequate approach towards data integration may be facilitated on that basis. It subsequently proposes a service-oriented, ontology-based mediation approach to data integration for an Internet of Things supporting autonomous cooperating logistics processes.

6.1 Introduction and Background

The concepts and technologies of the Internet of Things are rapidly becoming significant to challenges arising in the field of logistics. With today's globalised markets in a state of accelerating structural change, planning and control strategies need to be redefined, whilst traditional supply chains are evolving into complex networks of numerous stakeholders. The goods structure, logistics and structural effects identified by Aberle (2003) characterise these changes. The first describes a shift away from mass production towards a buyers' market, which creates a trend towards individual product customisation and, consequently, a noticeable increase in per-unit shipments. The second effect describes a shift towards road freight transport, which arises from the increasing demands for small shipments along with a high quality of service and due-date reliability. Finally, the structure effect indicates an individualisation of transport on the micro-logistics level. Co-operation is required between otherwise competing logistics service providers to satisfy today's customer requirements. These three effects lead to a dramatic increase in complexity and dynamics of transport logistics processes.

The core vision put forward by the Internet of Things, of networked, intelligent objects potentially capable of taking autonomous decisions based on decentral information processing, resonates strongly with research in the field of logistics aimed at addressing these effects. Autonomous cooperating logistics processes (Hülsmann et al. 2006) are a prominent example of such research. Here, "autonomous control" is understood as processes of decentralised decision-making in heterarchical structures. It anticipates interacting elements in non-deterministic systems, which possess the capability and possibility to render decisions independently (Böse and Windt 2007). Critical to this understanding is the decentralisation of decision-making responsibilities, in contrast to traditional, hierarchical process control. This approach is motivated by an expected improvement in robustness and increase in scalability of process control, amongst other effects.

An example of the convergence of the concepts of the Internet of Things and autonomous cooperating logistics processes can be found in the use of software agents to implement information processing and decision taking entities (Timm 2006) for logistics processes, as described in Trautmann (2007) and Jedermann et al. (Jedermann et al. 2008). In combination with an appropriate solution to mapping software agents as "digital counterparts" to physical logistics objects, using the auto-identification technology, middleware and standards of, for instance, EPCglobal (EPCglobal 2009) or ID@URI/Dialog (Främling et al. 2006), a foundation for an "Internet of Things for logistics" may be laid.

Research in autonomous cooperating logistics processes shows that different control problems arise from different applications of autonomous control, resulting in a wide spectrum of degrees of autonomy (Windt et al. 2008) with respective requirements towards the characteristics of the involved intelligent logistics objects as well as the underlying data processing, decision making and data integration strategies. This means that, in order for an Internet of Things to truly benefit the autonomous cooperating logistics processes on an operational level, the "things" therein not only need to be able to communicate with each other, but also to be suitably integrated into the overall logistics IT landscape.

The IT landscape in logistics is already a highly complex, distributed and heterogeneous one, even without taking autonomous cooperating processes into account. As shown in Figure 6.1, significant effort was, and still is, spent in order to achieve at least integration between systems of certain business partners by bridging the technological islands through specific ICT solutions (Hannus 1996). However, most of these solutions sooner or later become obsolete due to the continuous development of the islands reflected by a steadily decreasing sea level as well as the highly dynamic partnerships within today's enterprise networks. Instead of developing solutions for 1:1 relationships a general solution must be found, which allows a unique access to all relevant logistics data while accepting the diversity of existing systems and standards (Hans et al. 2008).

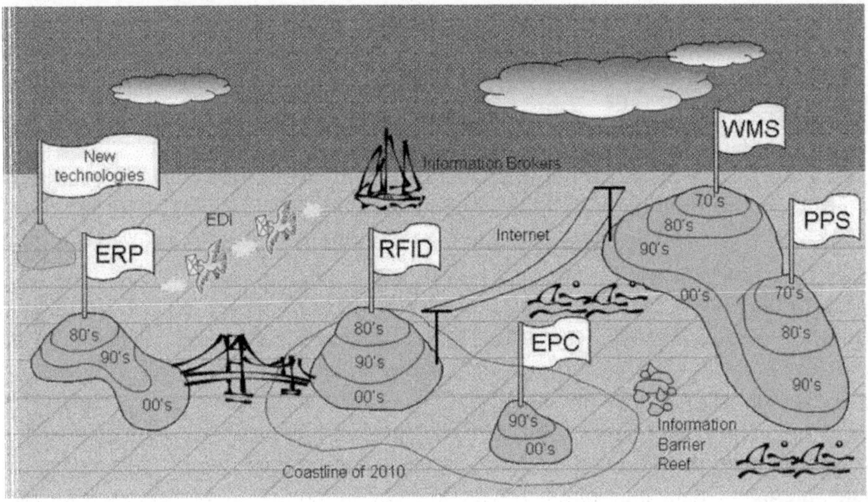

Fig. 6.1 The IT Landscape in Logistics – Bridging Islands (based on Hannus 1996)

This situation is exacerbated by developments in modern logistics, such as autonomous cooperating logistics processes and the Internet of Things. Both developments lead to the creation of "new islands" of technology development in the IT logistics landscape. Depending on the application, relevant data may be stored in heterogeneous enterprise systems, such as Warehouse Management Systems (WMS), Enterprise Resource Planning Systems (ERP) or disposition systems. At the same time, data from item-level tracking and tracing systems needs to be taken into account, in particular that pertaining to RFID. Data may also be generated and stored in systems embedded into logistics objects, such as trucks or containers, or be generated dynamically, for example by sensor networks monitoring the temperature of a refrigerated container. Whilst the specific requirements towards data integration differs according to the characteristics of each individual application of autonomous control, it can be said that, in general, digital counterparts representing individual logistics entities need to be able to access data relevant to their decision making processes, regardless which "island" that data may be located on.

The characteristics of the IT landscape underlying autonomous cooperating logistics processes outlined above, pose a number of challenges towards data integration. The heterogeneity of the data sources, their highly distributed nature, along with their availability make the application of traditional approaches problematic. The field of semantic data integration offers potential solutions to address these issues. This contribution aims to examine in what way an adequate approach towards data integration may be facilitated on that basis. It subsequently proposes a semantic approach to data integration for an Internet of Things supporting autonomous cooperating logistics processes, which combines ontology-based mediation with service-oriented integration.

6.2 State of the Art

The following sections outline the state-of-the-art in relevant areas of research. First, literature from the field of the Internet of Things is discussed in order to establish an understanding of the terminology used in the remainder of this contribution. Next, autonomous cooperating logistics processes are introduced and put into perspective with the perspective upon Internet of Things discussed in the previous section. Moving forward, approaches to data integration from two different areas are investigated. First of all, concepts and solutions for item-level information management for products and logistics objects are discussed. Secondly, approaches to traditional enterprise application integration are presented, which need to be combined with item-level information management approaches in order to facilitate intelligent data integration for an Internet of Things in logistics within the context of this contribution.

6.2.1 The Internet of Things

This section begins by clarifying the understanding of the Internet of Things used in the remainder of this paper, including a look at related developments in the field of Intelligent Products. It concludes outlining applications of the state of the art in the field of logistics.

Terminology

The term "Internet of Things" was first used by the Massachusetts Institute of Technology in the year 1999. Here, it was used in the sense of a networked system of autonomously interacting and self-organising objects and processes, which is expected to lead to a convergence of physical things with the digital world of the internet (Brand et al. 2009). This extrapolates the idea of the Internet - a global, interconnected network of computers – to describe a network of interconnected *things,* such as everyday objects, products, and environments. As such, the Internet of Things represents the common ground of a number of recent multi-disciplinary developments, such as Ambient Intelligence (Ducatel et al. 2001), Ubiquitous (Weiser 1991) and Pervasive Computing (Gupta et al. 2001), and Auto Identification (Cole and Engels 2002). At the heart of the concept lies the idea that objects - *things* - are capable of information processing, communication with each other and with their environment, and autonomous decision taking.

Intelligent Products

One example of quite research towards the realisation of intelligent objects, which exhibit the characteristics described above, is the field of "Intelligent Products". Intelligent Products are physical items, which may be transported, processed or used and comprise the ability to act in an intelligent manner. McFarlane et al. (McFarlane et al. 2003) define the Intelligent Product as

> "...a physical and information based representation of an item [...] which possesses a unique identification, is capable of communicating effectively with its environment, can retain or store data about itself, deploys a language to display its features, production requirements, etc., and is capable of participating in or making decisions relevant to its own destiny."

The degree of intelligence of an intelligent product may exhibit variations from simple data processing to complex pro-active behaviour. This is the focus of the definitions in McFarlane et al. (McFarlane et al. 2003) and Kärkainen et al. (Kärkainen et al. 2003b). Three dimensions of characterisation of Intelligent Products are suggested by Meyer et al. (Meyer et al. 2009): *Level of Intelligence*, *Location of Intelligence* and *Aggregation Level of Intelligence*. The first dimension describes whether the Intelligent Product exhibits information handling, problem notification or decision making capabilities. The second shows whether the intelligence is built into the object, or whether it is located in the network. Finally, the aggregation level describes whether the item itself is intelligent or whether intelligence is aggregated at container level.

The Internet of Things in Logistics

Concepts and technologies of the Internet of Things have previously been applied to problems in the field of logistics. For example, in the area of transport logistics, ten Hompel (2005) considers the autonomous transport of logistics objects from the sender to the delivery address, as an example of the Internet of Things. A further example is the discussion of the application of dynamic route planning algorithms in autonomous transport logistics networks (Berning and Vastag 2007). Besides basic, item-level tracking & tracing of goods along the supply chain and a general potential for the optimisation of processes (VDI/VDE Innovation + Technik GmbH 2008) as well as the improvement of Efficient Customer Response (ECR) (Gaßner and Bovenschulte 2009), the Internet of Things is of particular interest to the field of logistics.

The German national study QuinDILog, which focused on vocational qualification resulting from the implementation of the Internet of Things in logistics, identified a number of additional potentials of the Internet of Things for the field of logistics. For example, the granular, item-level documentation of supply chain events can allow for a greater transparency in contractual and legal matters (VDI/VDE Innovation + Technik GmbH 2008). Out-of-stock (OOS) situations may be avoided by automated positioning and warehouse management solutions (Gaßner and Bovenschulte 2009). Especially for critical goods, such as foods or medicine, quality assurance (Jedermann et al. 2008), product pedigree and history traceability can be enabled using the Internet of Things. Protection against product theft and plagiarism (Staake et al. 2005) based on unique identification and positioning technologies is another example. Last, but not least, completely new business models like fourth party logistics (4PL) may be developed on the basis of the Internet of Things (Schuldt et al. 2010).

Research in the field of Intelligent Products has also been applied to logistics. For instance, Kärkkiänen et al. (Kärkkiänen et al. 2003b) describe the application of the concept to supply network information management problems. Additional examples are the application of the Intelligent Products to the supply chain (Ventä 2007), to manufacturing control (McFarlane, et al. 2003), and to production, distribution, and warehouse management logistics (Wong, et al. 2002).

6.2.2 Autonomous Cooperating Logistics Processes

This section briefly introduces the research area of autonomous cooperating logistics processes. It furthermore presents the concept of intelligent logistics objects developed in that area of research. Subsequently, it puts the concept of intelligent logistics objects into perspective with the Internet of Things and intelligent products.

Terminology

In the context of this contribution, the term "Autonomous Control" is used following Böse and Windt (2007) to describe

> "...processes of decentralised decision-making in heterarchical structures. It presumes interacting elements in non-deterministic systems, which possess the capability and possibility to render decisions independently."

The research area of autonomous cooperating logistics processes (Scholz-Reiter et al. 2004) aims to meet today's logistics challenges, such as the goods structure, logistics and structural effects identified by Aberle (2003), by introducing autonomy and self-organisation into control, information processing and decision-making in logistics (Ehnert et al. 2006). The argumentation is that central control and planning of logistics processes has reached its limits in addressing these issues (Scholz-Reiter et al. 2004). Here, the term "autonomy" describes

> "...the capability of a system, process or an item to design its input-, throughput- and output-profiles as an anticipative or reactive answer to changing constraints of environmental parameters."

The application of autonomous control to logistics processes is expected to increase their robustness, flexibility, adaptability and reactivity to respond to changing business environments, requirements and to changing or partially conflicting objectives (Scholz-Reiter et al. 2004). A prominent characteristic of this understanding is the decentralisation of decision-making responsibilities in contrast to traditional, hierarchical process control. A dynamic heterarchy, in which otherwise passive logistics entities are equipped with the ability to process information, to render and execute decisions on their own, replaces the strict centralised top-down management of traditional logistics processes. Artificial agents are entrusted to act in their own "best interest" within the bounds of their operational, tactical or strategic (Timm 2006) autonomies. The motivation for this approach is, amongst others, an expected improved robustness and increased scalability of process control.

Intelligent Logistics Objects

The concept of an intelligent logistics object is inherent in the understanding of Autonomous Control in logistics systems proposed by Böse and Windt (2007). Here,

> "...autonomous control in logistic systems is characterized by the ability of logistic objects to process information, to render and to execute decisions on their own."

Logistics objects are defined in this context as both,

> "...material items (e.g. parts, machines or conveyors) and immaterial items (e.g. production orders) of a networked logistic system, which have the ability to interact with other logistic objects of the considered system."

In Scholz-Reiter et al. (Scholz-Reiter et al. 2007), the former are further differentiated as commodities and all types of resources, whilst constraining the immaterial logistics objects to orders.

According to this understanding, an intelligent logistics object is consequently either a material or immaterial logistics object which is capable of communicating and interacting with other logistics objects. It is a broader understanding than that of the Internet of Things which additionally encompasses autonomous objects without physical representations.

6.2.3 Item-level Information Management Approaches

Item-level information is that information, which is specific to individual logistics objects or products. It is created in all processes a logistics object is involved in. These include, for example, production logistics processes as they occur in the beginning-of-life (BOL) phase of the product lifecycle (Hong-Bae et al. 2007), distribution (Hribernik et al. 2009) and service logistics processes as occurring during the middle-of-life (MOL) phase, and reverse logistics processes, which take place in the object's end-of-life (EOL) phase (Schnatmeyer et al. 2005; Schnatmeyer 2008).

Item-level information management can be based on standards for product information modelling and exchange. Existing standards, such as those developed by the technical committee (e.g., ISO TC184/SC4), focus on product information and processes specific to BOL. The emerging standards ISO10303-239 (International Organization for Standardization 2009a), which defines Product Life Cycle

Support (PLCS) and ISO 15926 (International Organization for Standardization 2009b) are exceptions, which deal explicitly with item-specific product information. Although PLCS is under continuously development to widen its scope of application, it currently focuses on specific maintenance processes in MOL. ISO 15926 caters for the oil and gas production domain but contains generic parts, such as the ISO 15926 Part 2 Data model, which is also used by other initiatives. However, both standards remain restricted to particular domains or processes. Moreover, information standards only address the information transfer and interpretations issues in information management throughout the product life-cycle. Access to and consolidation of information is an issue to be solved for each inter-enterprise scenario.

Another prominent application domain is the area of shipment tracking. The tracking systems used by major forwarders or logistics service providers are perfectly engineered to suit situations where shipment is handled by a single organisation (Kärkkäinen et al. 2004). However, management of item-specific product information requires the support of multi-organisational networks, which is supported only by some approaches, i.e. the EPCglobal Architecture Framework, DIALOG, WWAI and the PROMISE Architecture. These approaches are introduced in the following sections in more detail.

EPCglobal Architecture Framework

The EPCglobal Architecture Framework (EPCglobal 2009) represents a collection of widely adopted industry standards in the field of auto-identification aimed at the coupling of information and material flows in retail logistics (cf. Figure 6.2). It includes tag protocol standards for the physical and logical requirements of RFID systems and encompasses standards for the definition of item-level, unique identification codes, the Electronic Product Code (EPC). Of foremost interest towards data integration are the EPC Information Services (EPCIS) (EPCglobal 2007) and Object Name Service (ONS) standards specified in the framework.

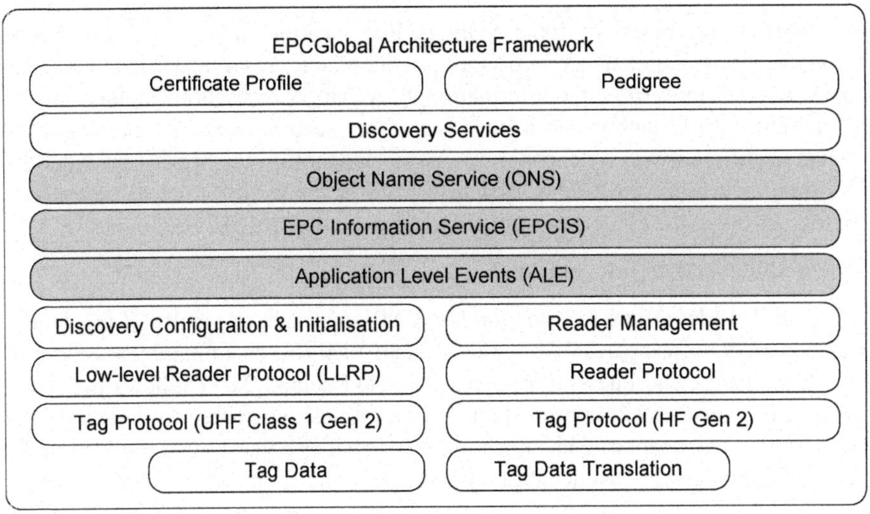

Fig. 6.2: The EPCglobal Architecture Framework (EPCglobal 2009)

EPCIS is a standard that defines interfaces for the sharing of data among trading partners. Its aim is primarily to enable supply chain participants to gain real-time visibility into the movement, location and disposition of assets, goods and services throughout the world (Soon and Ishii 2007). EPCIS can be leveraged to track individual physical objects and collect, store and act upon information about them. By providing a standard interface to that information, EPCIS enable cooperation partners to seamlessly query such information throughout supply chains. At the present time, information service discovery operates at the product class level, and not the item level.

ONS defines a mechanism by which authoritative metadata and services associated with EPC Identifiers may be located in the network. Its function is to transform the EPC stored, for example, on RFID-Tags, via their corresponding Identity URI encoding into URLs, which may respectively point to a Web Service or other information resource. Performance and security issues have moved EPGglobal to develop alternative "discovery services" (Meyer et al. 2009).

Dialog

The Dialog system developed at Helsinki University of Technology aims at solving the challenges of item level information management without the need of developing new standards for product coding. In the DIALOG approach (Kärkkäinen et al. 2003a) an ID@URI notation is used as product identifier, where the ID part identifies the product item at the URI. The uniqueness of a URL is guaranteed by the Domain Name System (DNS) infrastructure. For an ID@URI to be globally unique, the ID part should be unique for that URI. The URI part of the Dialog product code indicates the location of the tangible object's "agent". The agent is a background service running at the computer indicated by the URI. It offers various interfaces for functionalities like location updates, product information requests, maintenance information requests, etc. Each interface has its own characteristics concerning the information to exchange, restrictions on data security, authentication, authorisation etc. Therefore, security considerations can be treated in different ways, depending on how "dangerous" the service provided by each interface is for the Dialog system itself and for the information systems of companies using the Dialog system (Kärkkäinen et al. 2003b).

World Wide Article Information

The World Wide Article Information (WWAI) approach, originally of the Trackway company, now a part of Elisa, provides an XML-based communication protocol for exchange and querying of product-related data. WWAI follows the structured peer-to-peer (P2P) approach and utilises a hash algorithm to determine the node for the placement of data concerning a particular object, e.g. a product, in the network. This permits easy location of nodes potentially storing data for objects of interest. Furthermore, subscriptions can be specified for object IDs to automatically obtain new information on the objects. The main advantages of WWAI include an implementation of the protocol and the decentralised nature of the solution allowing for easy implementation and deployment. However, WWAI is a proprietary specification with currently little industry support. Furthermore, it is unclear how the approach addresses the critical issues of P2P networks, such as ensuring result quality and response time for queries, and limited querying capabilities confined to object ID exact matching (Do et al. 2006).

The PROMISE Architecture

The PROMISE Architecture is focussed on the concept of Product Embedded Information Devices (PEIDs). PEIDs (Jun et al. 2007) realise the concept of intelligent products and components acting as embedded information gathering devices linked to sensors, which are able to sense their environment and their condition wirelessly, for example via RFID or Universal Plug and Play (UPnP). PEIDs are categorised according to their capabilities with regards to data storage and data processing capabilities (cf. Table 6.1). In addition to the categories data storage and data processing shared with Böse and Windt (2007) the devices' ability to integrate sensors as well as their options for network connectivity are used to differentiate different types of PEIDs. The sole common denominator of all types of PEID is, however, that they contain a global, unique identifier. This fulfils the most basic requirement towards integrating information with an intelligent object. PEID information is communicated to backend systems via a message and event based middleware (PROMISE Data Services) using a standardised and XML based PROMISE Messaging Interface (PMI) (Främling and Nyman 2008).

Type	Identification	Data Storage	Sensors	Data Processing	Connectivity
Type 0	✓				Passive

Type 1	✓	✓			Passive
Type 2	✓	✓	(✓)	+	Wireless
Type 3	✓	✓	✓	++	Wireless
Type 4	✓	✓	✓	+++	Always

Table 6.1 PEID Classification (according to The PROMISE Consortium 2008)

The PMI (Kärkkainen et al. 2003b) is an XML-based standard communication protocol linking the nodes of a PROMISE architecture implementation. An information model is instanced for each individual product on the basis of a semantic object model (Cassina et al. 2008). This is transformed into a semi-structured, syntactical model to define the information items related to each individual product (Främling and Nyman 2008), which in turn defines the structure and data types of the PMI messages. It provides functionality for the access to and management of item-specific product data in an expressive and generic way, and specifies both, the syntax and semantics of a request and response pairs. Its main task is to represent item-level read requests and write commands from and to PEIDs and the other nodes in the PROMISE architecture. The InfoItem-Element is the central concept of PMI, representing the message payload used to fulfil queries. InfoItems use unique identifiers to address specific items and define both, application and item specific data types. PMI supports event, message and subscription based communication between all nodes and forms the backbone of the PROMISE architecture.

6.2.4 Enterprise Application Integration Approaches

Taking into consideration the IT landscape in logistics as outlined in this contribution, approaches towards enterprise application integration need to be taken into account. This section outlines prominent approaches relevant to the problem area. First, traditional data integration approaches are outlined, with tightly coupled, loosely coupled and object-oriented approaches being discussed. Service-oriented architecture is briefly discussed and, subsequently, the field of semantic mediation is introduced.

Traditional Data Integration Approaches

Whilst a tightly coupled approach can quickly be dismissed on grounds of its inflexibility, loosely coupled and object-oriented approaches cannot be adopted without critical analysis. An object-oriented approach generally provides good mechanisms for avoiding integration conflicts. However, when considering this approach, one must take into account that a single canonical model is required to describe the entire data model, which clearly restricts its flexibility and scalability. Each time a new stakeholder or data source enters the logistics system, the model needs to be extended. Depending on the dynamics of the logistics system, this may or may not be a disqualifying factor with regards to this approach. As the fluctuation of data sources, stakeholders and systems in a complex logistics system with any degree of autonomous control can be assumed to be high, an object-oriented approach to data integration is likely to be unsuitable. A loosely coupled approach requires detailed knowledge of each of the heterogeneous data sources to be able to be successfully employed. With regards to complex logistics systems, further analysis is required to determine whether this is feasible or not. The possibility of requiring highly flexible, and thus possibly not always pre-determinable, context data, for example from sensor networks, may prove to be an argument against this approach.

Service-oriented Architecture

A software architecture is described as service-oriented when it uses loosely coupled software services to provide functionality (Stojanovic et al. 2004). Here, logic is not packaged as individual programmes, but is distributed across an amount of independent services. The actual implementation details of these services by their provider are completely transparent to the consumer. The most popular implementations of service-oriented architecture (SOA) are carried out using Web Services (Thoben et al. 2003), which are built using the combination of the XML standards. Simple Object Access Protocol (SOAP) (cf. Gudgin et al. 2003), Web Service Definition Language (WSDL) (cf. Christensen et al. 2001) and Universal Description, Discovery and Integration (UDDI) (cf. Clement et al. 2004). These standards, in combination with the Hypertext Transfer Protocol (http), provide a system independent approach to the discovery, identification, provision and consumption of services according to SOA. However, a SOA may also be built using other technology, such as Common Object Request Broker Architecture (CORBA), Distributed Component Object Model (DCOM) or Enterprise Java Beans (Blanke et al. 2004).

Semantic Mediation

Besides the traditional approaches to data integration, a number of predominantly semantic approaches remain to be taken into account. Here, the main concepts constituting architecture of such data integration systems are mediators (Ullman 1997; Wache et al. 2001 and Wache 2003). In this approach, both syntactic and semantic descriptions of the data to be integrated are applied. The semantic mediator is capable of extracting knowledge regarding the data structures of the underlying data sources and, subsequently, transforming, decomposing and recomposing data requests according to that knowledge. The mediator relies on semantic descriptions of the data sources. In the case of autonomous logistics processes, this implies a wholly semantic modelling of the relevant logistics information and data across the distributed, heterogeneous sources, for which a number of approaches, such as ontologies, may be chosen. Here, extensive research is required to determine whether such semantic descriptions of logistics data are feasible and adequate to address the requirements of autonomous logistics processes.

6.3 Problem Analysis

In the following, this contribution concentrates on examining how decentral data storage may be facilitated to support an Internet of Things for the autonomous control of logistics processes. It examines the advantages of combining a service interface with a semantic approach to data integration for solving this data integration problem.

The understanding of an Internet of Things for logistics presented in this paper so far, contributes to the fulfilment of these criteria and, consequently, to an increase in the degree of autonomy of logistics processes. First, the Internet of Things displays high degrees of data processing decentrality with Things explicitly required to be capable of local data processing. Furthermore, things are expected to communicate with each other in order to coordinate their decisions.

Currently, the discussion of data integration from the perspective of the Internet of Things in logistics focuses mainly on the facilitation of information exchange between individual physical, intelligent objects. The reasoning behind this is to separate the operational level of control from the strategic in order to create more autonomous, robust and flexible operational systems. However, this is not always the most appropriate solution. Research in autonomous cooperating logistics processes shows that different control problems arise from different applications of autonomous control, resulting in a wide spectrum of degrees of autonomy (Windt et al. 2008) with respective requirements towards the characteristics of the involved intelligent logistics objects as well as the underlying data processing, decision making and integration strategies.

6.3.1 Logistics Systems Integration Targets

A market study of the current IT logistics system landscape was carried out by the authors in the context of CRC637 Autonomous Cooperating Logistics Processes in order to identify potential integration targets for intelligent logistics objects in autonomous cooperating logistics. The study covered 122 different IT systems on the market used in distribution, transport, retail, and warehouse logistics. It focused on identifying the foremost data exchange formats in the field by collecting information regarding the systems' interface capabilities. Additionally, database and ERP interoperability were examined alongside auto-ID support.

According to the study, the most prominent interface is EDIFACT EANCOM with 62% implementation, followed by SAP with 54%. Third is EANCOM XML with almost 39% use, closely followed by ebXML with almost 36%. To put this in perspective, almost 32% indicated the implementation of bespoke proprietary interfaces only.

The study reveals that a large number of different IT systems, data models and exchange formats are employed to support logistics processes. Whilst a significant share of the market can be addressed by EANCOM, SAP and ebXML, still almost one third of all systems exhibit proprietary interfaces.

6.3.2 Integrating Intelligent Logistics Objects

Dynamic data sources in an Internet of Things in logistics are foremost material intelligent logistics objects themselves. The characteristics of intelligent logistics objects are quite close to those defined by Meyer et al. (Meyer et al. 2009) in their classification of Intelligent Products. Furthermore, the PEID classification scheme in Table 6.1 gives an overview of the types of technology and interface used to realise intelligence embedded into products. This classification is a good indicator to the requirements towards data integration for such devices. The following sections have been derived from the classification scheme and discuss the identification of intelligent logistics objects, data storage and connectivity issues and sensors. A section dealing explicitly with immaterial logistics objects concludes this discussion.

Identification of Intelligent Logistics Objects
In order to be able to make decisions regarding the entities within an autonomous logistics system, a mapping between the individual entities and their descriptive data is imperative. The study mentioned in the previous section shows the most supported auto-identification technology by logistics systems today is RFID (65%) with EAN and EPC numbering schemes, both exhibiting about 60% support. Alternative approaches should, however, not be discounted. Among these are the approaches discussed in section 6.2 (state of the art*)*, such as ID@URI/Dialog and WWAI.

Data Storage and Connectivity
With a mapping between the individual entities and their descriptive data facilitated using auto-ID technology, as described above, data in back-end systems can be attributed to intelligent logistics objects by mapping an identifier to the data. This can, in principle, be applied to data stored on dynamic data sources as well. However, dynamic data sources imply added complication of not always being accessible, having data volume restrictions, and other hardware-related issues. Furthermore, the disparity of different implementations is a problem. From RFID via embedded systems to full-scale, integrated computing devices, such as On-board Units (OBUs), the scope of systems to be integrated is wide. A number of existing approaches exist to overcome these limitations. RFID middleware, such as the EPCglobal Architecture Framework, can be used to abstract from RFID hardware. For PEIDs, the PMI standard can be used in combination with the PROMISE development CorePAC, which is a hardware abstraction layer for different PEID

types. Other approaches in this area include the deployment of OSGi components, amongst a number of more proprietary solutions.

Sensors and Actuators

Sensors are of particular interest to autonomous cooperating logistics processes for their ability, for example, to monitor the condition of cargo. The integration problem, with regards to sensors and sensor networks, is similar to that of data storage above – sensors can simply be seen as a specific type of dynamic data source. On top of the approaches discussed above, such as PMI and OSGi, a number of standards for the description and communication of sensor data exist. Foremost amongst these is the work of the Open Geospatial Consortium, SensorML. SensorML provides standard models and an XML encoding for describing sensors and measurement processes.

Actuators are relevant to the field of autonomous cooperating logistics processes where intelligent logistics objects with capabilities for autonomous decision making are designed to directly act upon the physical logistics environment. Examples of these kinds of intelligent logistics objects include autonomous forklift trucks (Schuldt and Gottfried 2008) or intelligent production machines (de Souza et al 2008). Besides bespoke proprietry data exchange formats, a number of contributions towards the standardisation of interfaces towards actuators exist. Foremost are the contributions from OPC, ASAM-GDI and SAP. The OPC Unified Architecture (OPC UA) encompasses a comprehensive framework for the integration of automation technology including actuators. In contrast to the preceding standard OPC Data Access (OPC DA), the Unified Architecture makes use of a service-oriented approach instead of the Microsoft Distributed Component Object Model (DCOM) interface. The standard General Device Interface (GDI), proposed by the Association for Standardisation of Automation and Measurement Systems (ASAM), and the related Open Robot Resource Interface for the Network API (ORiN) aim to provide platform and framework independent access to devices, such as actuators, and has been adopted into the ISO standard 20242 on Industrial automation systems and integration (International Organization for Standardization 2009c). Finally, the SAP-driven SOCRADES initiative strives to establish a service-oriented integration architecture for manufacturing resources (de Souza et al 2008).

Immaterial Logistics Objects

In addition to the categories defined here, immaterial logistics objects need to be considered. This is, in fact, merely a formal amendment – immaterial logistics objects can be considered to be Intelligent Products with intelligence located in the network, but without a physical manifestation. Without having to resort to proprietary implementations, immaterial logistics objects may be handled using existing standards. For instance, orders, invoices and other data pertaining to this type of object can be interfaced using the relevant messages of the EDIFACT EANCOM standard data exchange format. Furthermore, as shown in Hribernik et al. (Hribernik et al. 2009), immaterial logistics objects may be identified and described by participating stakeholders using URIs in EPCIS events. Purchase orders may be mapped to physical entities via the BusinessTransactionID vocabulary that may point to an URI describing the transaction.

6.3.3 Summary of Data Integration Requirements

The following table (Table 6.2) summarises the previous section by listing the major integration targets for autonomous cooperating logistics processes. Four types of integration targets are differentiated between:

1. Logistics IT systems, describing IT systems in logistics, such as ERP, WMS, disposition and other "traditional" enterprise systems used in logistics
2. Intelligent material logistics objects – which relate to material intelligent logistics objects, which exhibit characteristics of the PEID classification scheme
3. Digital counterparts – these relate to the decision making components of intelligent logistics objects, whether located in the object or in the network
4. Sensors and actuators – relating to sensors, sensor networks and actuators, which fall outside of the previous categories

A specific category has neither been defined for immaterial logistics objects nor smart environments. If the former is "intelligent", it is merely a digital counterpart without a physical component. Consequently, the category "digital counterpart" suffices. Where a logistics object merely exists, e.g. as an order in an ERP system, the underlying logistics IT system is all that is needed to be taken into account.

With regards to the latter, intelligent logistics objects encompass not only goods but also logistics resources, such as machines, vehicles, transport nodes, etc. Consequently, data sources in a "smart environment", as suggested by ubiquitous computing and ambient intelligence, are simply specific instances thereof and need not be treated separately. Examples are a warehouse with a sensor network installed or a truck equipped with an on-board unit. There is obviously some overlap between the categories, especially between "intelligent material logistics objects" and "digital counterparts", depending on the implementation choices towards decision making and information processing. For example, it is quite possible to install software agents on OSGi components. These grey areas, however, do not affect the aim of identifying data integration requirements.

Integration Target	Type(s)	Interface/standard	Type of interface	Importance (0-5)
Logistics IT systems	General	EDIFACT EANCOM	Semi-structured text	●●●●●

		EANCOM XML	Semi-structured text	●●●
		ebXML	Semi-structured text	●●●
	SAP compliant	SAP RFC (Remote Function Call)	ABAP function interface (proprietary)	●●●●●
	Other	Bespoke proprietary	Misc. proprietary interfaces	●●●
Intelligent logistics objects	EPC compliant	EPCIS	Semi-structured text, service binding	●●●●●
	ID@URI compliant	Dialog	Services	●●●
	PEIDs	PMI	Semi-structured text, Service binding	●●●●
	OSGi-based	OSGi	Service	●●●
	Other	Bespoke proprietary	Misc. proprietary interfaces	●●●
Digital counterparts	Multi-agent based (e.g. JADE, PlaSMa, Dialog)	ACL (Agent Communication Language)	Agent language, onotlogy	●●●●
		Agent proxies	Services	●●●
		Dialog agent	Services	●●●
		EDIFACT EANCOM	Semi-structured text	●●
Sensors & actuators	Java-based	OSGi	Service	●●●
	OGC compliant	SensorML	Semi-structured text	●●●
	PEIDs	PMI	Semi-structured text, service binding	●●●●
	Other sensors	Bespoke proprietary formats	Mainly semi-structured text	●●●●●
	OPC	OPC DA	MS DCOM	●●●
		OPC XML DA	MS DCOM, Semi-structured text, service binding	●●●

	OPC AU	Services	●●●●●
General	GDI	Remote procedure calls, GDI data types	●●●
	ORiN API	Service, DCOM, semi-structured text	●●●●
Smart Embedded Devices in Manufacturing	SOCRADES	Services	●●●
Other actuators	Bespoke proprietary formats	Misc. proprietary interfaces	●●

Table 6.2 Major Integration Targets in Autonomous Cooperating Logistics Processes

With regards to logistics IT systems, EDIFACT EANCOM and SAP RFC are the most prominent targets. However, more than 30% of systems with proprietary interfaces cannot be neglected. Consequently, a data integration approach must be able to cope with both, semi-structured, standard data exchange formats as well as function interfaces, and be flexible enough to cope with arbitrary proprietary interfaces.

To integrate intelligent material logistics objects, the support of RFID middleware standards, such as the EPCglobal Framework Architecture, foremost EPCIS, is mandatory. In addition, a means to interfacing emerging standards for the integration of PEIDs and other embedded devices is necessary. PMI currently offers the most comprehensive and structured approach to this.

The field of digital counterparts is dominated by software agent technology. The PlaSMa platform is dedicated to the support of autonomous cooperating logistics processes and is, consequently, of highest priority. Other approaches favour service interfaces. The possibility of agent communication via EANCOM strengthens the need for EANCOM support, but is at the present time not widespread.

Sensor and sensor network integration is at the present time largely a case-by-case decision, with most interface using proprietary approaches. However, emerging standards, such as PMI or SensorML are increasing in importance and should not be neglected. Therefore, a data integration approach needs to be highly flexible towards sensor data sources. With regards to actuators, a promising contribution can be found in the Unified Architecture standards put forwards by OPC. A proposed data integration approach should also take into account the standards emanating from ISO 20242 and factory automation initiatives, such as SOCDRADES.

6.4 Solution Concept – A Service-oriented, Ontology-based Mediator

The following sections outline a solution concept for data integration for an Internet of Things for autonomous cooperating logistics processes. The concept describes a service-oriented, semantic approach to data integration which addresses the requirements outlined in the previous section. The concept consists of two main solution components – first is an ontology-based mediator, second a service interface layer defining logical views upon that mediator. These two components are described in the following sections.

6.4.1 Ontology-based Mediator

At the heart of the solution concept lies an ontology-based mediator component (Ullman 1997, Wache et al. 2001 and Wache 2003), which is capable of composing queries to any combination of relevant logistics data sources. It achieves this by semantic mediation. Each data source is fully described syntactically and semantically by an ontology, which can be mapped onto the others by the mediator. Wrapper components handle the transformation to and from the relevant data sources in a rule-based fashion.

The proposed system architecture illustrated in Figure 6.3 follows the traditional pattern of a semantic mediator - besides the actual mediator component, which possesses an ontology of autonomous cooperating logistics processes, the wrapper components each contain extension ontologies, which fully formalise the data sources they are responsible for as semantic descriptions. Heterogeneity conflicts are solved either by the mediator component itself or by the respective wrapper, depending on the type of conflict.

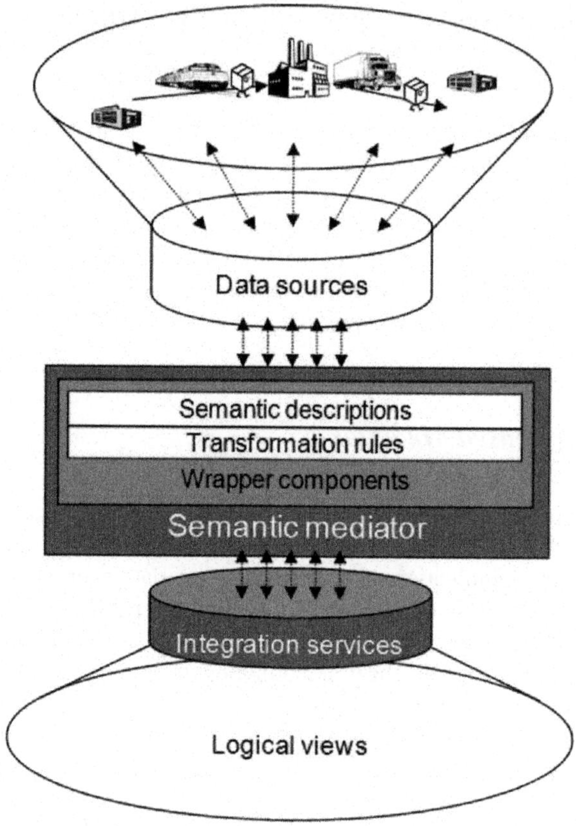

Fig. 6.3 Concept of a Service-oriented, Ontology-based Mediator

The following sections describe the components of the ontology-based mediator in more detail, on the basis of a prototypical implementation of the solution concept.

Semantic Descriptions of the Data Sources

The Web Ontology Language (OWL-DL) (Smith et al. 2004) was used for the specification of the ontology, which describes the individual data exchange formats. OWL-DL was chosen for three reasons: first of all, it is used in the multi-agent system for the description of the domain of autonomous controlled logistics. Secondly, it was judged to be adequately expressive to cover the semantic description of both, the standard exchange formats used in transport logistics and the overarching concepts of autonomous logistics processes. Finally, a number of Java libraries and reasoners are readily available for OWL-DL, which was expected to significantly accelerate the development of a prototypical implementation.

By implementing semantic descriptions and transformation rules for the major interfaces to IT systems in logistics and intelligent material logistics objects identified in Table 6.2, access to the majority of relevant data sources can be given. As a proof of concept, this was prototypically implemented as wrappers for both the EDIFACT EANCOM and EPCIS formats. Additional standards and proprietary data sources can be integrated easily by adding a new wrapper with the relevant semantic description and set of transformation rules, making the concept highly extensible. This approach also allows the service consumer to either easily integrate the required services into its own logistics IT landscape, or, for example, utilise thin clients to access a web-based GUI towards the cloud services.

Data Transformation in the Wrappers

The wrappers query data from the respective data sources and transform it via an internal format in order to enable the processing of data from heterogeneous data sources and formats. The transformation is carried out within the wrappers and is transparent to the actual mediator component. This allows for a complete abstraction from the data sources. Transformation in the wrappers is rule-based. A first prototypical realisation described and implemented these, using the business rule management system "Drools" (Drools Community 2009) (Drools - Business Logic Integration Platform). The use of Drools offered the possibility to react more quickly and flexibly to modifications to individual data sources. However, this approach proved slow and inaccurate in practice. A dedicated, generic algorithmic approach, which makes use of transformation rules stored in XML-Files, was specified and implemented. Should a change be needed, only the rule files would need to be updated. Modifications to the source code of the wrappers with subsequent recompilation and deployment can be avoided in this way.

Internal Query Interface

The query language "SPARQL" (Prud'hommeaux and Seaborne 2008) is used as the query language at the query interface of the system. It was specifically developed for querying ontologies and thus provides an adequate basis for queries to the semantic mediator. However, SPARQL only offers the possibility to query the system but not write to it. However, SPARQL alone doesn't fulfill all of the requirements, because agents representing autonomous objects also need to be able to create messages and data. To extend the functionality to support bidirectional queries, the "SPARQL Update" (Seaborne et al. 2008) language was used to extend the SPARQL query language. This allows for editing of ontologies with a similar syntax to SPARQL. A combination of both languages was specified and prototypically implemented as the query language of the semantic mediator.

Hardware Abstraction towards Dynamic Data Sources

The proposed concept also facilitates the direct integration of dynamic data sources used in logistics processes, such as RFID, sensors, sensor networks and other systems integrated into physical logistics objects. By abstracting from the physical interfaces towards these data sources, the semantic mediation approach may be applied in much the same way it is to static data sources. The abstraction layer is required to be able to provide a reliable interface, regardless of the physical accessibility of the dynamic data sources at any time. It is responsible for buffering, filtering and routing data to and from the respective data sources. It may consist of elements such as the FOSSTRAK (Floerkemeier et al. 2007), HAL towards EPC-compliant RFID, PMI (Främling and Nyman 2008) towards PEIDs (Jun et al. 2007) or OSGi towards sensor components (Ahn et al. 2006).

Interoperability with Existing Ontologies

The ontology used is a critical success factor of any semantic mediator. It has to reflect all the characteristics of the application domain and simultaneously has to be as simple and comprehensible as possible. Many existing onotologies may be taken into consideration for the semantic description of the entities in the given transport logistics scenario, such as those used in the fields of product lifecycle and data management, as exemplified by Terzi (2005), Tursi (2009) and Lee et al. (2009). However, none of these truly reflect the syntax and semantics of autonomous logistics processes whilst encompassing the syntax and semantics of standard logistics data exchange formats. Consequently, as a first step, a new ontology was designed based on both the application scenario and the top-level ontology of the multi-agent system, which describes basic concepts of autonomous logistics processes. It can be extended by incorporating additional ontologies into the system. One particularly interesting option is the alignment of the ontology with the PROMISE semantic object model, which already reflects many aspects of item-level information management of intelligent objects, albeit in the field of Product Lifecycle Management.

6.4.2 Service Interface Layer for Logical Views

A service layer is designed as the external query layer towards the semantic mediator. This design decision was made for three interrelated reasons. The first and decisive reason for a service-oriented interface layer lies in its ability to efficiently implement logical views upon the heterogeneous, distributed data made accessible by the semantic mediator. In this context, a logical view is a data model into which a subset, or even entirety, of the data, made available by the semantic mediator, may be transformed. In simple terms, the mediator pre-processes the queried data into that data model the recipient requires. The goal of the implementation of such logical views is to make the process of semantic mediation as transparent as possible to the consumers of the mediation service.

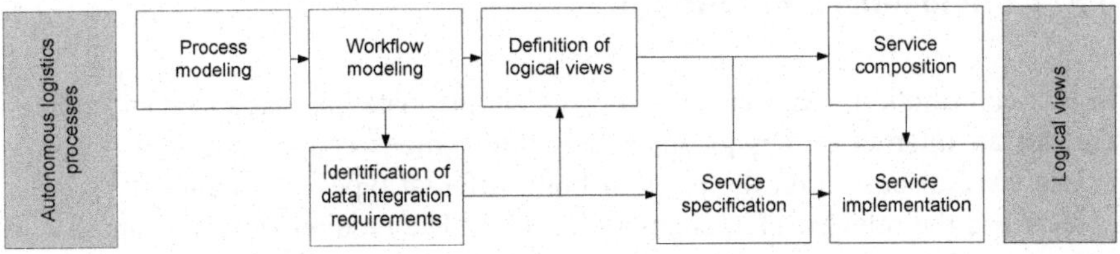

Fig. 6.4 Method for Defining Logical Views and Corresponding Service Compositions

Logical views upon mediated data may be designed effectively using the methods of service-oriented architecture. This also brings with it the distinct benefit of facilitating a process-oriented, model-driven approach to their design. The procedure sketched in Figure 6.4 may be applied to design logical views for specific autonomous cooperating logistics. The first step is to model the autonomous cooperating processes, e.g., using EPK. From that model, a workflow model may be derived in a language facilitating the model-driven design of SoA, such as Business Process Modelling Notation (BPMN). On the basis of such a model, data requirements may be identified using e.g., UML sequence diagrams. Both the workflow model and the data requirements can be used to derive a logical view, which can be modelled e.g., as UML class diagrams. In the next step, services fulfilling the data requirements may be specified. The services are then implemented and mapped to mediator queries, in this case to SPARQL queries. Service compositions may be designed based on the defined logical views in e.g., Business Process Execution Language (BPEL), which can be mapped across from the BPMN workflow model. This mapping may be guided by the defined logical views.

The second reason for designing a service interface is that a number of major integration targets already, at least optionally, define service interfaces to their data. Especially for the most relevant service bindings, such as EPCIS and PMI, the addition of native service interfaces towards the mediator has several benefits. One advantage is that systems supporting these service interfaces gain native access to all data sources integrated by the mediator. They consequently inherit all of the advantages of semantic mediation for autonomous cooperating logistics processes in their specific application domain. Such service interfaces can simply be designed as logical views, as described above – using this mechanism, the mediator might, for example, provide the logical view of EPCIS events upon a subset of the available data.

Finally, to define the proposed internal SPARQL query layer as the external interface would effectively create a further "proprietary data source" in the IT landscape in autonomous cooperating logistics processes.

6.5 Conclusions and Outlook

In this contribution, requirements towards an adequate approach to data integration for an Internet of Things supporting autonomous cooperating logistics processes were discussed on the basis of both different types of intelligent logistics objects and the relevant IT systems in logistics. The requirements indicate that in order to successfully provide intelligent logistics objects with the data they need for varying degrees of autonomous control, a number of data integration targets need to be addressed. For the operational integration of IT systems in logistics into an Internet of Things supporting autonomous cooperating logistics processes, foremost specific data exchange formats, such as EDIFACT EANCOM and ebXML need to be taken into account besides SAP RFC. However, due to the fact that almost one third of logistics systems do not comply with standard interfaces, the data integration needs to be flexible enough to efficiently cater for arbitrary proprietary interfaces.

Besides interfacing such enterprise systems, an adequate data integration mechanism is also required to cater for the integration of dynamic data sources, foremost material and immaterial intelligent logistics objects and sensors. Here, the necessity for catering for a plethora of different semi-structured and service-based interfaces defined the data integration problem. Furthermore, abstraction towards the hardware platforms of the data sources is identified as a further problem to be solved.

A service-oriented, ontology-based mediator is proposed as one approach meeting these integration requirements. Ontology-based mediation brings with it a number of advantages when tackling the identified integration issues. For one, the heterogeneous data sources need not be touched. By defining wrapper components containing semantic descriptions of the data sources along with transformation rules, data sources may be integrated in a flexible fashion. Access to dynamic data sources can be ensured using hardware abstraction towards both, physical intelligent logistics objects and sensor components. Interfaces to existing hardware abstraction middleware, such as defined by PROMISE or EPCglobal may also be integrated. Finally, a service interface layer is proposed to provide logical views for consumers of the mediation service. By leveraging the strengths of service-oriented architecture, logical views can be developed for individual consumers, according to a model-driven method. Logical views can consequently be designed on the basis of models of autonomous cooperating logistics processes, for both, participating IT systems and intelligent logistics objects.

Validation of the prototypical implementation against exemplary application scenarios has demonstrated the applicability of the semantic data integration to an Internet of Things in the field of autonomous cooperating logistics processes (Hribernik et al. 2009). The semantic mediator is proven capable of fulfilling bidirectional data integration in these scenarios. However, a number of issues remain to be tackled. Foremost is the better integration of sensors and sensor networks into the data integration approach. Here, work will be focused on the definition of a more comprehensive hardware abstraction layer for fulfilling all requirements towards the integration of dynamic data sources and sensors.

Future research will concentrate on leveraging the potential of using ontologies to semantically describe data sources. In order for the ontology-based mediator to react flexibly to changes in the autonomous cooperating process and IT landscape, methods of ontology learning may be applied to contribute to automating the currently manual process of data source description.

Acknowledgments

This research was supported by the German Research Foundation (DFG) as part of the Collaborative Research Centre 637 "Autonomous Cooperating Logistic Processes".

References

Aberle G (2003) Transportwirtschaft: einzelwirtschaftliche und gesamtwirtschaftliche Grundlagen, 4th edn. Oldenbourg, Munich

Ahn H, Oh H and Sung CO (2006) Towards Reliable OSGi Framework and Applications.SAC '06: Proceedings of the 2006 ACM symposium on Applied computing ACM, New York

Berning M, Vastag S (2007) Simulation selbststeuernder Transportnetze. In: Bullinger H-J, ten Hompel M (ed) Internet der Dinge. Springer, Berlin

Blanke K, Krafzig D, Slama D (2004) Enterprise SOA: Service Oriented Architecture Best Practices. Prentice Hall International

Böse F, Windt K (2007) Catalogue of Criteria for Autonomous Control in Logistics. In: Hülsmann M., Windt K. (eds) Understanding Autonomous Cooperation and Control in Logistics - The Impact on Management, Information and Communication and Material Flow. Springer, Berlin

Brand L, Hülser T, Grimm, V, Zweck A (2009) Internet der Dinge – Perspektiven für die Logisitk. Zukünftige Technologien Consulting

Cassina J, Taisch M, Potter D, Parlikad AKN (2008) Development of PROMISE architecture and PDKM semantic object model. 14th International Conference on Concurrent Enterprising, June 2008, Lisbon

Christensen El, Curbera F, Meredith G, Weerawarana S (2001) Web Services Description Language (WSDL) 1.1. W3C Recommendation, World Wide Web Consortium. http://www.w3.org/TR/wsdl.html. Accessed 06 September 2010.

Clement L, Hatley A, von Riegen C, Rogers T (eds) (2004) UDDI Version 3.0.2. http://uddi.org/pubs/uddi_v3.htm, OASIS Open. Accessed 06 September 2010.

Cole PH, Engels DW (2002) Auto ID - 21st Century Supply Chain Technology. Proceedings of AEEMA Cleaner Greener Smarter conference, October 2002

de Souza LMS, Spiess P, Köhler M, Guinard D, Karnouskos S, Savio D (2008) SOCRADES: A Web Service based Shop Floor Integration Infrastructure. Springer.

Do H-H, Anke J, Hackenbroich G (2006) Architecture Evaluation for Distributed Auto-ID Systems. Proceedings of the 17th International Conference on Database and Expert Systems Applications

Drools Community (2009) Drools Introduction and General User Guide 5.0.1 Final. JBoss Enterprise. http://downloads.jboss.com/drools/docs/5.0.1.26597.FINAL/drools-introduction/html_single/index.html. Accessed 07 December 2009

Ducatel K, Bogdanowicz M, Scapolo F, Leijten J, Burgelman J-C (2001) Scenarios for Ambient Intelligence in 2010. European Commission, Technical Report

Ehnert I, Arndt L, Mueller-Christ G (2006) A sustainable management framework for dilemma and boundaries in autonomous cooperating transport logistics processes. Int J Environ Sustainable Dev 5:355-371. doi: 10.1504/IJESD.2006.011555

EPCglobal Inc. (2007) EPCIS (Electronic Product Code Information Service). Frequently Asked Questions. EPCglobal Inc., Lawrenceville

EPCglobal Inc. (2009) The EPCglobal Architecture Framework, 1.3, Standard Specification. EPCGlobal Inc.

Floerkemeier C, Lampe M, Roduner C (2007) Facilitating RFID Development with the Accada Prototyping Platform. In: PerCom 2007. IEEE Computer Society, White Plains

Främling K, Ala-Risku T, Kärkkäinen M, Holmström J (2006) Agent-Based Model for Managing Composite Product Information. Comput Ind 57:72-81

Främling K, Nyman J (2008) Information architecture for intelligent products in the internet of things. In: Autere V, Bask A, Kovács G, Spens K, Tanskanen K (eds) Beyond Business Logistics. Proceedings of 20th NOFOMA logistic conference, Helsinki

Gaßner K, Bovenschulte M (2009) Internet der Dinge - Technologien im Anwendungsfeld RFID/Logistik. In: Botthof A, Bovenschulte M (eds) Das Internet der Dinge. Die Informatisierung der Arbeitswelt und des Alltags - Erläuterung einer neuen Basistechnologie. Hans-Böckler-Stiftung, Düsseldorf

Gudgin M, Hadley M, Mendelsohn N, Moreau J-J, Nielsen HF (2003) SOAP Version 1.2 Part 1: Messaging Framework, W3C Recommendation, World Wide Web Consortium. http://www.w3.org/TR/soap12-part1/. Accessed 06 September 2010.

Gupta SKS, Lee W-C, Purakayastha A, Srimani PK (2001) An Overview of Pervasive Computing. IEEE Pers Commun Mag 8:8-9

Hannus M (1996) Islands of automation in construction. In: Žiga Turk (ed) Construction on the information highway. CIB publication, University of Ljubljana

Hans C, Hribernik KA, Thoben K-D (2008) An Approach for the Integration of Data within Complex Logistics Systems. In: Haasis HD, Kreowski H-J, Scholz-Reiter B (eds) Dynamics in Logistics. First International Conference LDIC 2007 Proceedings. Springer, Heidelberg

Hong-Bae J, Kiritsis D, Xirouchakis P (2007) Research Issues on Closed-loop PLM. Comp in Ind 58:855-868. doi: 10.1016/j.compind.2007.04.001

Hribernik KA, Hans C, Thoben K-D (2009) The Application of the EPCglobal Framework Architecture to Autonomous Control in Logistics. Proceedings of the 2nd International Conference on Dynamics in Logistics. Springer, Berlin, Heidelberg

Hülsmann M, Windt K, Wycisk C, Philipp T, Grapp J, Böse F (2006) Identification, Evaluation and Measuring of Autonomous Cooperation in Supply Networks and other Logistic Systems. In: Baltacioglu T (ed) Proceedings of the 4th International Logistics and Supply Chain Congress, Izmir, Turkey

International Organization for Standardization (2009a) ISO/DIS 10303-239:2005: Industrial automation systems and integration – Product data representation and exchange – Part 239: Application protocol: Product life cycle support. International Organization for Standardization, Geneva, Switzerland.

International Organization for Standardization (2009b) Industrial automation systems and integration – Integration of life-cycle data for process plants including oil and gas production facilities – Part 2: Data model. International Organization for Standardization, Geneva, Switzerland.

International Organization for Standardization (2009c) ISO/DIS ISO 20242-1:2005: Industrial automation systems and integration -- Service interface for testing applications -- Part 1: Overview. International Organization for Standardization, Geneva, Switzerland.

Jedermann R, Antunez LJ, Lang W, Lorenz M, Gehrke JD, Herzog O (2008) Dynamic Decision making on Embedded Platforms in Transport Logistics. In: Haasis HD, Kreowski HJ, Scholz-

Reiter B (eds) Dynamics in Logistics. First International Conference LDIC 2007. Springer, Berlin Heidelberg

Lee J, Chae H, Kim C-H, Kim K (2009). Design of product ontology architecture for collaborative enterprises. Expert Syst App 36:2300-2309. doi:10.1016/j.eswa.2007.12.042

Jun H-B, Shin J-H, Kiritsis D, Xirouchakis P (2007) System architecture for closed-loop PLM. Int J Comput Integr Manuf 20:684-698. doi: 10.1080/09511920701566624

Kärkkäinen M, Främling K, Ala-Risku T (2003a) The product centric approach: a solution to supply network information management problems? Comput Ind 52:147-159. doi: 10.1016/S0166-3615(03)00086-1

Kärkkäinen M, Holmström J, Främling K, Artto K (2003b) Intelligent products - a step towards a more effective project delivery chain. Comput Ind 50:141-151. doi: 10.1016/S0166-3615(02)00116-1

Kärkkäinen M, Ala-Risku T, Främling K (2004) Efficient tracking for short-term multi-company networks. Int J Phys Distrib Logist Manag 34:545-564. doi: 10.1108/09600030410552249

McFarlane D, Sarma S, Chirn JL, Wong CY, Ashton K (2003) Auto ID systems and intelligent manufacturing control. Eng Appl Artif Intell 16:365-376. doi:10.1016/S0952-1976(03)00077-0

Meyer GG, Främling K, Holmström J (2009) Intelligent Products: A Survey. Comput Ind 60:137-148. doi:10.1016/j.compind.2008.12.005

Prud'hommeaux E, Seaborne A (2008) SPARQL Query Language for RDF. http://www.w3.org/TR/rdf-sparql-query/. Accessed 15 July 2009

Seaborne A, Manjunath G, Bizer C, Breslin J, Das S, Harris S, et al. (2008). SPARQL Update. A language for updating RDF graphs. http://www.w3.org/Submission/SPARQL-Update/. Accessed 28 July 2009

Schnatmeyer M, Schumacher J, Thoben K-D (2005) EOL Information Management for Tracking and Tracing of Products. 18th International Conference on Production Research (ICPR-18) Proceedings. Salerno

Schnatmeyer M (2008) RFID-basierte Nachverfolgung logistischer Einheiten in der Kreislaufwirtschaft. PhD Thesis, University of Bremen

Scholz-Reiter B, Windt K, Freitag M (2004) Autonomous logistic processes: New demands and first approaches. In: Monostori L (ed) Proc. 37th CIRP International Seminar on Manufacturing Systems. Hungarian Academy of Science, Budapest

Scholz-Reiter B, Kolditz J, Hildebrandt T (2007) Specifying adaptive business processes within the production logistics domain – a new modelling concept and its challenges. In: Hülsmann M, Windt K (eds) Understanding Autonomous Cooperation & Control in Logistics – The Impact on Management, Information and Communication and Material Flow. Springer, Berlin

Schuldt A, Gottfried B (2008) Selbststeuerung in der Intralogistik: Kognitive räumliche Repräsentationen für autonome Fahrzeuge. Ind Manag 24:41-44

Schuldt A, Hribernik KA, Gehrke JD, Thoben K-D, Herzog O (2010) Cloud Computing for Autonomous Control in Logistics. 40th Annual Conference of the German Society for Computer Science. Gesellschaft für Informatik

Smith MK, Welty C, McGuinness DL (2004) OWL Web Ontology Language Guide. http://www.w3.org/TR/owl-guide/. Accessed 15 July 2009

Soon TJ, Ishii S-I (2007) EPCIS and Its Applications. Synthesis Journal, 109-124. Information Technology Standards Committee. IDA Singaport & SPRING Singapore. http://www.itsc.org.sg/pdf/synthesis07/Five_EPCIS.pdf. Accessed 06 September 2010

Staake T, Thiesse F, Fleisch E (2005) Extending the EPC network: the potential of RFID in anti-counterfeiting, Proceedings of the 2005 ACM symposium on Applied computing. Santa Fe

Stojanovic ZA, Dahanayake NW, Sol HG (2004) Modeling and Design of Service-Oriented Architecture. Proceedings of the 2004 IEEE International Conference on Systems, Man and Cybernetics: Impacts of Emergence Cybernetics and Human-Machine Systems. IEEE Computer Society

ten Hompel M (2005) Das Internet der Dinge: Status, Perspektive, Aspekte der aktuellen RFID-Entwicklung. Dortmunder Gespräche 2005. Fraunhofer Symposium RFID, Dortmund.

Terzi S (2005). Elements of Product Lifecycle Management: Definitions, Open Issues and Reference Models. PhD Thesis, University Henri Poincaré Nancy 1 and Politecnico di Milano

The PROMISE Consortium (2008) PROMISE Architecture Series Volume 1: Architecture Overview. Finland: Promise Innovation Oy

Thoben K-D, Hribernik KA, Kirisci P, Eschenbächer J (2003) Web Services to support Collaborative Business in Manufacturing Networks. In: Weber F, Pawar K S, Thoben K-D (eds) Enterprise Engineering in the Networked Economy. Proceedings of the 9th. International Conference on Concurrent Enterprising, Espoo

Timm IJ (2006) Strategic Management of Autonomous Software Systems: Overview Article. Technical Report 35. University of Bremen, Centre for Computing and Communication Technologies (TZI)

Trautmann A (2007) Mulitagentensysteme im Internet der Dinge – Konzepte und Realisierung. In: Bullinger, H-J, ten Hompel M (eds) Internet der Dinge. Springer, Berlin

Tursi A (2009) Ontology-approach for product-driven interoperability of enterprise production systems. PhD Thesis, University Henri Poincaré Nancy 1 and Politecnico die Bari

Ullman JD (1997) Information integration using logical views. In: Afrati F N, Kolaitis P, (eds) Proceedings of the 6th International Conference on Database Theory (ICDT'97). Delphi, Greece

VDI/VDE Innovation + Technik GmbH (2008) Intelligente Logistiknetze mit RFID: Praxisnahe Informationen für Hersteller, Anwender und Dienstleister. Bundesministerium für Wirtschaft und Technologie, Berlin

Ventä O (2007) Intelligent Products and Systems: Technology Theme - Final Report. VTT, Espoo

Wache H (2003) Semantische Mediation für heterogene Informationsquellen. Akademische Verlagsgesellschaft Aka, Berlin

Wache H, Vögele T, Visser U, Stuckenschmidt H, Schuster G, Neumann H, Hübner S (2001) Ontology-Based Integration of Information - A Survey of Existing Approaches. In: IJCAI 2001WS on Ontologies and Information Sharing. Seattle.

Weiser M (1991) The Computer for the Twenty-First Century. Sci Am 265:94-104

Windt K, Philipp T, Böse F (2008) Complexity Cube for the Characterization of Complex Production Systems. Int J Comp Integr Manuf 21:195-200. doi: 10.1080/09511920701607725

Wong CY, McFarlane D, Zaharudin AA, Agarwal V (2002) The Intelligent Product Driven Supply Chain. Proceedings of IEEE International Conference on Systems, Man and Cybernetics. Hammamet, Tunisia

7 Resource Management in the Internet of Things: Clustering, Synchronisation and Software Agents

Tomás Sánchez López[1], Alexandra Brintrup[2], Marc-André Isenberg[3], Jeanette Mansfeld[3]

[1] Department of Engineering, University of Cambridge

[2] Said Business School, University of Oxford

[3] BIBA - Bremer Institut für Produktion und Logistik, University of Bremen

Abstract. The objects of the Internet of Things will be empowered by embedded devices whose constrained resources will need to be managed efficiently. It is envisioned that these devices will be able to form ad-hoc networks, and that the connection from these networks to the Internet of Things infrastructure will not always be possible. In this chapter we propose the use of clustering, software agents and synchronisation techniques in order to overcome the challenges of managing the resources of the Internet of Things objects. We argue that clustering will be beneficial to reduce the energy expenditure and improve the scalability and robustness of the object networks. Software agents will aide in the automation of task, both for the objects and the Internet of Things users. Finally, synchronisations techniques will be necessary to address the various challenges of harmonising plenty of copies of object data with potentially partially disconnected Internet of Things architecture components.

7.1 Introduction

Despite the many technical and operational questions arising from the Internet of Things concept and the various interpretations of what the Internet of Things is and what promises it will deliver, it appears there is general consensus that the Internet of Things will empower users and objects to share information in a seamless, automated manner. In this context, the Internet of Things promises a new generation of the Internet, in which global connectivity moves towards everyday objects and things, radically widening the scope of Internet-based applications. On these premises, the management of an escalating number of connected devices, together with a movement towards their increasing autonomy and relatively limited capabilities, pose a number of challenges that are yet to be explored.

This chapter will investigate a number of techniques aimed at addressing the challenges arising from the increasing number of connected objects, such as limited computation and energy, unreliable wireless channels and the impossibility of ubiquitous network access, repetitive and mundane user interactions, given the complexity of the architecture. Within this scope, three major interconnected topics will be explored: First, the grouping of objects into *clusters* in order to overcome scalability, energy efficiency and robustness issues, secondly, the use of *software agents* to represent and manage objects and users, moving part of the complexity to the architecture and providing a bridge between the users and the things, and, thirdly, techniques for bidirectional *synchronisation* of object knowledge in order to support operations and provide resilience in situations having only intermittent or unreliable network connectivity.

Hence, this chapter attends to represent a useful contribution on the implications of the Internet of Things vision, putting emphasis in actual problems and functional needs that will arise from a future architecture that today is little more than abstract ideas. The remaining chapter is organised as follows: Section 7.2 includes a literature review about the current state of research as well as related research areas in terms of chapter scope. Section 7.3 presents general assumptions as well as a definition about the Internet-connected objects underlying this chapter. Sections 7.4, 7.5 and 7.6 refer to the three interconnected topics, namely clustering, software agents and synchronisation, and illustrate their possible adoption within an Internet of Things. Concluding the chapter section 7.7 summarises the presented concepts and gives an outlook about the expected consideration of the described concepts within the development of the Internet of Things.

7.2 Background and Related Work

7.2.1 Clustering

Clustering is a popular method of organising wireless network topologies, in which a few nodes, the cluster heads (CH), are elected as representatives to route the traffic originated in the entire network. The clustering of intelligent computing devices has been widely researched in the fields of Wireless Sensor Networks (WSN) and Mobile Ad-hoc NETworks (MANET), although aiming at different objectives. The main objective of a MANET is network reliability and the accessibility of nodes. This is realised by building meshed networks without central authorities. Each node is connected to several other nodes, which always allow alternative communication routes from one node to another. Due to the functionality of in-network routing, all nodes act as routers with their own routing tables; this causes high activity rates of the nodes with the corresponding energy consumption. On the other hand, the clustering approaches of WSN are more hierarchical, using CHs as decentralised authorities for realising mostly star or tree topologies. WSNs vary in their objectives: there are existing approaches aiming to fault-tolerance, load-balancing, energy consumption, increased connectivity and reduced packet delay. While MANETs are generally built to handle objects in dynamic environments, WSN are traditionally used to cluster more or less static nodes. Although the mobility rates within clusters of autonomous objects within the Internet of Things is envisioned to be higher than the traditional mobility inside WSNs and MANETs, the research on those approaches is a valuable basis for the development of energy-efficient clustering methods for autonomous objects. For a better understanding of the requirements and challenges of the clustering of objects within the Internet of Things, this section will review the literature of WSN and MANETs in this area.

We would first like to compare those ad-hoc wireless clustering protocols that consider both mobility and energy-efficiency. The properties that we would like to compare are listed below. It is important to note that this comparison does not pretend to be an exhaustive listing of properties, but just those aspects that we consider most important. The studied properties are the following:

- Type: If the protocol is specifically for WSNs or for more general MANETs.
- Controlled variable CH period: CHs may be elected for periodic or aperiodic time intervals. Aperiodic intervals provide more flexibility since they can better manage the extra resources that the CH will use. We only consider those aperiodic intervals that can be effectively controlled, listing their variable that is used to compute them.
- CH election according to node conditions: If a node is elected according to its own conditions. The type of condition is also listed. Node's condition influencing its election as CH is generally a beneficial strategy since it provides first hand decision information.
- Synchronisation: If the nodes need synchronisation for either electing the CH or operating inside the cluster. Synchronisation among network nodes is costly and must be avoided when possible.
- Global cluster information: If the cluster nodes need to store information about all the cluster members in order to perform the CH election or to operate. Global information implies poor scalability with the number of network nodes.
- Multi-hop routing: Sometimes the clustering protocol may lead to developing a routing mechanism to exchange information among network nodes. Multi-hop routing mechanisms are beneficial because they can route communication packets between two nodes that are not directly connected.
- CH election complexity: An estimation of the complexity to elect a new CH. In general, the lower the complexity, the more efficient is the proposed algorithm.

Table 7.1 shows a summary of the comparison. We found five MANET clustering protocols that explicitly consider node's energy as a factor for CH election. We also found two WSN protocols that consider not only energy but also mobility. By mobility we mean not only that nodes may move inside the network, but also that the addition of new nodes and the removal or death of nodes is also considered. The small number of related work found, suggests that it is not common for MANET clustering protocols to focus on CH energy efficiency, and is also not common for WSN clustering protocols to consider mobility. Judging by the distribution of protocol types, the latter group seems rarer than the former.
All the listed protocols consider node's residual energy in order to elect the CH, although some of them use also other factors. Controlling the period that a node will be a CH is quite uncommon. Only MoCoSo has a variable CH period that is calculated upon the residual energy of the node (Sánchez López et al. 2008). Also, only MoCoSo and Onodera and Miyazaki do not require any global information while still providing multi-hop routing (Sánchez López et al. 2008, Onodera and Miyazaki 2008). MoCoSo integrates a new hierarchical routing mechanism, called Sequence Chain, that uses the addresses of the nodes to perform nearly zero cost routing along the addressing tree. Although a similar technique is employed by Onodera and Miyazaki, their protocols do not actually implement a clustering mechanism, but rather a tree formation algorithm in which parents are chosen and reconfigured according to their residual energy.

Protocol	Type	Variable CH Period	Conditioned election	Synchronisation	Global information	Multi-hop	Complexity
DMAC (Basagni 1999)	MANET	No	Weight	No	Yes	No	O(n)
WCA (Chatterjee et al. 2002)	MANET	No	Weight	Yes	Yes	No	O(d+m+1) *
LIDAR (Gavalas et al. 2006)	MANET	Mobility	Energy+	No	Yes	No	O(n)
ANDA (Chiasserini et al. 2004)	MANET	No	Energy	Yes	Yes	No	O(nxc) **
Wu et al. 2001	MANET	No	Energy+	No	Yes	Yes	O(v+N[x]) ***
Liu and Lin 2005	WSN	No	Energy	No	Yes	No	O(n)
Onodera & Miyazaki 2008	WSN	No	Energy	No	No	Yes	O(n)
MoCoSo (Sanchez Lopez et al 2008)	WSN	Yes	Energy	No	No	Yes	O(y) ****

* d: number of direct neighbours; m: number of messages regarding the cluster-related status
** c: number of CHs
*** N[x]: number of neighbours of node x; v: total number of vertex in the network graph
**** y is the number of nodes that answer a CH election messages, y <= n

Table 7.1 Energy Considering MANET Clustering Protocols and Mobile WSN Protocols

We would also like to compare all the clustering protocols in WSN that, while not supporting mobility, they consider energy-efficiency in the election of the CHs. The reason for including this comparison is the novelty of these protocols in their use of residual energy as a CH election variable, which represents the current state of the art in WSN clustering. They also serve as the proof that mobility in WSN is hardly considered.

Table 7.2 shows a summary of our comparison, following a similar column distribution as Table 7.1. MoCoSo meets most of the desirable requirements for an energy efficient clustering protocol (Sánchez López et al. 2008). The most important advantage over all the other protocols is the abnegation of global cluster information. For example, LEACH, and all the protocols that derive from it, need to synchronise their communication with the CH, which can only be done by knowing all the cluster members. EDAC, being the only comparable protocol to MoCoSo in terms of variable CH period, needs to store and update in the CH the residual energy values of all the cluster members in order to choose a successor. Every node in GESC needs to store a graph of all the cluster members in order to elect the CH. Finally, in HEED, every sensor node needs also to store a list of "candidate" CHs every time that a cluster election is triggered.

Protocol	Variable CH Period	Conditioned election	Synchro-nisation	Global in-formation	Multi-hop	Complex-ity
LEACH (Heinzelman et al. 2002)	No	None	Yes	Yes	Yes	O(n)
(Liang & Yu 2005)	No	Energy	Yes	Yes	Yes	O(n)
EECS (Ye et al. 2005)	No	Prob + Energy	Yes	Yes	Yes	O(n)
EDAC (Wang et al. 2004)	Energy	Energy	Yes	Yes	Yes	O(n)
HEED (Younis & Fahmy 2004)	No	Energy	No	Yes	No	Nit x O(n) *
GESC (Dimokas et al. 2007)	No	Significance	No	Yes	No	O(n x u) + O(n) **
MoCoSo (Sanchez Lopez et al. 2008)	Energy	Energy	No	No	Yes	O(y) ***

* Nit is the number of iterations defined beforehand
** u is the number of edges of the graph formed by the cluster nodes
*** y is the number of nodes that answer a CH election messages, y <= n

Table 7.2 Comparison of WSN Protocols

7.2.2 Software Agents

Agent Based Systems are an evolving software paradigm that strives to create software that can possess human characteristics, such as autonomy, adaptability, sociality, judiciousness, mobility and reactivity. Commonly cited definitions of computational agents found in literature are:

- Intelligent agents are software programs that continuously perform three functions: perception of dynamic conditions in the environment; reasoning to interpret perceptions, solve problems, draw inferences, and determine actions. (Hayes-Roth 1995)
- Autonomous agents are computational systems that inhabit some complex dynamic environment, sense and act autonomously in this environment, and by doing so, realise a set of goals or tasks for which they are designed. (Maes 1995)

From the above definitions, it can be gathered that for a software entity to be named an agent, it should maintain the following properties:

- Autonomy, which suggests that agents should operate without the direct intervention of external forces, and control over their actions and internal state
- A description of the current state of its environment; in order for it to perceive the state it is in. In the case that the environment consists of other agents, the agent needs to have "social-ability, i.e., an interaction protocol and language. An agent's social-ability might be collaborative, competitive or even antagonistic.
- Reactivity (reflex based agent) and/or proactivity (goal/utility based agent), meaning that they should respond to changes in the environment and exhibit goal-directed behaviour by taking the initiative and planning to reach its goals.
- Knowledge of how the agent's actions affect its environment, in order for reactivity and proactivity to happen.

Agents might also "learn" to improve their behaviour using feedback from its performance, evolve or self-replicate depending on the needs of a particular application. Agent design has been the focus of much debate in the study of artificial intelligence over the years. A good review of agent design is given in Russell and Norvig (2003).

Although there have been no attempts for agent integration within the Internet of Things, to date, software agents can greatly enhance the functionality of the core Internet of Things architecture in two ways: Firstly, user centric agents can enable the automation of user queries and alert users to any changes in specific items or trails. Users, if they wish so, can offload monitoring duties to a user agent and customise alerts to be sent to them. Secondly, product centric agents can enable the concept of intelligent autonomous products (i.e., things) to be integrated with the Internet of Things, and help bring the intelligent product concept alive, by enhancing services that the Internet of Things can offer to its users.

To date, we have seen many examples of Intelligent Products, which, at their highest level of intelligence, are physical objects coupled with computational software agents to pursue their goals. More formally, an Intelligent Product, defined by Wong et al. (2002) is the coupling of a product and an information based representation that (1) possesses a unique identification, (2) is capable of communicating effectively with its environment, (3) can retain or store data about itself, (4) deploys a language to display its features and requirements, and (5) is capable of participating in, or making decisions relevant to, its own destiny (Wong et al. 2002). For a recent review of intelligent product definitions please see Holmstöm et al. (2009). Although there has been many variations on this definition, and debates on what we expect from an intelligent product, the last decade saw examples of autonomous products that manufacture themselves (Bussmann and Sieverding 2001) and monitor themselves, ordering maintenance when needed (Brintrup et al. 2010). More primitive "intelligent" products have encompassed other parts of the product lifecycle, such as retail, service, and recycling, where products had no autonomy, but users gave decisions upon them using a combination of sensory data and decision support software. For a detailed review of the intelligent product research landscape, please see Brintrup et al. (2008).

We envisage that the connection to the Internet of Things will be the next step for intelligent product research, as the Internet of Things offers a powerful platform to connect products with other products and service providers. Using the Internet of Things, products can send updates on their status and service requests to their stakeholders. Intelligent products then can move into a mode where they autonomously and continuously look for ways to bring leverage to their owners and producers by maximising their lives in service. They optimise their production, configuration, search for replacement parts as well as find suppliers and negotiate with them. They can minimise their carbon footprint. They can promote themselves, advertise new services and alert users for service upgrades. When necessary, they can cooperate with other products to place batch orders, and compete with other products to acquire rare parts. They can report any faults to their producers and recycle themselves at the end of their lives. To enable this vision, there has to be a seamless, scalable, and lightweight integration of software agents within the Internet of Things.

7.2.3 Data Synchronisation

Understanding the Internet of Things as a new generation of the Internet, where more and more objects and things will be connected globally, it is to expect that the amount of data and information created and exchanged in this context will extremely increase. The data distributed in the Internet of Things can be stored in the objects themselves or in heterogeneous online repositories, and might exist in connected and/or (partially) disconnected environments. In order to maintain a coherent cross-infrastructure view of the object information, the synchronisation of data across the architecture components is necessary. Due to the complexity and pervasiveness of the Internet of Things architecture, it is envisioned that this synchronisation will be a big challenge, so services, such as data access on demand and data consistency, are provided.

Once the requirements of data synchronisation in the Internet of Things have been analysed, it is easy to find many similarities with distributed database systems. Bell and Grimson describe a distributed database as a logically integrated collection of shared data, which is physically distributed across the nodes of a computer (Bell and Grimson 1992). In the case of the Internet of Things, these data will be additionally distributed across autonomous and heterogeneous objects, adding even more complexity to the system. There has been a lot of research in distributed database systems during the last three decades. We believe that its results offer a valuable base for developing synchronisation requirements for the Internet of Things. The following requirements for distribute databases are outlined by Bell and Grimson (1992):

- Data Handling,
- Query Optimisation,
- Concurrency Control,
- Recovery,
- Integrity and Security.

Two additional requirements are added by Öszu (1999):

- Transaction Management,
- Replication Protocols.

These requirements need to be met in order to support efficient, secure and consistent data synchronisation in distributed databases, and can be set as key requirements for data synchronisation in the Internet of Things as well.

While older approaches designing distributed database systems, Bell and Grimson propose the use of a central instance (similar to a distributed database management system) to coordinate database activities, new approaches apply mobile agents without a specific master node. Such agents support distributed transactions and security tasks (Assis Silva and Krause 1997, Niemi et al. 2007, Krivokapic 1997). Assis Silva and Krause (1997) describe the agent-based concept as:

- Very suitable for supporting transactions processing in massively distributed environments,
- Very suitable for supporting activities in dynamically changing environments,
- Providing an adequate support for mobile devices,
- Fulfilling coordination requirements of different types of application.

Regarding the data itself, its synchronisation involves different types of information in order to ensure data consistency:

- Object data: information describing an object,
- Security data: information supporting access control to an objects information,
- Event data: information about an objects history.

Knowledge about the structure, syntax and semantic of object information is required to filter data that has to be synchronised. There are different approaches and standards providing such knowledge. Table 7.3 gives an overview about the related work in this area. It is not intended to be complete, but to give a summary of work that may support data synchronisation in heterogeneous, distributed environments, such as the ones found in an Internet of Things architecture.

Due to the distributed locations of object information in the Internet of Things, the network availability between all information resources is of a special interest. Suzuki and Harrison (2006) show different scenarios describing possible operations on RFID tags in connected and disconnected environments and the corresponding synchronisation operations required to update a central database managing tag data. The authors also introduce a proposal for a Data Synchronisation Protocol. Pátkai and MacFarlane (2006) also show a classification of data synchronisation scenarios.

Reference	Description	Related data
Bonuccelli et al. 2007	Clock synchronisation and global time	Event data
Cilia et al. 2004	Concept-based approach to provide content information	Semantic
Grummt 2010	Requirements for Item Information Services and in Discovery Services	All data and semantic
Canard and Coisel 2008	Scheme for key synchronisation supporting RFID authentication	Security data
Ray et al. 2000	Model of Semantic correctness	Semantic

Table 7.3 Related work in the area of synchronisation

As mentioned above, common approaches require a stable or at least partial network connection. The Internet of Things is envisaged to contain distributed heterogeneous databases, applications and services, which might be always connected, partially connected or even permanently disconnected. All of these components will potentially receive new or updated object information, and, therefore, a new approach to secure data consistency in the Internet of Things has to be researched.

7.3 Assumptions and Definitions

The Internet of Things advocates the extension of the Internet infrastructure that we know today towards the inclusion of objects or *things* as information producers. For the sake of clarity, we could define these objects as manufactured items, whose information and state is relevant to some service or application that is connected to this Internet of Things and, therefore, to the users that make use of those services. Some manufactured objects may require power or certain computation or communication to fulfil their primary purpose. This is the case with electrical appliances or computing equipment. These objects might evolve to support the electronic hardware and software necessary to gain access to the Internet of Things infrastructure. Other objects, whose traditional purpose doesn't require power, computation or communication of any kind, will not be able to produce any information. Therefore, there is a need for devices that can be attached (and eventually embed) to them and produce information on their behalf. The capabilities of those devices will greatly influence the level of participation of those objects in the Internet of Things, the same way that the evolution of an electrical appliance towards the Internet of Things connectivity will influence its level of participation on it. For the rest of our discussion, however, we will assume that the participation level of any Internet of Things object is based at least in the following capabilities:

- A unique identity
- Ability to sense and store their condition. Condition is the status of an object obtained by interpreting the output of sensor transducers associated with it
- Ability to make their information (be it identification, condition or other attributes) available to external entities
- Ability to communicate with other objects
- Ability to take decisions about thereselves and their interactions with other objects

Since the Internet of Things enablement of objects without any previous power, computation or communication capabilities, is specially challenging, the rest of the chapter focuses on the challenges arising from them, although many of the discussion here can be applied to any Internet of Things object. We will therefore assume that the communication capabilities of the devices that represent the objects are realised over the air using radio signals. Therefore, all the protocols that will be discussed in this chapter assume wireless communications implemented by the devices that represent the objects to which they are attached. In the context of wireless networking, an independent computing agent is generally called a "node". For this reason, we will call the devices that represent the Internet of Things objects "nodes".

The Internet of Things architecture needs to be supported by an infrastructure that connects all the architectural components. This infrastructure would have the current Internet as its core backbone, as the Internet is the most pervasive global computer networking infrastructure available today. The devices attached to the objects mentioned above would need to connect to the infrastructure in some way. Some argue that on an Internet of Things, the things themselves need to be connected directly to the Internet. However, although the IETF and other global organisations are working on embedded Internet protocol stacks, such as the 6LoWPAN or the ROLL, a generic Internet of Things architecture should not require these kinds of capabilities (Kushalnagar et al. 2007, Vasseur et al. 2010). A good reason for this is the already existing great number of legacy networking protocols that cannot be adapted to work directly on top of IP stacks (e.g. mobile phones, proprietary WSN systems). Another reason is that many of the low cost devices that could create a really pervasive Internet of Things cannot support even the lightest of the proposed embedded Internet protocols (e.g. RFID tags, low cost WSN). In this chapter, we assume that local networks of objects can communicate with the Internet of Things infrastructure transparently, either directly with the support of IPs, or via gateways that can translate legacy protocols to the ones used on the Internet. Many times we will refer to "infrastructure gateways", meaning the computing devices that serve as bridges of local networks to the infrastructure. Those bridges may or may not provide translation services.

Following the definition of object and their characteristics above, along this chapter we will also assume that objects can create networks with other objects. We will also refer to the clustering capabilities of these networks of objects, where clustering is a particular mechanism for organising the objects into networks. Generally speaking, a network may contain several clusters, and certain elected members of those clusters communicate among each other creating a certain hierarchy. It would also be possible to create further clusters with these elected members, creating a double clustering network architecture. For example, an elected member of the cluster elected members could be chosen to communicate with the infrastructure gateway, creating a single elected representative for the whole network and, therefore, for all its clusters and objects. For simplicity, in this chapter we will focus in a single clustered network, and will use the terms "cluster" and "network" interchangeably. This assumption does not limit the discussion, as the same concepts could be applied if several layers of clustering would be considered.

Finally, in an Internet of Things context, the words "objects", "things" or "products" are often used interchangeably. In this chapter, we will use any of the aforementioned words to refer to the "things" of the Internet of Things. Conversely, the words "intelligent" and "smart" are used extensively in the same context to denote the capabilities of those things to process information and to make informed decisions that influence the objects life and that of its surroundings. We will use any combination of those words to refer to emphasise the computation and reasoning capabilities of the Internet of Things objects.

7.4 Clustering for Scalability

7.4.1 Clustering Principles in an Internet of Things Architecture

Objects such as goods, product parts, assembly machinery, logistics and transportation items (e.g., pallets, containers or vehicles), warehouses, retailer's facilities or end-user assets are eligible for condition monitoring and can provide valuable information for themselves or other objects in their vicinity. In order to monitor their condition, embedded devices with wireless communication capabilities could be attached to them, becoming a part of the object, the same way a barcode sticker is part of the vast majority of today's products.

WSN are excellent candidates for becoming the devices attached to the objects of the Internet of Things, because many of its principles of operation address the Internet of Things requirements. These requirements include the clustering needs and the assumptions presented in section 7.3. Nevertheless, there are a number of differences between the "traditional" WSN and the devices that we propose will represent the Internet of Things objects. The main differences include the lack of a standardised unique identification scheme, the assumption of static deployments, the assumption of centralised base stations and the inflexible topologies that WSN are usually constructed upon.

Common WSN features include multi-hop communication, cooperative applications and events triggered inside the network. These are characteristics of active (as opposed to passive) networking. A clustering design for the Internet of Things requires the use of active networking to create collaborative, multi-hop and always-dynamic interactions among objects, which are equipped with wireless embedded devices as mentioned in section 7.3. This strategy extends the paradigm on object information gathering, since now it is possible not only to communicate the status of more than one object at the same time, but also to trigger the reporting of information in a bottom-up approach with no need for external control (i.e., there is no need for readers to initiate the reporting process, as is the case in passive RFID). What is more the ability to forward messages inside the same networks and provides the opportunity for distant objects, which were previously unable to reach the reader in a single hop, to report their status information to the system.

Active networking also creates the possibility of extending the information of an object by using other nearby object's information to enrich its status. To maximise the potential of these attributes, it would be beneficial to design a data structure model which organises the information of all the objects. The maintenance of such structure would be rooted on events originated at the active network itself, providing a real-time repository of network and object information. This structure would also provide the basis for sharing real-time object information among several information consumers. The type of information stored in this online repositories, as well as its synchronisation with the objects' real-time data, is a topic that we address in section 7.6.

One of the most important limitations of the devices attached to objects is power: since the devices are not powered by readers but by batteries, every action in which the device is involved, such as sensing or using the wireless transceiver, consumes part of its energy. For this reason, the protocols that manage node communication and networking must be carefully considered, since these devices are expected to function for months or years with the same battery charge. Clustering can be used to manage the power of the devices that represent the objects in the Internet of Things networks. In essence, clustering extends the network lifetime by electing a representative network member, or CH, which collects all the communication within the network and forwards it to the outside (so-called data aggregation). CHs consume more energy than the rest of the network members and their role must be periodically rotated in order to avoid the premature exhaustion of their battery power. The rotation of each CH is realised through an election process which computes the "best" candidate for the next period. This election may consider the particular *static* capabilities of each node (e.g. longer radio range, computing power), as well as its current *dynamic* status. One of the most important dynamic attributes of a node is its current residual energy. The election of the best candidate is paired with the decision of how long it will remain as the new CH. In the same way, the election itself is based on dynamic and static information about a node and the time that a node will have. The role of CH can be static (i.e. always the same time) or dynamic (i.e. a different time for each CH election), depending on the attributes of this specific node.

The election mechanism should be a distributed decision via collaborative messaging among all the nodes of the cluster. Centralised solutions cannot provide the scalability features that the Internet of Things should encourage, specifically when networks and clusters may be formed of hundreds or thousands of nodes. The election mechanism should also be dynamic, in the sense that changes on the network (e.g. the election of a new CH) should not be limited to static events, such as the exhaustion of an CH's representation period, but should also be triggered by unpredictable changes, such as the addition of new objects to the group or the removal of a group of objects that were part of a particular cluster. The need for the support of dynamic operation is a result of the differences between WSN and Internet of Things architecture outlined above, especially due to the mobility of things.

In this section, so far, we have outlined how the principles of clustering can provide benefits regarding the management of resources inside Internet of Things object networks. However, clustering is a general strategy with many dimensions, and the protocols that manage the clustering mechanisms have to be tailored to the specific challenges and needs of the Internet of Things architecture. The rest of this section is dedicated to provide a number of design guidelines based on this early evaluation.

7.4.2 The Role of Context

The clustering of autonomous objects into groups requires similarities between the participating objects, which sufficiently confine the cluster groups from each other. For the detection of such similarities the autonomous objects always require the best available up-to-date information to arrive at a substantiated and aim-oriented clustering decision. That information can arise from two different sources: out of the physical environment of the object (e.g. other objects, infrastructure gateways, environmental parameters) and/or by connecting spatially separated resources (e.g. central databases) across the Internet of Things. Due to the possibility of disconnected environments, in which object networks may temporally loose connection to the Internet of Things infrastructure, the lack of knowledge about the objects environment and its surrounding situation as well as the systemic objective of robustness, a central clustering authority is impractical; clustering decisions require the objects' direct involvement and depend on the objects' own information, especially the objects' context and their capability of context awareness. In the context of ubiquitous and pervasive computing, there are several definitions for the term of context (awareness) (Crowley et al. 2002, Dey 2000, Schilit et al. 1994, Brown et al. 1997, Ryan et al. 1998). We will use the following definitions by Dey (2000) and Schilit et al. (1994) to set the basis for the rest of our discussion about context:

> "Context is any information that can be used to characterise the situation of an entity. An entity is a person, place, or object that is considered relevant to the interaction between a user and an application, including the user and applications themselves." (Dey 2000)

> "A system is context-aware if it uses context to provide relevant information and/or services to the user, where relevancy depends on the user's task." (Dey 2000)

> "Such context-aware software adapts according to the location of use, the collection of nearby people, hosts, and accessible devices, as well as to changes to such things over time. A system with these capabilities can examine the computing environment and react to changes to the environment." (Schilit et al. 1994)

The usage of environmental knowledge enables contextual clustering, which does not necessarily depend on a comparison of predefined characteristics or attributes of the participating objects (e.g. shipment destination); on the contrary, it is based on the object's situational information and status. As an example, consider the clustering process of objects depending on the residual energy of the surrounding objects, the duration of the proximity between them (i.e., neighbourhood) or the degree of similarity between the objects' tasks. Hence, contextual clustering offers an access to self-categorised object groups over the Internet of Things; it is driven by a contextual rather than a process perspective (e.g., process oriented package flows). The utilisation of situation-dependent attributes enables a precise and useful clustering and allows a human-like understanding of the objects' situations. However, a disadvantage of a pure application of the contextual clustering conceptualisation is the uncertainty about the rate of change of the context, namely, the time-dependent validity of the context and the differentiation between the long-term and short-term validity of the context. This time, dependency could create problems with a high rate of changes on the context surrounding a particular group of objects, triggering a high number of re-clustering processes, which will incur in the use of too many network resources. Additionally, there is the challenge of the context awareness itself, since it would be desirable for the objects to possess a generic context analysing power not limited to specific parameters (e.g. fuzzy logic, complex event processing). This would also results in high demands in terms of the computing power and, consequently, in high energy consumption. A compromise could be a hybrid clustering approach, using the objects' characteristics as well as the predefined objects' context. Hybrid clustering could reduce the re-clustering effort while partly benefiting from situation-dependent clustering advantages. In dynamic and mobile scenarios, such as those encountered in the Internet of Things, the role of context for clustering should be considered, since it would bring significant contribution for precise and efficient clustering.

7.4.3 Design Guidelines

The development of clustering algorithms for physical objects involves a number of components and protocols. Some of them are necessary for the logical infrastructure within the clusters and serve as the basis for other services and applications to build upon. These necessary components, which are fundamental for an efficient clustering process, include the CH election process, a suitable addressing scheme and an efficient routing procedure.

CH Election

This chapter proposes for the Internet of Things device networks to utilise clustering for power management and, therefore, for scalability. Although a detailed description of clustering and its benefits was presented in section 7.4.1, let's recall that the main objective of clustering is to extend the object network lifetime by electing a representative network member which collects all the communication within the network and forwards it to the outside. This section outlines the CH election mechanism that an Internet of Things device, representing an object, would take as part of a clustered network of devices. As an important and powerful feature of this design, CHs are elected according to their residual energy.

Not every device would be eligible at any time to become a CH. It is possible for certain devices to be in range with an infrastructure gateway while some other remain "hidden" or out of range. Apart from the energy efficiency requirements, the CH election procedure should also avoid choosing a CH which is hidden while another CH from the network is in range of an infrastructure gateway. To address this issue, infrastructure gateways could send advertisement packets to announce their presence. Only CHs that receive an advertisement would participate in the CH election process.

Let T_{CH} be the duration that a node will have, once the CH role has been elected. T_{CH} could be calculated as a function of the node's residual energy:

T_{CH}= C x Residual Energy, where C is a constant

In each CH, a node would calculate its own proposed T_{CH} according to the above equation, and would send its proposal to the rest of the cluster members. A consensus decision would be made, and the selected node would become the new CH for the time T_{CH}. Although the factors that would elect a new CH could vary, a straightforward decision would select the proposal with the highest computer T_{CH}, since this would minimise the number of CH election procedures and, therefore, conserve more energy over time. A random delay, also function of the node's residual energy could be introduced to avoid collisions in the wireless channel when a big number of nodes are in the same cluster.

A CH election procedure would start when any of the following situations occur:

- T_{CH} expires
- The current CH cannot communicate with any infrastructure gateway
- A CH, whom did not participate in the previous election and has more residual energy than the current CH, receives an advertisement packet from a gateway
- A CH cannot communicate with the current CH before T_{CH} expires
- A new object is added to the CH's network.

According to this, if a network loses its CH and no CH receives an advertisement packet from a gateway, the election process would not start again to choose a new CH. This situation is undesirable, because the condition information of the associated objects may still be useful locally (e.g., for storing it in the node's memory for a later synchronisation – see section 7.6). Moreover, it is not practical to reject new associations due to temporal disconnections. To avoid this problem, the CH election procedure could be started by any node which runs for more than a certain amount of time without being able to communicate with its CH. When the connection with the information infrastructure is re-established, networks with a CH selected in this way would start a regular CH election procedure again.

Cluster Membership

The above mechanism to select a CH would take place inside a cluster of nodes. But how do the nodes decide to become part of the same cluster?

In order for objects and networks to find other objects and networks, one or several nodes could send periodic discovery broadcast packets. Nodes receiving these packets would process the packet information and decide to become part of the same network or cluster by sending back a response packet. The results would be communicated to all the network members and a new CH election process would begin. This process could involve several objects and networks at the same time. We could call the process "association" by which multiple objects and networks join together to form a unique cluster.

```
Association request reception
-----------------------------------

Input: List of association requests
Output: Association response, update to local network

if (list of requests NOT null)
        HAL = list of requests from networks with highest number of nodes;
        if (I am in HAL)
            #Nodes: #Nodes + #Nodes in the list of association requests;
            Initiate CH election
            Send association response to the list of association requests;
            Update my network;
        end
end

Association response reception
-----------------------------------

Input: Association response
Output: Update to local network

According to the association response...
        change my network ID;
        change my address;
        update #Nodes;

Update my network;
```

Fig. 7.1 Algorithm for Global Knowledge by Localised Association Procedures

As presented earlier, the objective of object clustering is not only to manage the energy resources of the network, but also to form collaborative groups of objects that share a common situation or purpose. For this reason, we could propose an association procedure which considers aspects of the object's nature in order to filter and classify potential object interactions before they occur. This association procedure would therefore have two phases: The first phase, in which association requests would be organised and filtered, and a second phase, in which the final association procedure would take place, and which would include the update of the cluster attributes, such as the election of a new CH or the re-factoring of routing addresses.

Association requests could include information not only about static attributes of the objects (e.g. ID, address, network), but also dynamic and contextual attributes. We could divide these attributes into two groups, regarding the need for their presence in order for the analysis of the requests to proceed. *Mandatory* attributes would need to be held by both, the request sender and the receiver, while *optional* attributes would not be a pre-condition for the association procedure to continue, but rather they would add up to the decision on membership of the requesting object. At the same time, due to the heterogeneity of the Internet of Things, attributes coming from different objects might not always be totally compatible, and a certain degree of fuzziness in the extraction and comparison of association request attributes would be necessary.

While the comparison of static and dynamic attributes could be relatively easy to implement, contextual information is more difficult to compare. For objects involved in different contextual situations, the contextual attributes might mean different things. For example, for objects involved in a shipment, the shipment's destination would be the most important contextual attribute. For objects stored in a warehouse, the delivery date could be more important than the destination. This kind of contextual clustering could be realised, for example, by defining priority decision rules for different types of autonomous objects depending on possible locations. However, this methodology would just take the 'meta-context' into account; but there would be as many contexts as objects are present in the situation, deduced from the individual object perspectives. Even if the membership decisions could be limited to the object's location, the 'meta-context' of the situation, the combination of present objects, their internal status and the integration of context into membership decisions would still be a challenge for the computing capabilities of embedded devices. We could envision that this first phase of the clustering mechanism would grow in complexity as the capabilities of the object's embedded devices would increase, from just location-based prioritisation to the analysis of complex statistical data and rules. However, it would also be necessary to evaluate the complexities of such algorithms against the quality of the resulting clusters, since any complex processing would have important repercussions on the energy expenditure, the scalability and the robustness of the resulting clusters.

In wireless networks, a network identifier is usually selected to distinguish a particular network from the others. If clusters of objects are built in a meaningful and contextually rich manner, this identifier could also provide a useful hint on the 'theme' of the cluster. The objects from the Internet of Things are likely to have unique identifiers which follow a meaningful encoding. Examples of this encoding can be found in the Electronic Product Code set of standards (Armenio et al. 2009). When a cluster is updated with the addition of new members, decisions should be made on which identifier would represent the resulting cluster. This decision could be taken in a second phase of the association procedure. Another important task of this phase would be the assignment of local addresses for communication and routing. Some guidelines on the design of an addressing scheme will be given later in this section.

Many clustering mechanisms, as well as other network-wide operations, may require the knowledge of the number of nodes of the network, or the number of nodes in particular parts of the network, such as an address branch for hierarchical addressing. Often, this knowledge is considered as a 'global' attribute, since only a 'global viewer' would have access to the information of every single node in the network. In distributed networks, such as the one that we are describing in this chapter, it is however possible to set mechanisms in place that will keep every node updated of global attributes with little extra processing. In the case of the number of nodes of the network, the association procedure could keep track of them with the assumption that every network starts up with a single node, and that successive association procedures are build up to create large clusters and networks. When a particular attribute needs to be equal network-wide (e.g., the network ID), this global knowledge could be very useful, for example, to decide which party in an association keeps its attributes and which one will have to update them. Figure 7.1 presents a simple algorithm demonstrating how this global knowledge can be obtained from localised association procedures. This algorithm can be made more complex by taking into consideration issues such as:

- Calculation of addresses
- Hierarchical node structures (e.g., which node will become the parent and which the children)
- Calculation of the number of nodes in specific hierarchy branches to decide, for example, which branch will have to re-assign its addresses
- Changes on the direction of the parent-children relationships due to network mergers.

A disassociation procedure would need to be undertaken if an object or group of objects leave the network. This process would need to update all the attributes presented before, such as the number of nodes per network or the addresses of the nodes. The algorithms would also need to include provisions for re-merging a network whose routing structure was broken, the selection of a new network identifier if the previous one is not representative enough after the disassociation, the election of a new CH if the previous CH left the network, etc.

Addressing and Routing

Dynamic address allocation is a common problem for wireless ad-hoc networks where mobile nodes constantly join and leave the network. Unlike wired networks, which lack of strong power and infrastructure constraints, mobile networks must optimise their operation and keep the connectivity even when unexpected topology changes occur. Wireless Sensor Networks pose additional challenges due to their especially scarce resources.

Due to the complexity of fitting a full protocol and application set into the sensor nodes, WSN dynamic addressing has been somehow left aside in favour of other research areas, such as routing, MAC layer design, synchronisation protocols, etc. However, the recent interest in the community to design networks that can adapt to the environment is forcing to reconsider all aspects of WSN autoconfiguration and maintenance, including dynamic addressing.

One of the main challenges in the mobile object networks of the Internet of Things is the support for dynamic associations of objects. The purpose of considering dynamic association of objects is to provide a flexible solution in which the network members do not need to be known in advance. In this way, a great variety of applications can fit into the architecture without the obstruction of an inflexible design. The networks resulting from object interactions are likely to be relatively small: a group of boxes or pallets in a freight, robots in an assembly line, containers in a cargo bay, parts of complex assets, such as machinery or vehicles, etc. However, it is also likely that the shadowing effect provoked by objects in the environment would prevent every object to reach the network CH in a single hop. Furthermore, larger networks are also viable, such as those in big warehouses or retail shops. For these reasons, we would like to consider a multi-hop addressing and routing protocol, simple enough to adjust to the object's embedded devices constraints but fulfilling all the requirements of the Internet of Things networks. We devise the following requirements for an Internet of Things addressing scheme:

- Addresses must be unique inside a network
- Addresses must be reused when objects leave
- Addressing must be dynamic. Address assignation should be fully distributed
- Addressing must be scalable
- Support for network merge and split should be provided
- The protocol overhead must be minimised

To meet these requirements, the following list summarises the ideal properties that an addressing scheme for the Internet of Things devices should have:

- *Hierarchical*: The nodes involved in this addressing scheme are the devices that represent objects. Nodes receive addresses organised in a tree structure. When two or more objects associate, the object that computes an address becomes a parent. Hence, senders of association requests become children.
- *Distributed and unique:* Each node should be only responsible for assigning addresses to its children. The address a child receives should be derived from its parent address in a way that makes that address unique for the network.
- *Scalable:* If a node leaves a group, its address should become automatically available for any other node joining with the same parent. Moreover, the ad-

dresses should not be limited in size by the scheme, but rather increase their size as the network becomes bigger. Network merges should also be supported by reassigning the addresses of the network with the smallest number of nodes.
- *Low overhead:* A parent should only need to know its immediate children to assign addresses in a unique manner. The addressing scheme should provide routing along the tree with nearly no cost. A node could know how many hops it is away from any destination by just analysing the address, and parents could route packets following the tree by just comparing its address with the packet's destination address. Hierarchical assignation of addresses should allow parents to provide shortcuts to the destination in an equally simple way by routing packets to neighbours that are closer to the destination than following the tree.
- *Extensible:* The addressing scheme should provide mechanisms to allow an unlimited number of children per parent even if the original limits for address space assignation per parent have been reached.

An addressing scheme that fulfils all of these properties was proposed by Sánchez López et al. (2010) as part of a distributed protocol for the management of resources inside Smart Object networks (Sánchez López et al. 2008).

7.5 Software Agents for Object Representation

The Internet of Things envisages a global architecture in which objects become first class citizens of the Internet, and therefore they are not only able to report their status to human users via computing infrastructures, but are also able to communicate with each other and other Internet of Things components in order to influence their own destiny. Although the Internet of Things can potentially serve any type of objects, commercial products and assets constitute one of the main drivers for its conceptualisation. On this note, we would like to discuss the influence of products in the Internet of Things vision, and use this discussion to introduce the role of software agents for object representation. We start this discussion by listing the communication scenarios that might typically occur in a product lifecycle:

Product to product communication:

- Products that request service, asking other products for batch orders
- Problem co-diagnosis, where products of the same firm consult each other for undiagnosed failure modes

Product to supplier communication:

- Products asking service provision (including recycling, maintenance, scrappage, and logistics) from suppliers with a given service request, time, and price
- Manufacturer communicating product upgrades or recalls to products

- Products communicating performance data to manufacturer

Product to user communication:

- Product performance and actions
- Product location and state
- Upgrades, promotions and additional services

The Internet of Things may benefit from the automation of the above communications and actions between product and user (suppliers, manufacturers and owners). Computational agents provide us with a suitable abstraction as well as practical tools to carry out this task. Agents could be used to take on mundane monitoring tasks from a human user, such as the monitoring of a set of products travelling across a supply chain, the arrival of products on a specific location, the divergence of a product from a pre-specified path, the health, expiry or maintenance actions of a product and so on. The human user shall be able to pre-configure queries and subscribe to alerts on these queries. In addition to monitoring users, we might encounter the need for service providing agents in the Internet of Things. These could be organisational agents that offer maintenance or logistics to products, for example.

The Internet of Things would require the automation of data gathering and analysis on the "things" by making the things themselves responsible through the intelligent product paradigm. Hence we need to examine ways in which objects can characterise themselves, communicate with others and automate their actions. One suggestion might be a Service Oriented Architecture (SOA), which refers to a design paradigm where various services can be loosely coupled and accessed via "Web Service protocols", such as XML, SOAP, WSDL, etc. This method promotes a service view rather than a product based view. On the other hand, academic literature and industrial frontrunners in the area argue that a product centric view is an intuitive one that distributes risk and reduces bottlenecks (Brintrup et al. 2010). Given that multiple organisations and objects will use the Internet of Things architecture throughout a product's lifecycle, it is important that a scalable and interoperable architecture is proposed, which reduces reliance on centralised databases and processing. An important point here is to aim for a generic architecture that can be used in as many product lifecycle scenarios as possible. Having solely an SOA-based architecture could bias the architecture towards sensor nodes that require high processing power and memory in order to wrap messages that are compatible with web services. This would have a cost implication and, therefore, bias the use of the architecture to complex high value products, and is consequently unfavourable.

Agent based systems and associated technologies, such as object-oriented, peer-to-peer and service-oriented architectures, have matured to the point where intelligent products can leverage their potential. An agent oriented viewpoint provides an intuitive encapsulation to the intelligent product, while being practical. The approach allows complex decision making without having to go through many architectural layers. Recent advancements in agent-based open software, such as Open Source Cougaar and JADE (COUGAAR 2010 and JADE 2010), point to synergetic environments where SOA principles work in harmony with those of agent-based systems. The Internet of Things shall therefore aim to provide an architecture that will make use of the strengths of both, SOA and agent-based systems. Intelligent reasoning can then occur at each level of the system to reduce overall system load and increase responsiveness, while SOA principles, such as modularity, reuse and abstraction, will be exploited.

The current thinking in the area points to the use of an architecture that will allow agent characterisation using the Ontology Web Language, agent definition via re-usable XML based plugins, a discovery service for finding the requested service, followed by a one-to-one interaction between the provider and client, similar to SOA. For instance, finding supplier user agents through yellow pages and then negotiating with them on a one-to-one basis may well follow this procedure. There may also be instances where this exact procedure is not required, for example, when products need to communicate to one another to arrange batch orders, negotiate scheduling, or learn from one another for co-diagnosis. There may also be complex decision making at the object level, such as deciding the next step of production or which sub-components to recycle. Having agents representing objects on the network and processing these decisions would make the architecture work faster and be more applicable to a large number of scenarios. Since we might potentially have millions of product agents on the Internet of Things, the product characterisation, data storage and communication protocols shall be as lightweight as possible. Here, clustering can play the important role of increasing scalability, and the same concepts applied to the clustering of physical devices can be extended for clustering of agents. In fact, with appropriate and accurate-enough information, many of the clustering burdens could be relegated to the agents themselves, which would reside in parts of the architecture with much more readily available computing resources than those of the embedded devices in the objects themselves. Decisions taken by the agent system would then be synchronised with the physical world, providing a balance between scalability, resource management and real-time information.

7.6 Data Synchronisation

7.6.1 Types of Network Architectures

Data synchronisation in the Internet of Things depends on the availability of connectivity within the network architecture in which the objects are moving. There are three types of architectures that have to be examined:

- Connected architectures (Internet),
- Partitioned architectures (Intranet, Extranet),
- Disconnected architectures (local resources).

Connected Architectures
The Internet is a globally distributed network of computers carrying many services and information resources. It is assumed that resources are always connected to share and update information. Objects moving in a connected architecture do not need a special synchronisation method, because object information can be updated at any place inside the network in real-time (see Figure 7.2).

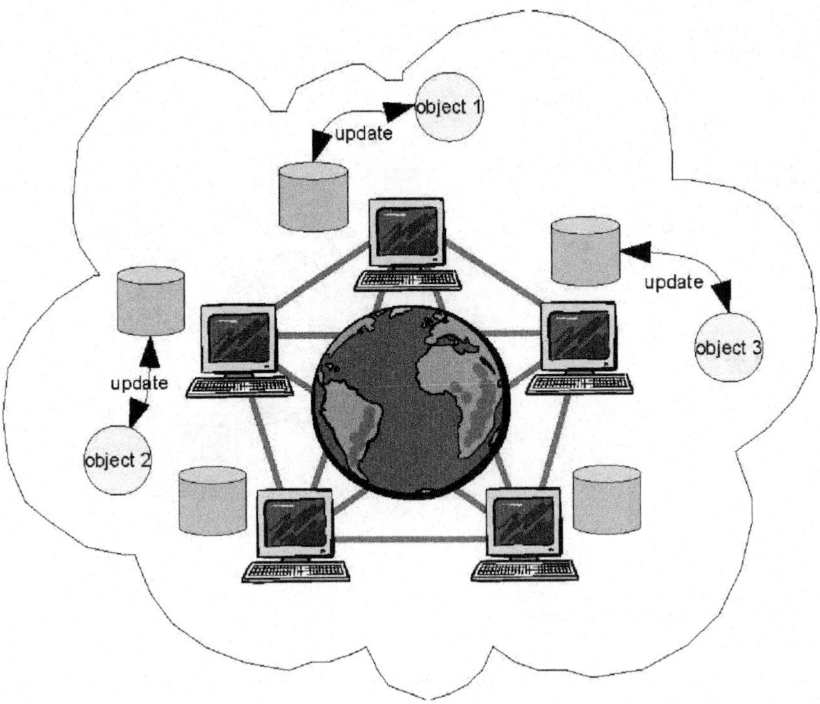

Fig. 7.2 Internet Architecture

The EPCglobal Architecture, as a proposal for implementation of the Internet of Things supporting the logistic supply chain, is based on this connected architecture (Armenio et al. 2009). Objects may be identified by the Electronic Product Code while related information is stored in distributed repositories. Capturing and accessing object information requires a stable network connection.

Partitioned Architectures

Intranet and Extranet may be seen as types of partitioned networks. They base on the same technology as the Internet, but they are only reachable inside a closed area (Intranet) or with a special authentication (Extranet). Beside partitioned networks, there are partially disconnected devices, such as mobile devices, which are disconnected while updating object information (e.g., for maintenance). Data will be updated at the mobile device first and has to be synchronised to related network information resources when the mobile device will be connected again. Synchronisation is similar to those supporting objects moving between partitioned networks. But there is a new complexity, because a mobile object may be connected to the network before data is synchronised between the mobile device and the network resources (see Figure 7.3).

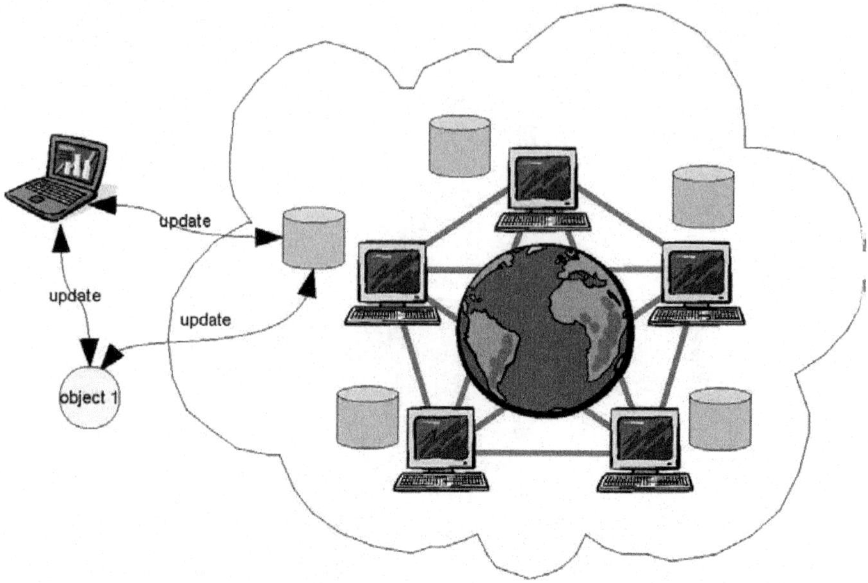

Fig. 7.3 Partitioned Architecture

Objects moving in partitioned networks may act the same way as objects inside the Internet: Objects move within the physical world while information is generated and exchanged within information networks. Information about an object might be generated while it has no reliable network connectivity (e.g. accumulation of sensor data). Therefore, it is necessary to have a robust mechanism to synchronise such information updates to the network when connectivity is available, in order that the data can be available within a single company and also shareable with other organisations. Other challenges for data synchronisation to be considered, are synchronising data (e.g., configuration instructions) from the network to an object that only has intermittent connectivity, or interpreting the current 'state' of an object correctly, when some of the event information that should contribute to the state of determination arrives late, out of sequence, or not at all.

Disconnected Architectures

Many things that interact with the Internet of Things may not have permanent reliable connectivity to communication networks. This may be caused by missing technical resources (e.g., in remote areas) as well as technical restrictions that may not allow Internet connectivity (e.g., in dangerous production areas). Nevertheless, there might be objects that have to be maintained inside such areas. Those things would need information about their life cycle without having permanent reliable Internet connectivity. By contrast to their special needs, many of today's Internet-based applications require a stable network infrastructure and routinely store object information into networked repositories.

A permanent disconnected architecture has to distinguish between objects and other resources without network availability. Objects may be disconnected, because they are unmovable or only moving in environments without network availability. Mobile devices could be used to exchange information between disconnected objects and other network resources. In the case of disconnected applications or repositories, information could be exchanged via mobile objects. Both cases are shown in Figure 7.4.

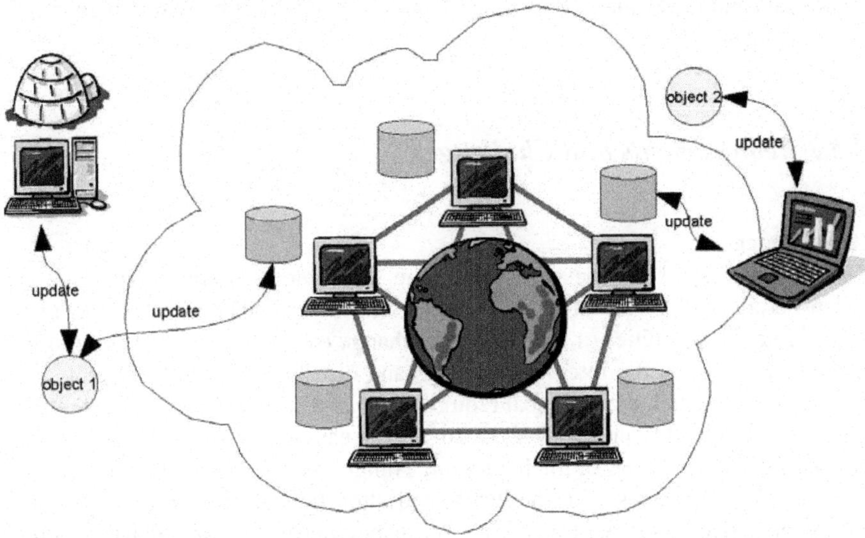

Fig. 7.4 Disconnected Architecture

Interfaces

The Internet of Things has to handle the information exchange in heterogeneous distributed networks with characteristics of all architectures described above. Therefore, it is important to identify all the interfaces for information exchange and to define synchronisation mechanisms to assure data consistency and security. Table 7.4 shows a summary of possible types of information exchange and the related time of synchronisation considering all architectures.

Architecture	Types of information exchange	Time of synchronisation
Internet	Real-time between object and connected resources	Real-time
Partitioned networks	By objects	Next connection
Partially disconnected	By objects and mobile devices	Next connection
Permanently disconnected (object)	By mobile devices	Next connection

Permanently disconnected (local resources)	By objects	Next connection

Table 7.4 Types of Information Exchange

Synchronisation in the Internet of Things has to handle these different types of information exchange. The autonomy of objects and the possibility to change data in (partially) disconnected environments causes difficulties with the serialisation of object data and the availability of consistent data at any location within the Internet of Things.

7.6.2 Requirements and Challenges

Requirements

The whole Internet can be seen as a huge distributed heterogeneous database (Niemi et al. 2007). The Internet of Things will increase that enormous database and add new possibilities of information exchange (see Table 7.4). Because of the similarities between the Internet of Things and distributed database systems it is necessary to research the technical requirements of distributed databases (see section 7.2.3, Bell and Grimson 1992, Öszu 1999, Leanvitt 2010) considering the architectural characteristics of the Internet of Things (see section 7.6.1).

Data Handling deals with the distributed allocation of data to the nodes of a computer network and the transformation of heterogeneous data. It has to solve problems resulting from the distribution of data as well as defining a common database description to make heterogeneous data understandable. Common distributed databases use a global master represented by a distributed database management system to organise data allocation. Most of them support relational databases using SQL to query data. There are different approaches to implement object-oriented databases, but they have not the acceptance of relational databases. Only during the last few years non-relational databases have reached increased popularity (Leanvitt 2010). NoSQL databases use different technologies for data handling. The most popular types are key-value stores, column-oriented databases and document-based stores. A well-known implementation is Bigtable by Google using the column-oriented approach (Chang et al. 2006). The main advantage of NoSQL databases is their better scalability, but they do not ensure consistency. *Data Handling in the Internet of Things* will add the problem of (partially) disconnected environments and autonomous objects. It remains unclear if it will be possible to keep only one global instance of object information and to manage distributed data, because the availability of data access cannot be assured permanently. Clustering will be an approach to support distributed data allocation without permanent network availability (see section 7.4). Data transformation will aim for harmonising existing databases like relational or XML databases and object data, which is expected to be stored in a differing way. Object data will need efficient approaches for data management, because storage capacity will be limited. There is a need to define a common global language to understand all information resources within the Internet of Things.

Query Optimisation is necessary to provide efficient access to distributed databases which are affected by huge data resources and the possibility of unstable connectivity. Structural details of information access should be hidden from the user. The most common language for receiving and manipulation data is SQL.

Although SQL is very powerful and well standardised, it is focused on relational databases. There are also proposals for querying other resources, such as SPARQL, which is used for the Semantic Web, or the different implementations of OQL for querying object-oriented databases.

Query Optimisation in the Internet of Things has to support querying of data in heterogeneous structures and should provide methods to handle (partially) disconnected architectures. The structure of object data has to be examined considering query mechanisms as well as scalability.

Transaction Management has to organise the correct execution of database transactions, which are series of actions that have to be processed as single indivisible units. A transaction has four important properties:

- Atomicity (executing a transaction as a single unit),
- Consistency (transforming a database from one consistent state to another),
- Independence (providing execution independent from another transaction),
- Durability (making transaction results persistent in the database).

Those properties are known as ACID properties. There has to be a good concurrency control as well as a recovery system to provide them as a whole. While common relational databases follow the ACID approach, NoSQL databases implement a set of weaker properties named BASE (Basically Available, Soft-state, Eventual consistency). *Transaction Management in the Internet of Things* will not be able to implement all ACID properties, although they are essential to assure data consistency. It can be expected that they will not be achieved at any time and any place. Providing consistency and independence will require intelligent approaches of synchronisation, due to the data manipulation needs in (partially) disconnected environments. Software agents can be suitable to bridge this gap (see section 7.5, Assis Silva and Krause 1997). Furthermore, transaction methods of NoSQL databases have to be examined. The CAP theorem is also worth mentioning here. It states that it is impossible to provide consistency, availability and partition tolerance in a distributed system at the same time (Gilbert and Lynch 2002). Providing consistency in the Internet of Things will cause similar problems, because of the partitioned architectures.

Concurrency Control comprises different methods to ensure transaction management in distributed databases. It is concerned with scheduling and serialisation of transactions and offers different techniques of concurrency control. A transaction consists of a sequence of reads and writes. The entire sequence of reads and writes by all concurrent transactions in a database is a local schedule, while such a sequence affecting distributed databases is a global schedule. Ordering all reads and writes of a schedule in a way that they can be processed sequentially one after the other means serialisation. To support concurrency control, Bell and Grimson (1992) distinguish three techniques:

- Locking methods,
- Timestamp methods,
- Optimistic methods.

Concurrency Control in the Internet of Things has to manage large amounts of data resources, which are distributed among common databases with permanent network availability, data resources in (partially) disconnected environments and autonomous objects. Therefore, scheduling and serialisation will find a new complexity in that context. Object information could be updated in different locations that may not be connected with each other. If those locations are partially disconnected, synchronisation would be possible during the next connection at an unknown time. In the meantime, object information could have changed again, which may cause problems for serialisation. If we assume that objects are moving, it could be suitable to support data consistency with a local object-oriented data management. As mentioned before, this solution could be supported by software agents. On the other side, there might be stationary objects which could not be supported by that solution. Instead, these objects would require a synchronisation of a partially disconnected database. The traditional mechanisms of concurrency control would also reach their limits.

Locking methods are not suitable in the context of the Internet of Things. Assuming that objects are working autonomously and data resources can exist in partially disconnected environments, the usage of locking methods could cause a large number of deadlocks that would prevent further synchronisation. Considering that updating object information could be possible at different times and in partially disconnected environments, the usage of a *timestamp method* could be suitable. Serialisation could be realised by timestamps of data updates. Nevertheless, the usage of such a method would require a global reference time to work correctly. Bonuccelli proposed the enforcement of a global clock and described it as a difficult task (Bonuccelli et al. 2007). Finally, *optimistic methods* are based on the premise that conflict is rare. Considering the disordered movement of objects in the Internet of Things, these methods could be hardly considered suitable in this context.

Recovery refers to the ability to ensure the consistency of data resources in case of unpredictable failures of hardware or software components. Like concurrency control, recovery is tightly linked to transaction management, because a transaction is the smallest recovery unit. Considering the ACID properties of a transaction, concurrency control ensures consistency and independence, while recovery provides durability and atomicity (Bell and Grimson 1992). Recovery requires a history of transactions to identify the point where to restart transactions after the occurrence of an error. Such a history might be implemented by a log file. Corresponding to transaction management and concurrency control it has to be distinguished between local and global transactions in distributed systems. *Recovery in the Internet of Things* has to solve similar problems to the concurrency control. Especially working in (partially) disconnected environments may cause problems. The usage of an object-oriented history could be suitable to ensure the consistency of object data. Supporting recovery methods by a history would also require a global time reference.

Integrity and Security aim to avoid the corruption of data resources. *Integrity* tries to ensure the logical correctness of data. There might be local and global constraints to avoid the storage of incorrect data. *Security* mechanisms protect data resources from the access of unauthorised users. Depending on their special needs, users may have different views on the data resources. On this basis, they would need to provide identification and get authentication in order to access the associated data. Encryption of data is an additional approach to protect data if an unauthorised user gets access to a secured data resource.

Integrity and Security in the Internet of Things are of great importance, because many different users will request and manipulate many heterogeneous data resources. There is a need to guarantee that no data will be manipulated or destroyed by unauthorised users. Working in (partially) disconnected environments will be of special interest, because constraints and authorisation rules have to be known locally in cases where no network will be available.

Replication is the process of storing redundant data resources with the aim to support recovery methods and data availability in distributed environments. Redundant copies of the same data resource require methods to assure consistency among those copies. Öszu (1999) mentions the one-copy equivalence as an important consistency criterion. It requires that the value of all copies should be identical after a transaction. A typical replication control protocol is the Read-Once/Write-All (ROWA) protocol. *Replication in the Internet of Things* is not expected to be a suitable approach. As discussed earlier, there is a potential for various non-homogeneous copies of object data across the architecture, due to the manipulation of data in disconnected environments and the usage of autonomous objects. It is expected that data might not be in the same state at every point in time. It will therefore not be possible to support one-copy equivalence. Nevertheless, experiences from replicated databases could be applied to use different copies of object data (although not always in the same state) to improve data availability in partially disconnected environments as well as recovery in the Internet of Things.

Challenges

From the perspective of data management, the Internet of Things can be considered as a heterogeneous distributed database with the following special architectural characteristics (see section 7.6.1):

- The existence of (partially) disconnected environments
- The handling of (mobile) autonomous objects

Following this idea, the Internet of Things will need a data management strategy providing the same tasks as the data management of distributed environments, but considering these special needs. Section 7.6.2 gave a summary of all those tasks and identifies the special requirements in the Internet of Things. As a result, the following topics would need further investigation before data synchronisation can be introduced in an Internet of Things architecture:

- Advantages and disadvantages of global and shared data management in (partially) disconnected environments,
- Defining a global language fitting the needs of common data structures and object-associated data structures,
- Methods for transaction management, concurrency control and recovery in environments not being able to achieve the ACID properties,
- Implementation of a global reference time in (partially) disconnected environments,
- Guarantee of data integrity and security in (partially) disconnected environments.

We can therefore conclude that finding intelligent solutions to overcome these challenges is a precondition for providing efficient and secure synchronisation in the Internet of Things.

7.7 Summary and Conclusion

Among the many challenges in the realisation of the Internet of Things vision, many times the management of the resources of the embedded devices that will power the Internet of Things objects is overlooked. In this chapter, we have discussed three techniques that will assist these constrained devices to empower the Internet of Things services for extended periods of time, while providing the objects with enhanced capabilities that positively influence the collection of object information.

The clustering of the Internet of Things objects will support the networking of autonomous intelligent objects by influencing their lifetime, scalability and robustness. By considering both, the energy of the devices as well as their context, the presented techniques can not only ensure that the objects will be able to produce information for longer periods of times, but also that only those objects involved in the same contextual situation will cooperate and share information. The use of software agents can add up to these benefits by taking representation of both, the physical objects and the Internet of Things users, in situations requiring automation and objective decision making. Examples of these situations are the monitoring of products in supply chains, the management of procurement processes for objects representing commercial products or the execution of periodic queries for object information in networked databases. At the same time, software agents can aid the clustering processes by moving part of the decision making process to the architecture side, reducing the burden of the embedded devices attached to the Internet of Things objects. Finally, object data synchronisation will be needed in order to cope with (partially) disconnected environments where objects are not permanently connected to the Internet of Things infrastructure. While many lessons can be learned from the research in distributed databases, the unique requirements of the Internet of Things will call for a new set of solutions in areas such as data integrity, transaction management, concurrency control or a global language for object information management.

References

Armenio F, Barthel H et al. (2009) The EPCglobal Architecture Framework. http://www.epcglobalinc.org/standards/architecture/architecture_1_3-framework-20090319.pdf. Accessed 11 June 2010

Assis Silva FM, Krause S (1997) A distributed Transaction Model Based on Mobile Agents. In: Rothermel K, Popescu-Zeletin R (eds) Proceedings of the First International Workshop on Mobile Agents. Springer, Berlin-Heidelberg

Basagni S (1999) Distributed Clustering for Ad Hoc Networks. Proceedings of the 1999 International Symposium on Parallel Architectures, Algorithms and Networks, Fremantel

Bell D, Grimson J (1992) Distributed Database Systems. Addison Wesley Publishers Ltd.

Bonuccelli M, Ciuffoletti A, Clo M, Pelagatti S (2007) Scheduling and Synchronization in Distributed Systems. http://citeseerx.ist.psu.edu/viewdoc/summary?doi=10.1.1.39.509. Accessed 9 June 2010

Brintrup A, Ranasinghe D, McFarlane D, Parlikad A (2008) A review of the intelligent product across the product lifecycle. Proceedings of the 5th International Conference on Product Lifecycle Management, Seoul

Brintrup A, McFarlane D, Owens K (2010) Will intelligent assets take off? Towards self-serving aircraft assets. IEEE Intell Syst. doi:10.1109/MIS.2009.89 (In Press)

Brown PJ, Bovey JD, Chen X (1997) Context-Aware Applications: From the Laboratory to the Marketplace. IEEE IEEE Pers Commun. doi:10.1109/98.626984

Bussmann S, Sieverding J (2001) Holonic control of an engine assembly plant-an industrial evaluation. Proceedings of the 2001 IEEE Systems, Man, and Cybernetics Conference, Tucson

Canard S, Coisel I (2008) Data Synchronization in Privacy-Preserving RFID Authentication Schemes. Proceedings of 4th Workshop on RFID Security, Budapest

Chang F, Dean J et al. (2006) Bigtable: A Distributed Storage System for Structured Data. Proceedings of the 7th Conference on USENIX Symposium on Operating Systems Design and Implementation - Volume 7 (Seattle, WA, November 06 - 08, 2006). USENIX Association, Berkeley

Chatterjee M, Das SK, Turgut D (2002) WCA: A Weighted Clustering Algorithm for Mobile Ad Hoc Networks. Clust Comput. doi:10.1023/A:1013941929408

Chiasserini CF, Chlamtac I, Monti P, Nucci A (2004) An energy-efficient method for nodes assignment in cluster-based Ad Hoc networks. Wirel Netw. doi:10.1023/B:WINE.0000023857.83211.3c

Cilia M, Antollini C, Bornhövd A, Buchmann A (2004) Dealing with Heterogeneous Data in Pub/Sub Systems: The Concept-Based Approach. Third international workshop on distributed event-based systems DEBS '04, Edingburgh

COUGAAR (2010) An Open-Source Agent Architecture for Large-Scale, Distributed Multi-Agent Systems. http://www.cougaar.org/. Accessed 20 June 2010

Crowley JL, Coutaz J, Rey G, Reignier P (2002) Perceptual Components for Context Aware Computing. Proceedings of the UBICOMP 2002, Goteborg

Dey A (2000) Providing Architectural Support for Building Context-Aware Applications. Dissertation, Georgia Tech

Dimokas N, Katsaros D, Manolopoulos Y (2007) Node Clustering in Wireless Sensor Networks by Considering Structural Characteristics of the Network Graph. Proceedings of the International Conference on Information Technology 2007, Las Vegas

Gavalas D, Pantziou G, Konstantopoulos C, Mamalis B (2006) Lowest-ID with Adaptive ID Reassignment: A Novel Mobile Ad-Hoc Networks Clustering Algorithm. Proceedings of the 1st International Symposium on Wireless Pervasive Computing, Phuket

Gilbert S, Lynch N (2002), Brewer's Conjecture and the Feasibility of Consistent, Available, Partition-Tolerant Web Services. doi: 10.1145/564585.564601

Grummt EO (2010) Secure Distributed Item-Level Discovery Service Using Secret Sharing. http://www.faqs.org/patents/app/20100031369. Accessed 20 June 2010.

Hayes-Roth B (1995) An architecture for adaptive intelligent systems. ARTIF INTELL. doi:10.1016/0004-3702(94)00004-K

Heinzelman WB, Chandrakasan AP, Balakrishnan H (2002) An Application-Specific Protocol Architecture for Wireless Microsensor Networks. IEEE Trans Wireless Commun. doi: 10.1109/TWC.2002.804190

Holmstöm J, Kajosaari R, Främling K, Langius K (2009) Roadmap to tracking based business and intelligent products. Comp Ind 60: 229-233. doi:10.1016/j.compind.2008.12.006

JADE (2010) Java Agent Development Framework. http://jade.tilab.com/. Accessed 20 June 2010

Krivokapic N (1997) Synchronization in Distributed Object Systems. Proceedings of BTW'1997, pp.332-341

Kushalnagar N, Montenegro G, Schumacher C (2007) IPv6 over Low-Power Wireless Personal Area Networks (6LoWPANs): Overview, Assumptions, Problem Statement, and Goals. http://citeseerx.ist.psu.edu/viewdoc/summary?doi=10.1.1.73.4790. Accessed 20 June 2010

Leanvitt N (2010) Will NoSQL Databases Live Up to Their Promise?, Comp. doi: 10.1109/MC.2010.58

Liang Y, Yu H (2005) Energy Adaptive Cluster-Head Selection for Wireless Sensor Networks. Proceedings of the 6th International Conference on Parallel and Distributed Computing, Applications and Technologies, Dalian

Liu JS, Lin CHR (2005) Energy-efficiency clustering protocol in wireless sensor networks. J Adhoc Netw. doi:10.1016/j.adhoc.2003.09.012

Maes P (1995) Artificial Life Meets Entertainment: Life like Autonomous Agents. CACM. doi: 10.1145/219717.219808

Niemi T, Niinimäki M, Sivunen V (2007) Integrating Distributed Heterogeneous Databases and Distributed Grid Computing, http://citeseerx.ist.psu.edu/viewdoc/summary?doi=10.1.1.2.1963. Accessed 9 June 2010

Onodera K, Miyazaki T (2008) An Autonomous Algorithm for Construction of Energy-conscious Communication Tree in Wireless Sensor Networks. Proceedings of the 22nd International Conference on Advanced Information Networking and Applications – Workshops. IEEE Computer Society, Washington

Öszu MT (1999) Distributed Databases. http://citeseerx.ist.psu.edu/viewdoc/summary?doi=10.1.1.33.2276. Accessed 9 June 2010

Pátkai B, McFarlane D (2006) RFID-based Sensor Integration in Aerospace. http://www.aero-id.org/research_reports/AEROID-CAM-009-Sensors.pdf. Accessed 9 June 2010

Ray I, Ammann P, Jajodia S (2000) Using semantic correctness in multidatabases to achieve local autonomy, distribute coordination and maintain global integrity. Inf Sci. doi:10.1016/S0020-0255(00)00062-1

Stuart Russell S, Norvig P (2003) Artificial Intelligence: A Modern Approach. 2nd Edition, Prentice Hall

Ryan NS, Pascoe J, Morse DR (1998) Enhanced Reality Fieldwork: the Contextaware Archaeological Assisstant. http://www.cs.ukc.ac.uk/projects/mobicomp/Fieldwork/Papers/CAA97/ERFldwk.html. Accessed 26 May 2010

Sánchez López T, Kim D, Canepa GH, Koumadi K (2008) Integrating Wireless Sensors and RFID Tags into Energy-Efficient and Dynamic Context Networks. Comput J. doi:10.1093/comjnl/bxn036

Sánchez López, T. Huerta Canepa, G. (2010) Distributed and Dynamic Addressing Mechanism for Wireless Sensor Networks. *Submitted to Int J Distrib Sens Netw.Will be published in November 2010.*

Schilit WN, Adams NI, Want R (1994) Context-aware Computing Applications. Proceedings of the 1st International Workshop on Mobile Computing Systems and Applications, Santa Cruz

Suzuki S, Harrison M (2006) Data Synchronization Specification. http://www.autoidlabs.org/single-view/dir/article/6/265/page.html. Accessed 9 June 2010

Vasseur JP et al. (2010) Routing Over Low power and Lossy networks (roll). http://datatracker.ietf.org/wg/roll/charter/. Accessed 20 June 2010

Wang Y, Zhao Q, Zheng D (2004) Energy-Driven Adaptive Clustering Data Collection Protocol in Wireless Sensor Networks. Proceedings of the International Conference on Intelligent Mechatronics and Automation, Chengdu

Wong CY, McFarlane D, Zaharudin A, Agarwal V (2002) The intelligent product driven supply chain. Proceedings of the 2002 IEEE International Conference on Systems, Man and Cybernetics, Hammanet

Wu J, Gaol M, Stojmenvic I (2001) On Calculating Power-Aware Connected Dominating Sets for Efficient Routing in Ad Hoc Wireless Networks. In: Ni LM, Valero M (eds) International Conference on Parallel Processing: 3-7 September 2001 Valencia, Spain. IEEE Press

Ye M, Li C, Chen G, Wu J (2005) EECS: An Energy Efficient Clustering Scheme in Wireless Sensor Networks. Proceedings of the International Profesional Communication Conference 2005, Limerick

Younis O, Fahmy S (2004) HEED: A Hybrid, Energy-Efficient, Distributed Clustering Approach for Ad Hoc Sensor Networks. IEEE Trans. Mobile Comput. doi:10.1109/TMC.2004.41

8 The Role of the Internet of Things for Increased Autonomy and Agility in Collaborative Production Environments

Marc-André Isenberg, Dirk Werthmann, Ernesto Morales-Kluge, Bernd Scholz-Reiter

University of Bremen, Planning and Control of Production Systems, Germany,
E-Mail: ise@biba.uni-bremen.de

Abstract. This chapter discusses the contribution of the Internet of Things for providing a fine-grained information infrastructure within collaborative production environments. Such infrastructure makes up-to-date information available to autonomous objects to render contextual decisions that evolve elemental agility. A technical discussion illustrates the feasibility of autonomous objects as well as their possible involvement in the Internet of Things. Additionally, a demonstrator is described, which exemplifies the effects of autonomous objects on agile processes within the automotive industry. Concluding, the chapter relevant research questions in the field of the Internet of Things and collaborative production environments are specified.

8.1 Introduction

Market structures underlie a continuous process of change, caused by innovations of enterprises, technical improvements, new market participants, amendments, or changes in society's values. The concerned enterprises need to react to these changes and need to adapt their services and products in a quick and adequate manner in accordance with the new market conditions. The rate of market changes grew steadily over the previous decades, especially due to the improvement as well as the development of existing and new information and communication technologies (ICT). Fulfilling the demands of the market is a bigger challenge than ever before – within very small time intervals a market can change fundamentally (Pavlou and El Sawy 2005). The automotive industry can serve as an example: within a few years the demand for powerful and fast cars has decreased significantly, while the customers' sensitivity for ecological fuel-saving cars has increased noticeably – that was much faster than the leading automakers had expected and taken into account within their product ranges and corporate structures (Zalubowski 2008). Automotive supply chains are typical collaborative production environments, in which different companies participate in producing a final product. Hence, varying the product range or changing the corporate structure of a central supply chain member, like automakers, always affects the processes and structures of its supply chain partners. The transmission of the need for changes from partner to partner and the detection of the individual modifications in processes and structures require a long time. Thus, an improvement of the network-wide reaction time offers a high potential to save and to strengthen the market position. To reduce the time gap between the detection of change and the necessary adjustments in production, it is necessary to find technologies and methods which enable the production networks to react autonomously for developing an agile supply chain coordination design.

Potential technologies that can confer this behaviour to collaborative production environments are the Internet of Things and autonomous objects (Uckelmann et al. 2010). The Internet of Things underlies different definitions, but mainly the term describes the increasing interconnectedness of electronic devices, using a common information infrastructure. Sometimes the Internet of Things is depicted as an unclear vision of unknown technologies; but it is neither a future vision, which is far away like a utopia, nor a technological breakthrough, like the invention of the radio or television was; it is a realistic prediction of the future convergence of present technological developments, the infrastructural expansion and the general trend of ubiquitous online accessibility. The future Internet of Things will use protocols and algorithms which will be based on those we use in the Internet today; it will just extend the capabilities of a more extensive machine-to-machine and human-to-machine communication, resulting in a higher number of specialised communication participants within the Internet. Autonomous objects are objects which are equipped with intelligence (small central processing units (CPUs) and algorithms) to be capable of making contextual routing decisions or handling activities. Both, the Internet of Things concepts and the autonomous objects, are complementary. The Internet of Things acts as an infrastructure and helps to realise the new systemic characteristics of autonomy and agility by providing an object-oriented information architecture for precise real-time data and ubiquitous Internet access. When applying the described concepts consequently, several existing paradigms of production environments are affected. Paradigms like "Just in Time" are no longer applicable, for instance. Most of the existing paradigms have in common that they require deterministic environments. Introducing a high degree of freedom leads to non deterministic environments and, as a result, state-of-the-art methods have to be extended. This contribution investigates the suitability and cooperation of the Internet of Things and autonomous objects for autonomic and agile processes in collaborative production environments.

This chapter is structured as follows: Section 8.2 gives an overview of recent demands of networked enterprises. Following, section 8.3 explains the fundamental concepts of agility and autonomy. After that, section 8.4 demonstrates the suitability of the Internet of Things for implementing autonomic and agile production processes in networked enterprises. Section 8.5 describes the technical requirements for fulfilling the new demands in production logistics. A prototypical example of a production scenario in which autonomic products are administrating themselves is given in section 8.6. Section 8.7 derives fundamental research questions and challenges which arise from the development and potential integration of the Internet of Things. Finally, in section 8.8 a conclusion and an outlook summarise the results.

8.2 Emerging Challenges of Networked Enterprises

The basic challenges of enterprises are well defined by Porters *Five Forces*; those are the rivalry within an industry, the bargaining power of suppliers, the bargaining power of customers, the threat of substitute products and the threat of new entrants (Porter 1979). All together create the market environment, which shows an inherent trend of increasing dynamically, caused by a more comprehensive use of ICT (more extensive and precise information) as well as of the extension of the worldwide infrastructure (more efficient and reliable transports/material flows).

Even if markets work by the law of supply and demand, the market participants influence the market by using many different strategic measures to exploit this law for their own objectives. They challenge their rivals through product innovations, strategic partnerships, procedural efficiency, pricing policies, acquisitions or in exploring new market opportunities (Morgan and Strong 1998). Particularly enterprise networks, where the individual enterprises concentrate on their core competences, are established to evolve synergies and to strengthen their market positions. However, the more extensive an enterprise network is, the more complex are control, synchronisation, fault recovery and reorganisation of the overall process flow. If the networks are collaborative production environments with a physical exchange of objects, the process flow is divided into an informational and a material flow, resulting in combined management, which is even more difficult.

Resuming the example of automotive industry from section 8.1, which also consists of collaborative production environments, it can be noted that there are still other components apart from the rivalry of enterprises which can affect the market conditions. For example, automakers also have to regard changes in customer preferences as well as in legal frameworks of different countries/markets. Due to the financial crisis of 2008 and high fuel prices, the customer preferences for powerful cars slid, while the concerned automakers did not offer alternatives (Zalubowski 2008). The situation was intensified by the European Union, which had enacted more rigorous emission regulations for passenger cars (European Union 2007), as well as by car scrappage schemes of some countries that were bound to the purchase of low-emission cars (ACEA 2009). The product range of the concerned automakers did not meet the new demands of the market, which resulted in considerable losses in their sales figures and profits and, consequently, partly lead to bankruptcies (Isidore 2009). Even if the development portfolios of the automakers had contained fuel-saving cars, since their supply chains contain a high number of partners, a fast adoption of new car models, as a reaction to the new customer demands, would probably not have had the necessary celerity in terms of restructuring of material and information flows. Additionally, customers increasingly demand the possibility to influence the design of their ordered cars by customising the cars' configurations. The automakers fulfil this demand by variant management, which allows the usage of different types of the same component in production processes, regarding the individual custom order. This also affects possible process sequences and constitutes an enormous challenge in terms of storage and scheduling. The demonstrator of the Collaborative Research Centre (CRC) 637[89] in section 8.6 shows an innovative approach to this challenge.

Another reason why enterprises have to ensure their structural agility is the necessary preservation of their potential compatibility to other enterprises apart from their actual partners. In supply chains, a dominating partner often defines the standards for information and material exchange and beats down the prices as a result of his absolute monopsony. The exclusive focus on the structures and specifications of the main customer lead to dependencies in the enterprise processes and can be detrimental, if the enterprise wants to cooperate with new partners, whose specifications differ from that one of the main customer. Furthermore, incorporating a new partner into a supply chain, which is dominated by one company, also requires its adaptation in structures and processes; this can reduce the attractiveness of its dedication. Additionally, markets are not longer limited by country or continental borders. Globalisation has created an international competition of the cheapest production sites as well as very efficient global logistics providers so that distances do not play the same important role as before.

Taking these market developments into account, it can be summarised that enterprises have to observe and to forecast the market in much more detail than before – due to global markets and to the more extensive enterprise rivalries, the frequency of changes within the markets have increased. Additionally, focusing on the main customer leads to dependencies in structures and market activities. These conditions cause a general new need for agile enterprise structures to strengthen the systemic characteristic of agility for ensuring a fast and adequate process adaptation to new market conditions. For implementing agile enterprise processes and satisfying the demand for customised products it is not sufficient to have agile strategies and structures; it is also necessary to give the operational processes the capability of being agile. One possibility of creating agile processes is given by the concept of Autonomous Control, which defines autonomous objects that are able to make their own decisions on the basis of certain information. For the purpose of agile and autonomous processes, the enterprises have to integrate high-density informational and control networks, which provide extensive real-time

[89] www.sfb637.uni-bremen.de

data and enable fine-grained controlling and objective specification for management. This feature can be served by the Internet of Things.

8.3 Fundamental Concepts of Agility and Autonomy

This section gives a detailed description of agility and autonomy, which will be the arising challenges of modern production and logistical systems.

8.3.1 Agility

While the general linguistic usage relates the term "agility" to the ease of movement or the human behaviour of being quick, light, nimble or mentally alert, agile manufacturing implies a lot more. A general definition of agility in manufacturing, out of a market oriented view, is made by Bessant et al.:

> "Agility in manufacturing involves being able to respond quickly and effectively to the current configuration of market demand, and also to be proactive in developing and retaining markets in the face of extensive competitive forces." (Bessant et al. 2001)

Another definition by Katayama and Bennett is also market oriented, but more focussed on the company's capabilities as well as on the customer requirements:

> "Agility relates to the interface between the company and the market. Essentially it is a set of abilities for meeting widely varied customer requirements in terms of price, specification, quality, quantity and delivery."(Katayama and Bennett 1999)

A more comprehensive definition is given by Yusuf et al., who have researched the drivers, concepts and attributes of agile manufacturing. They perceive the concept of agility as a system with input factors, operating mechanisms and outputs, and define it as follows:

> "Agility is the successful exploration of competitive bases (speed, flexibility, innovation, proactivity, quality and profitability) through the integration of reconfigurable resources and best practices in a knowledge-rich environment to provide customer-driven products and services in a fast changing market environment." (Yusuf et al. 1999)

Furthermore, the agility can be classified into three levels: macro, micro and elemental agility. While elemental agility refers to individual resources like people, parts or machinery, micro agility denotes the enterprise perspective (Goldman et al. 1995). Macro agility extends this consideration to enterprise networks and is suggested by Yusuf et al. For receiving an agile collaborative production environment, it is important to note that these levels cannot be optimised separately, but they are built on each other and they need to be optimised in a harmonic way (Yusuf et al. 1999).

In order to develop the systemic capability of agility, an organisation has to pursue the four core concepts of agile manufacturing, which emanate from a strategic management perspective (Yusuf et al. 1999; Katayama and Bennett 1999; Gunasekaran 1998). Figure 8.1 illustrates these concepts; afterwards the core concepts will be described briefly.

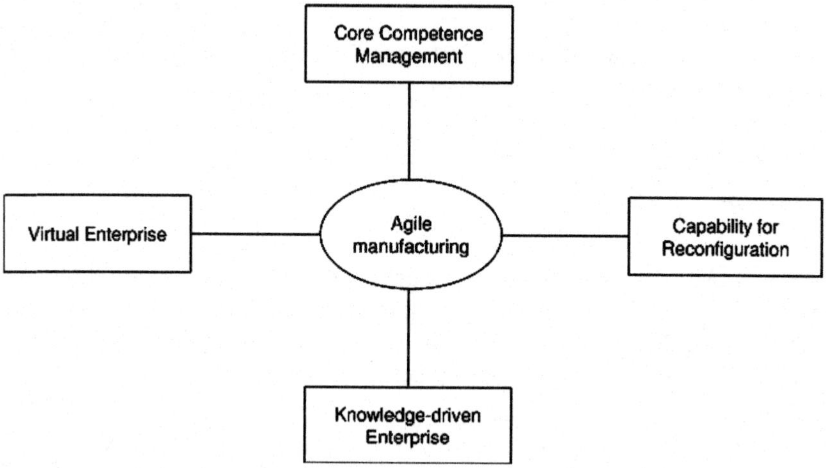

Fig. 8.1 Core Concepts of Agile Manufacturing (Yusuf et al. 1999)

Core Competence Management comprises all measurements and methods for saving, intensifying and developing a company's core competences. Such competences are all "*...skills that enable a firm to deliver a fundamental customer benefit*" (Prahalad and Hamel 1990). They are also called the collective knowledge of an organisation, which mainly means the technological and organisational skills as well as the know-how of the employees about manufacturing techniques, project management, communication, product development, etc. The firm's core competences should be particularly strong compared to other firms in the same industry.

A detailed understanding of the core competences enables organisations to interact in a *Virtual Enterprise*. Within virtual enterprises, organisations complement each other by providing competences which are necessary for their partner's production, but missed or outmatched by their counterpart. Participating organisations are still legally independent and their respective employees, who work for the virtual enterprise, are still placed in the organisations premises; the co-operation and communication within virtual enterprises take place by using modern ICT, mainly the Internet. Hence, usage of the Internet of Things offers an improvement of the potential co-operation, since it prolongs the informational range into the partner organisation. Virtual enterprises are often temporary organisations, which are built for a special purpose. Due to low effort for building a virtual enterprise, the possibility of an uncomplicated and flexible concentration of high qualified employees as well as the purposeful construction give organisations an extensive potential for creating agile structures.

The *Capability of Reconfiguration* is probably the most intuitive one in conjunction with the term agile manufacturing. It contains the structural and operational flexibility to shift the enterprise's focus and realign its business to the changed market environment. Additionally, it capacitates the enterprise to lead the way in competition (Yusuf et al. 1999). Realising this competence is a two-step process with a top-down approach. It starts with the development of a strategic architecture that features a corporate wide map of core skills (Prahalad and Hamel 1990), which enables the management to react fast to a necessary change by a quick identification of the corresponding elements. The second step concerns the implementation of modern ICT into the operational processes for achieving operational flexibility; only with executors, who have access to reliable real-time information and who receive their commands without delay, the *Capability of Reconfiguration* can be achieved.

The concept of the *Knowledge-driven enterprise* is based on the increasing acceptance of inimitable knowledge and information as the source of corporate success. The owners of this knowledge and information are mainly the employees. They generate the enterprise's success by transferring their collective knowledge into saleable products. To enable an agile enterprise, it is necessary to build up a knowledge-rich workforce, which is able to react quickly and adequate to the need for change in structures and products (Yusuf et al. 1999). For that purpose the enterprises have to integrate a knowledge management, which tasks are the

- prevention of knowledge loss by employee turnover,
- extension of the collective knowledge through further education,
- execution of workshops for cross-generation knowledge transfer as well as
- the arrangement of structured knowledge documentation.

The management of an enterprise can use these four concepts as a tool set for developing agile strategies (Gunasekaran 1998). Due to the difficult prediction of change they have to be very generic; not until the occurrence of change they will be instanced and specified by the individual parameters of the situation.

There are two comprehensive suggestions for determining the agility level reached by an enterprise. The first is from Yusuf et al., who categorised enterprises with agile attributes which are grouped in decision domains (Yusuf et al. 1999). Exemplary decision domains are

- competence (attributes: multi-venturing capabilities, developed business practice difficult to copy),

- technology (attributes: technology awareness, leadership in the use of current technologies, skill and knowledge enhancing technologies, flexible production technology),

- partnership (attributes: rapid partnership formation, strategic relationship with customers, close relationship with suppliers, trust-based relationship with customers/suppliers) or

- market (attributes: new product introduction, customer-driven innovations, customer satisfaction, response to changing market requirements).

The second suggestion is from Gunasekaran. He defined agility enablers as well as corresponding metrics (Gunasekaran 1998). The enablers are

- virtual enterprise formation tools/metrics;

- physically distributed teams and manufacturing;

- rapid partnership formation tools/metrics;

- concurrent engineering;

- integrated product/production/business information systems;

- rapid prototyping tools and

- electronic commerce.

The objective of this explanation was to describe the concept of agility in manufacturing process as well as to depict potential starting-points for a supportable concept integration of the Internet of Things.

8.3.2 Autonomous Control

The idea of Autonomous Control of making decisions and solving problems on the locality and within the environment where the objects are pending (the context) enables robust logistic processes. This has a significant impact on generating the elemental agility, which is immanent for achieving the integrated system agility.

The CRC 637 "Autonomous Cooperating Logistic Processes – A paradigm Shift and its Limitations" at the University of Bremen, has been analysing Autonomous Control since 2004. The CRC 637 defines the Autonomous Control as follows:

> "Autonomous Control describes processes of decentralized decision-making in heterarchical structures. It presumes interacting elements in non-deterministic systems, which possess the capability and possibility to render decisions independently. The objective of Autonomous Control is the achievement of increased robustness and positive emergence of the total system due to distributed and flexible coping with dynamics and complexity." (Hülsmann and Windt 2007)

The constituents of the definition can be divided into characteristics (autonomy, heterarchy, decentralised decision-making, interaction and non-deterministic system behaviour) and objectives (increased robustness and positive emergence). All of them are also discussed by Hülsmann, Windt and Böse (Hülsmann and Windt 2007, Böse and Windt 2007). For a better understanding of Autonomous Control, there will be a summary in the following section.

Characteristics

Autonomy describes the ability to make own decisions, independent from any external influence (Probst 1987). For autonomous objects those are mainly decisions which result in a physical handling activity of themselves. This can be a logistical decision, like a route or a transport mean or a production decision about the sequence of individual process steps like milling before drilling or vice versa.

Heterarchy is a form of a system in which its elements are theoretically on the same level of power and authority and where there is no entity which permanently dominates the others (Probst 1992). The elements have just a few relationships of superordination and subordination, so that controlling mechanisms are mostly executed by the elements themselves. That means that all system elements have the same organisational chance to take part in the interactions of the system (Hejl 1990). Thus, a heterarchical system design is closely related to the concept of Autonomous Control: while Autonomous Control describes the object behaviour, the heterarchy describes the characteristic of a system which is formed by the behaviour of the associated elements.

A heterarchical system with autonomous objects implicates *decentralised decision-making*. That means that the decisions on the operative level are not taken by a central co-ordinating instance, but the decision power is transferred to the elements themselves. They take contextual decisions between alternative actions on the basis of the environmental conditions or available information in line with their instructions and the predefined systemic objectives (Frese 1998). The capability of decision competence requires the presence of appropriate algorithms and methods.

Autonomous objects need exchange with their environment (e.g., other objects or sensors) for gathering crucial information, sending status messages and triggering actions. For these reasons they have to be capable of interacting with other system elements for co-operation and co-ordination. The *interaction* activity is the successful contact between systems or their elements, which is either intended by the object itself or induced by receiving a request from another object.

Non-deterministic system behaviour describes the unpredictability of a system's output despite having definite input variables, information about the system state as well as knowledge about the systemic transformation rules. Rerunning the system with identical input variables can cause different output results (Pugachev and Sinitsy 2002).

A prerequisite for achieving autonomous objects is to implement or to enhance some kind of intelligence into objects. The concept of the Intelligent Product pursues this approach and enhances products of today by adding competencies to them. Requirements of Intelligent Products are often verbalised as high level requirements and reflect the demand of autonomous products. McFarlane and Wong describe the Intelligent Product as a physical and information based representation of an item (McFarlane et al. 2003, Wong et al. 2002), which:

- possesses a unique identification;
- is capable of communicating effectively with its environment;
- can retain or store data about itself;
- deploys a language to display its features, production, requirements, etc. and
- is capable of participating in or making decisions relevant to its own destiny.

Definitions from Kärkkäinen et al. (2003) and Ventä (2007) reflect very similar properties of an Intelligent Product. They differ in the perspective from which they look at the Intelligent Product. While Ventäs point of view is a technical and systemic one, Kärkkäinen has a logistics focus. Based in this focus, he describes the Intelligent Product in a supply chain.

Similar to this, the Internet of Things concept formulates its requirements on intelligent items, which is highly congruent to the above mentioned description of Intelligent Products. There are key functionalities that are required to enable the interaction between items (Fleisch and Thiesse 2008):

- Identification: Objects in the Internet of Things are precisely identifiable by a defined scheme.
- Communication and Cooperation: Objects are capable to interact with each other or with resources across the Internet.
- Sensors: Objects can collect information about their environment.
- Storage: The object has an information storage that stores information about the object's history or/and its future.
- Actuating elements: Objects in the Internet of Things are capable to act on their own without having a super ordinate entity.

- User Interface: Adapted metaphors of usage have to be made available by the object.

Having in mind the definitions of the Intelligent Product as well as the aforementioned research field of Autonomous Control, in section 8.6 we will come up with an implemented application of these concepts.

Objectives

As Hülsmann and Windt explains, the Autonomous Control has two main objectives: increased robustness and positive emergence of the overall system.

The objective of *increased robustness* is based on the assumption that autonomous objects can react much faster to unforeseen events than higher planning and controlling instances (see also section 8.3.1 in terms of necessary agility in manufacturing). The direct concernment within the situation enables the objects to calculate their position and to react in a contextual manner, still in line with their instructions. If all incidents were solved by a higher instance, decisions and instructions would take longer to reach the executors and the system resources would have to handle a higher fluctuation in the number of objects.

Positive emergence means that the sum of the individual and context dependent decisions, which are made by autonomous objects, gain a better achievement of the total system objectives than it can be explained by the behaviour of every single element (e.g. lower delivery times and higher adherence to delivery dates) (Böse and Windt 2007).

Degree of Autonomous Control

The concept of Autonomous Control can be used in different degrees of intensity. An intuitive criterion for dividing the behaviour of logistical systems into those of a higher and those of a lower level of Autonomous Control is the proportion between autonomous controlled and conventional managed objects. But this is a very abstract criterion, due to the difficulty of deciding whether an object is autonomous or not. For a detailed categorisation of the system's level of Autonomous Control, Böse and Windt defined an extensive catalogue of criteria in the form of a morphological scheme, which tries to determine the property values of the overall system and the range of the object capabilities. Exemplary criteria are the organisational structure, the location of storage, the interaction ability or the resource flexibility (Böse and Windt 2007).

The consequences of the different degrees of Autonomous Control on the achievement of the logistic targets are not proportional. That means a higher level of Autonomous Control does not automatically lead to a better achievement of the logistic targets; additionally, they also depend on the complexity of the logistics systems. Philipp et al. showed that the concept of Autonomous Control helps to achieve the logistics targets within complex systems, but this support just endures until a specific level of Autonomous Control. If that limit is exceeded, the achievement of logistic targets will decrease markedly, as the system behaviour will increasingly resemble a state of anarchy (Philipp et al. 2007).

The correlation between the characteristics and objectives of Autonomous Control as well as their impact on agility are summarised in figure 8.2.

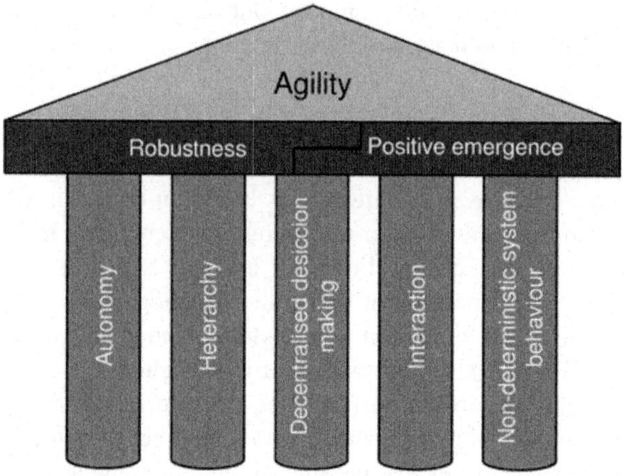

Fig. 8.2 Correlation between Characteristics and Objectives of Autonomous Control

8.4 Enabling Autonomy and Agility by the Internet of Things

Assuming that the runs of individual processes in a collaborative production environment are built and structured in an agile way (e.g., people are trained to switch between different activities, or changeability of process sequences is possible, etc.) and that the processes have a high level of ICT integration, the Internet of Things can constitute an agility enabler by providing the necessary communication infrastructure (compare to section 8.3.1). This impact occurs on two different ways: the managerial and the operational way. Both are shown in figure 8.3.

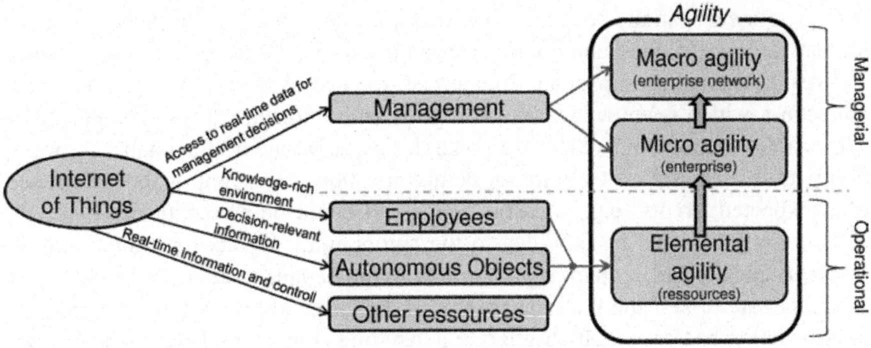

Fig. 8.3 Ways of Impact of the Internet of Things onto the Systems Agility

The managerial way takes effect about the human as a high level decision maker. By providing extensive real-time data out of the working environment through the Internet of Things, filter mechanisms can aggregate the relevant information and observe thresholds for critical processes, so that responsible persons receive their individual management status views with important exception messages. If necessary, the manager can take counteractive measures by using the Internet of Things as an instruction bearer. Due to the possibility of multidirectional communication, the instructions can reach all people and objects that are connected to the Internet of Things. An example for an unexpected event with market evidence is a product recall. After an enterprise has discovered a significant deficit within the product quality, it has to initiate several measurements in terms of its production processes. First of all, it is necessary to stop the current production of the defective product. While the development department remedies the products deficit, the management identifies the necessary changes within the production processes and adjusts the capacities for the reparation of the returning products. All these steps are supported by the infrastructure of the Internet of Things. If the product recall concerns an enterprise network, the Internet of Things demonstrates another advantage: standardised interfaces and protocols; the measurements in terms of the production processes would be the same. In this way the Internet of Things offers a kind of high-level controlling with shorter delays and thus enables faster reactions to unexpected events as well as quicker executions. Hence, agility on the level of enterprises and enterprise networks can be achieved.

The operational way of generating an agile system by the usage of the Internet of Things is through autonomous objects, operative employees and Internet connected resources. Autonomous objects are characterised by making their own decisions. Those decisions can just result from a reliable, accurate and actual information basis. If decision-relevant information is not in the proximity of the autonomous object, it can use the Internet of Things as an information provider to contact spatially distributed databases, objects and resources, which are connected to the Internet of Things architecture. The equipment of the objects with a long-range communication module (e.g., mobile communications) would offer another way to connect spatially separated sources of information, but it is still too expensive in terms of money and energy consumption in relation to the object values and realisable energy capacities. In contrast, the use of the Internet of Things architecture, which is embedded in the objects environment, causes adequate costs and enables autonomous objects to make decisions on the best available information with less energy expenditure for communication. For example, the occurrence of unexpected events (e.g., missing deliveries) could be automatically detected, the information could be provided to the autonomous objects for a situation dependent decision (e.g., changing their processing sequence due to missing parts) by using enterprises' intranets. There is also the usage of the Internet of Things as an instruction bearer, e.g., by triggering a handling activity on the autonomous object itself, like discharging it on a conveyer.

In this situation, operative employees work in knowledge-rich production environments, which are characterised by a high ICT integration within the working processes. The connection of the work stations to the Internet of Things achieves that the employees receive all the necessary information on their screens, regarding those product components which are relevant to their job. For example, these can be job lists with the product components' status (e.g., position, degree of completion), the next object individual work step (important in job shop manufacturing with a high number of product variants) or safety instructions. If a fast reaction to market changes is required, the management can send decisions concerning production changes directly to the employees' work station; employees, who are trained for changeability, can execute the new instructions instantly in their sequence of work. Internet connected resources are able to provide information about their environment based on a time interval, a threshold or by receiving a request. A good example is warehouse management. Within a collaborative production environment each enterprise owns warehouses and all of them depend on each other. The Internet of Things enables a permanent, network wide stock management and reduces the uncertainty about available goods; reliable data can be used for a better synchronisation of the processes as well as for a reduction of overstocking and understocking for cost savings, thus potentially reducing the bullwhip effect. Another example is the tracking and tracing of items like tools or products. The Internet connected resources do not influence the agility directly, but they provide a data base of reliable and actual information, thus enabling faster and more adequate reaction to a need for change.

On the operational side the individual non-human resources, which are able to act autonomously, the trained employees as well as the information transparency, provided by Internet connected resources, generate the elemental agility.

The influence of the Internet of Things on the agility of collaborative production environments can be summarised as an allocation of a close meshed information and control network, which obtains decision-relevant data of the participating objects and humans in real-time and offers an infrastructure for the quick and direct transmission of production instructions by determining the need for change.

8.5 Technical Requirements for Satisfying the New Demands in Production Logistics

Using the Internet of Things for optimising collaborative production environments with an increased autonomy and agility requires the development of hardware and software. Two main challenges can be derived from this need. First: the whole Internet of Things needs the ability to handle data from sensors, real-time location systems and other pervasive technologies available in the future (Thiesse et al. 2009). Second: the economical development and production of software for agents and hardware, such as sensors, actuators or dynamic material handling equipment. These challenges need to be coped in order to enter the available data in the Internet of Things and to make them accessible to objects for rendering their own decisions (Fleisch et al. 2005).

Bridging the gap between the real and the virtual world, the Internet of Things is the technological requirement for achieving an autonomous and agile collaborative production environment. In this environment all objects (whether humans or machines) can communicate with each other without the need of deliberate manual interaction. Currently, the Radio Frequency Identification (RFID) technology is penetrating different businesses and thus blazes a trail for the communication between objects and companies (Fleisch et al. 2005).

The Internet of Things is more than communication, the Internet of Things goes beyond communication; it equips the individual object with intelligence. This intelligence can be placed on the object itself or by representing the object in an IT-infrastructure, which can be near or far away from the object and which is linked to the object permanently or temporarily by the Internet of Things. By using available standard RFID technology, the object can only be identified and linked with information about the environmental conditions at the locations of the RFID-interrogators. Until now, the storage and processing of the data, which the object generates during its lifecycle, could be realised by the usage of IT-infrastructure, which is not physically linked to the object. The current Electronic Product Code (EPC) network architecture is designed to store the information an object collects during its whole lifecycle in IT-systems, which are placed at the supply chain partners, who have handled the object (EPCglobal 2009).

8.5.1 The Evolution from the RFID-based EPC Network to an Agent-based Internet of Things

For having a fully capable Internet of Things, every object should have the capability to process data in order to handle available information and to make decisions based on them, if necessary. This could be done by implementing Software agents and Multi-Agent Systems (MAS), which are common for implementing autonomous and interacting software systems. Agents are autonomous decision makers acting on behalf of physical objects implemented as a software program running in an MAS environment. These agents have sensors for perceiving their environment and actuators to act based on the results of reasoning processes. Moreover, agents are able to communicate with each other in an MAS to coordinate their actions. This results in an even better target achievement of each agent. Based on Knirsch and Timm (1999), agents are situated in an environment, act autonomously, and are able to sense and to react to changes.

In most conventional test scenarios the objects' agents are just running on server platforms. That implies the physically linkage of the agents to the objects, because embedded systems do not have enough computational power for running agent platforms. In the majority of scenarios the objects are connected to their intelligence, which is provided by the agent programs, by attaching unique identifiers, like RFID or barcode tags. These unique identifiers are detected when passing an RFID interrogator. The aim of a future Internet of Things is having the agents physically linked to most of the objects - especially the ones where it makes economical or strategic sense, like valuable goods or production relevant components. Moreover, Jedermann and Lang explain that it could be cheaper to attach intelligence to the object, instead of having much communication between the information technology infrastructure and the object (Jedermann and Lang 2008). To achieve this level of autonomy, computational hardware needs to become smaller and inexpensive, so that even everyday objects could be equipped with intelligence. Based on Moore's law, Mattern says that in the long term almost every object could be equipped with intelligence to run agents on its embedded system (Moore 1998, Mattern 2005). Not only the computational power is important when thousands of intelligent objects are produced every day, even economical and ecological aspects need to be considered. One step towards the production of cheaper and more environmental friendly tags offers the development of chips based on polymer electronics technology (CERP-IoT 2009).

Energy Supply for Embedded Devices

In addition to the capability of decentralised computational power, energy supply is an important topic. Unfortunately, the development of energy supply could not keep up with the development in processing technologies (Mattern 2005). Nevertheless, the ongoing miniaturisation of integrated circuits (ICs) and the software development have led to their reduced energy consumption by constant computational power. But there is still a need for innovative concepts regarding the energy supply. That does not necessarily mean energy storage; it also means to do research on approaches like energy harvesting. Sources for energy harvesting could be for example: vibration/motion, temperature difference, light or RF (electromagnetic waves). Combined with temporary storages and ultra low power micro control units these techniques can lead to a new ubiquitous sensor generation for the Internet of Things (Raju 2008).

Sensors for Collecting Information

Getting towards the future Internet of autonomous and agile Things, objects need to have information concerning their present situation. The agents running on IT-systems, which are physically or remotely linked to the objects, need this information for their decision making processes. Typical sensors could detect light, acceleration, temperature, their location or humidity (Mattern 2005). Currently, research is investigating new small sensors to analyse even liquids or gases. An example is the miniaturised gas chromatography system for detecting volatile organic compounds developed at the Institute for Microsensors, -actuators and -systems (IMSAS)[90] at the University of Bremen (Stürmann et al. 2005). Information about fresh products is collected by these small sensors by analysing the air which surrounds the products. This could be used by the object agents for calculating dynamic best-before dates. Based on that, approaches like "First Expires First Out" in supply chains could be implemented in new warehousing concepts for reducing losses due to a bad quality of perishable goods (Jedermann and Lang 2008).

One of the most important pieces of information, which is decision relevant for intelligent objects, is their location. The reason for this is that the majority of the objects will be mobile so that the objects themselves need to know about it (e.g., if the logistical object has reached its destination). Aside from the absolute position, the relative position might also be of interest. An example is the transport of fresh fruits: bananas should not be stored next to apples because this would lead to a fast ripening of the bananas. Currently, positioning systems are big, expensive, have high energy consumption and do not have the required accuracy. This is going to change (Mattern 2005). Location-sensing techniques like triangulation, proximity or scene analysis can be combined with different transmitting technologies based on radio, infrared light, magnetism, ultrasound or vision for designing location-aware objects (Hightower and Borriello 2001).

Communication

When the Internet of Things will consist of billions of objects (COMMISSION OF THE EUROPEAN COMMUNITIES 2009), this will result in extensive network traffic and will need a high number of network addresses. The communication needs of intelligent things could not be handled completely by common communication technologies: on the one hand, wired networks need further development for handling the increased network traffic over long distances. This will boost for example, the change from copper wires to fibre optics for long distance communication. On the other hand, the majority of objects will be mobile within the Internet of Things, so that wireless technologies for short- and long-distance communication need to be extended and developed further. Hence, technologies like ultra wide band (UWB), universal mobile communication system (UMTS), Zigbee or Long Term Evolution (LTE) will have to be even more common. But there will be

---

[90] www.imsas.uni-bremen.de

also a demand for special technologies, which consist of cheap components that are using low level protocols or need very low energy to bridge the gap between broad band networks and sensor networks. In order to handle all intelligent objects, being present in the Internet of Things, an address protocol with a wide address space is needed. An option is presented by the Internet Protocol version 6 standard (IPv6) for Internet communication (CERP-IoT 2009), which allows 2^{128} addresses.

Task Specific Degrees of Object Capabilities

In the future Internet of Things the technological equipment of (autonomous) objects will differ in its functional range, depending on the context in which the object acts as well as on its tasks. In the following, two opposed scenarios will be described, nevertheless, a lot more scenarios are possible in between these extremes:

In the first scenario the object presents a swap body with valuable goods inside, which carries a transponder for its identification, a general packet radio service (GPRS) module for long-distance communication, an embedded micro-controller for the agent platform as well as sensors for temperature and position measurement. This full capable object can act autonomously in offline cases and has an agent replica hosted in the Internet of Things.

The second scenario includes a box, which is used for the transport of low-cost items in automotive industry. This box is just equipped with a transponder, which enables its identification for the company´s load carrier management. An agent could be realised on a server within the Internet of Things, but not on the object itself. For receiving corresponding information, it would be conceivable to use the EPC network to find out more about the identified object. Equipment with only identification technologies could make sense, especially for linking low price products to the Internet of Things or for locations without a permanent network access.

This subsection explained the technological needs to enable an Internet of Things in collaborative production environments. An overview about the hardware status, necessary improvements and developments is illustrated in Table 8.1.

	Available technologies	Necessary improvements	Technologies in development
Identification	Barcode, OCR, RFID	RFID needs suitability for daily use	Polymer electronic (RFID)
Short distance communication	Zigbee, Bluetooth, WLAN, UWB, RFID	Reducing energy consumption, Simplify communication protocols	Bluetooth low energy, 6lowpan
Long distance communication	Copper wires, Fibre optics, GSM, UMTS	Expand the availability of broad band accesses	LTE, Smart Grid
CPU	Standard Silicon Technology e.g. ARM, Low voltage CPUs	Reducing energy consumption, Improving environmental friendliness	Photonic computing, Polymer electronic

Energy supply	Batteries (Lithium etc.), Capacitors	Higher energy density, Improving environmental friendliness, Extending life time, Fast battery charging	Energy harvesting, Polymer electronic, Novel batteries (Lithium-metal-air battery), Resonant energy transfer
Sensors	Small variety of miniaturised sensor types, e.g. temperature, acceleration	Developing more different sensors, Integration into embedded systems	Miniaturised sensors for e.g. gases and enzymes

Table 8.1 Capabilities of Autonomous Objects and Their Realisation by Technologies

8.5.2 Agents for the Behaviour of Objects

After describing the hardware which is needed to fulfil the needs of the Internet of Things, the development status of software technology will be illustrated:

For implementing the Internet of Things in collaborative production environments with increased autonomy and agility, it is necessary to go beyond standard centralised software architectures. To attend to this challenge, agent technology offers a promising approach.

Different agent platforms are available for programming agents, representing the objects in the Internet of Things. Two platforms which are used at the CRC 637 are the well known JavaAgentDEvelopment framework (JADE) as well as the Open Services Gateway initiative framework (OSGi) (Jedermann and Lang 2008). They are theoretically usable, but a wider usage of agents within the Internet of Things needs improved software for the setting up of agents. In a world of billions of Internet connected objects, many people, including the consumers, want to have the possibility to design and develop their agents; hence, the enhancement of the user-friendliness for agent software is very important (Mahmoud and Yu 2006).

Based on Uckelmann et al. different options are possible for implementing the agents in the Internet of Things (Uckelmann et al. 2010); the basic idea is using the EPC network as a standardised and well accepted structure. The already existing network needs to be extended by the following abilities: the capturing of dynamic data, the autonomous processing of data and the integration of intelligent material handling systems. These ideas are still not implemented. For a realisation the following questions need to be answered amongst others:

Where will the software agents work?
This is mainly influenced by the available techniques and whether it makes sense in an economical way to attach intelligence to an object. Based on these premises it is possible to integrate an agent on central IT-systems at the objects manufacturer or on an embedded system, which is physically linked to the object. In-

between these contrasts, the agent could be implemented on a local IT-system, which is named Internet of Things Information System (IoT IS). This IoT IS, which is described by Uckelmann, is hosted at logistical objects like warehouses or trucks and provides the computation capability for the agents of the logistical objects inside these objects. This enhanced system, based on the already existing standards of the EPC network, consist of the Query Interface with data synchronisation capabilities, the Repository that stores data as well as decisions, software-agents and preference sets, and the Capturing Application. This makes sure that the generated data is exchanged with the global Internet of Things in a correct and secure way (Uckelmann et al. 2010). Every option has its pros and cons. The main differences are whether the object needs network connectivity as well as the price and complexity of the hardware. In Table 8.2 possible locations for hosting agents are compared with each other.

Location of process capability	Technical requirements at the objects	Pros	Cons
Central IT-system	Object has a unique ID	No computational power at the object needed, Permanent access to agents, Cheap tags sufficient	No offline decision possible, Increased network traffic
Local IoT IS	Object has a unique ID	No computational power at the object needed, Cheap tags are sufficient, Agent is nearby the object	Agent has to transfer itself to the next IoT IS's
Embedded system	Object has a micro controller, Sensors and network connection	Offline decisions are possible, Low network traffic	Complex embedded hardware is needed, Synchronisation with agent replicas

Table 8.2 Comparison of Possible Hosting Locations in the Internet of Things for Agents

Additionally, figure 8.4 illustrates the two dimensions of an object's intelligence as well as the physical distance between the object and the location of its intelligence and shows some exemplary technology equipments. The rising level of intelligence by an increasing distance is due to the higher computational power of central IT and the ability of using more complex algorithms and heuristics.

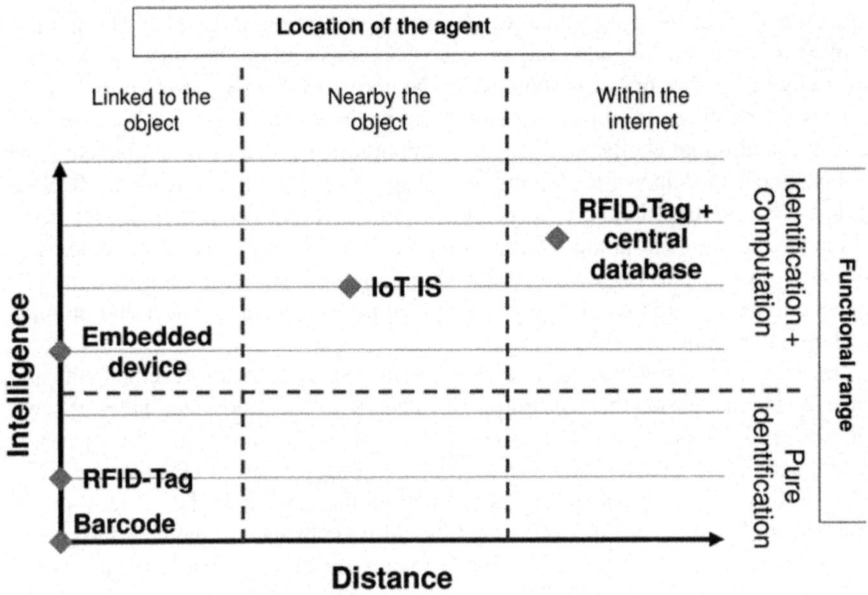

Fig. 8.4 An Object's Intelligence as well as Its Proximity to the Object

Who will be responsible for the agents on the organisational level?
Additionally, the localisation of agents implies questions of data security, product lifecycle responsibility and other social requirements. It is obvious to leave the agent in the sphere of the manufacturer. This solution includes the following advantages:

- It enables the manufacturer to collect a multitude of data about the lifecycle of the product, which could be used for product development.
- Moreover, the manufacturer can offer more services by using the data about the products available through the Internet of Things.
- The users of the product do not need to think about the agent's hosts.
- The manufacturer can easily refinance the costs for hosting the agent by including these costs into the price.

There are still some open issues, like the influence on the customer relationship, legal protection or the already mentioned data security.

Agent Replicas
The exclusive running of the agent on one IT-system or on the object itself seems impractical. There will be objects, which need both, an agent located on the object or in an IoT IS next to the object (depends on the functional range of the

object) and an agent replica continuously connected to the Internet of Things. The replica is needed if the agent has to make decisions without having network connectivity (e.g., if an object is standing in the corner of a warehouse without Wi-Fi coverage and needs to make sure being just in time at the customer's). This will result in additional challenges. The most important one is how to implement a reliable synchronisation of the agents being responsible for the same object (Uckelmann et al. 2010). Making objects capable of rendering decisions without being connected to an agent in the Internet of Things needs some rules which determine the authority of the agent next to the object to come up with decisions. This is needed to make sure that both agents responsible for the same object do not make different decisions.

In terms of the software, there are still some other necessary developments and open questions, beside the necessary enhancement of the user-friendliness of agent platforms, the agent location and the authorisation between the object's agent and its replica. These are, for example, the energy consumption of embedded devices (i.e., so that they consume energy depending on the context of the devices) or decision algorithms. A lot of applications should be improved in the near future, but there is still a long way to implement software agents on every level of product environments, especially in which real time control is needed.

8.6 Application Field: Automotive Tail-lights – Intelligent Product

The high complexity of logistics networks makes it more and more difficult to meet the demands of logistics. Having the right product at the right time at the right place – this is becoming very challenging with conventional planning and control methods. Thus, aspects such as agility, flexibility, proactivity and adaptability are in the focal point of the current research. This is done by applying concepts of decentralisation and autonomy on the logistics decision-making process.

These concepts that concentrate mainly on the methodology, require an information infrastructure they can rely on. The Internet of Things concept is deemed to act as an enabling infrastructure for distributing the information to items and logistics objects. As a future autonomous logistics object, the Intelligent Product is being introduced. The Intelligent Product will be presented in a production logistics scenario and is capable of acting autonomously through an assembly process. This assembly process is part of a production scenario designed to investigate the applicability in the domain of production logistics. The scenario illustrates an autonomous assembly system for an automotive tail-light. The whole equipment is tailored to communicate over the Internet, and, as a consequence, we can imagine extending this scenario in order to consider more than one location of the supply chain.

This section reflects an ongoing work on implementing Autonomous Control methods on logistics systems, specifically in production logistics, where the Intelligent Product plays a central role. The previously mentioned CRC 637 'Autonomous Cooperating Logistic Processes – A Paradigm Shift and its Limitations' hosts this work in a technical subproject.

8.6.1 Assembly Scenario

A production scenario is being implemented for investigating the applicability of Autonomous Control methods in the domain of production logistics. The scenario illustrates an autonomous assembly system for an automotive tail-light. The autonomous aspects refer to the decision-making and all related processes of transport of the components, etc.; the assembly itself is still designed to be a manual task.

The assembly scenario is originally designed to be a flow shop system that does not allow any flexibility within the sequence of processes. Today, automotive tail-lights are produced with variant types to meet the customer configuration demands. Due to this fact, variant flow shop systems evolved from the inflexible systems. However, these systems are still controlled centrally with a limited and predefined space of variants that are determined and scheduled beforehand. This realistic scenario was taken as a starting point to derive the introduced scenario with Autonomous Control by implementing variant types of the finished product which have to be chosen by the product itself. Figure 8.5 shows the assembly process and the related parts.

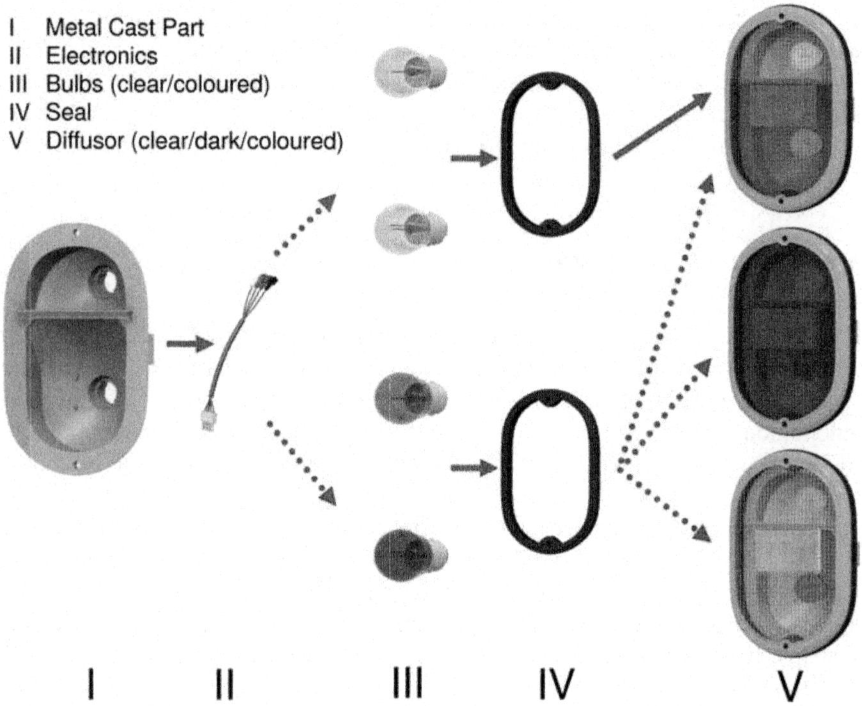

Fig. 8.5 Assembly Process of the Tail-light

8.6.2 Layout

The scenario consists of six stations; five of them are assembly stations, while one station is implemented as an input/output station to insert the semi-finished parts and also to take out the assembled/finished products. The assembly process consists of four stages, which are depicted in figure 8.5. The process starts with the insertion of the semi-finished metal-cast part into the material-flow system (compare to figure 8.6). The implemented assembly stations correspond to the five-stage assembly process and are designed to assemble bulbs (coloured and clear), seals and three types of diffusers.

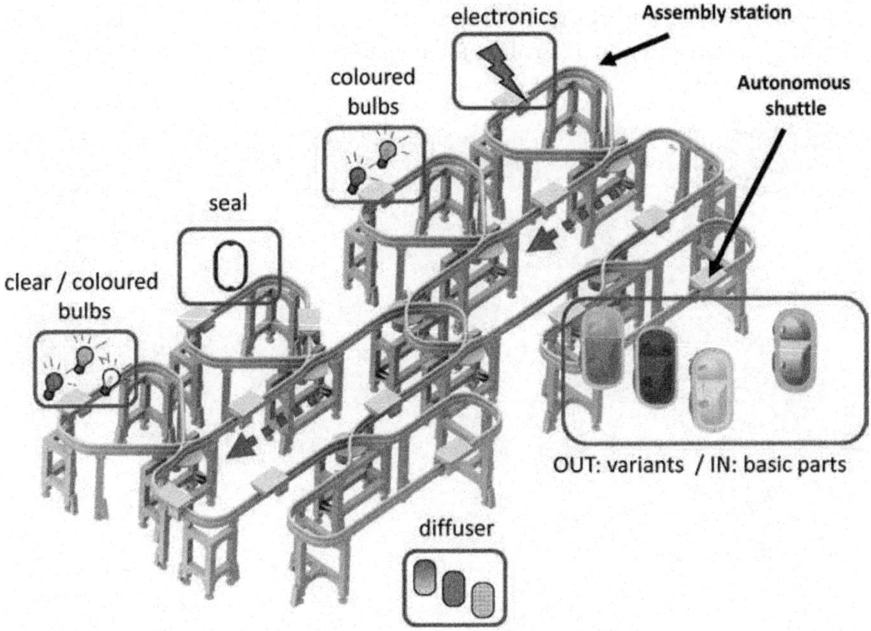

Fig. 8.6 Assembly/Production Scenario (Morales Kluge and Pille 2010)

To allow Autonomous Control, potential flexibilities have to be enhanced to the assembly process. This is realised by allowing the metal-cast parts with in-process-embedded RFID tags (Morales Kluge and Pille 2010) (basic structure for the automotive tail-lights) to choose which type variant to target. The variants require specific parts during the production process.

There are logical constraints that exclude products to choose the next assembly processes by chance. The currently available and feasible variant as well as the scheduling to the next assembly step is determined by the implemented decision methods.

8.6.3 The System

The actual set-up of the assembly scenario at the shop-floor of the Bremer Institut für Produktion und Logistik GmbH (BIBA)[91] allows the product to act flexible and to change the planned route by using the system integrated monorail-switches that offer multiple paths (compare to figure 8.7). The product has the ability to remain on the main line or to deviate to a bypass.

[91] www.biba.uni-bremen.de

The monorail system works with self-propelling shuttles with a mounting capable of carrying loads of up to 12kg. It is a modular system and gives the freedom of future extensions.

Fig. 8.7 The Monorail System (compare to Morales Kluge et al. 2010)

Fig. 8.8 Shuttles with Intelligent Products (Morales Kluge et al. 2010)

8.6.4 Technological Prerequisites

Hardware Abstraction Layer

The probably utmost important and especially relevant requirement of Autonomous Control is the ability of individual logistics entities to access to contextual as well as environmental data. Thus, the ability to understand and to compute the data from information sources is the prerequisite to build local decision-making systems (Hans et al. 2008). For this purpose a "Hardware Abstraction Layer" (HAL) was used, which was developed for having a structured access to nearly every hardware component of the system. It represents a layer that accesses hardware through the IP protocol, thus every brick of hardware had to be enabled to communicate over IP, beforehand. The HAL also takes into account the findings from the perspective of data integration. This facet goes beyond the usual HAL concepts but becomes necessary in this heterogonous context. Hans et al. as well as Hribernik et al. examined this from the point of view of data-integration in autonomous logistics networks (Hans et al. 2008; Hribernik et al. 2009). It also gives freedom in terms of future extensions of the system.

Metal Cast RFID

An automotive tail-light was tailored to be the Intelligent Product for the implementation scenario which has the feature of having an integrated 125 kHz RFID tag, enabling the identification of each item. Today's automotive parts are not equipped with material inherent Auto-ID Systems. This means that the tag is being inserted while producing the tail-light in a casting process. Morales Kluge and Pille describe the objectives of this approach (Morales Kluge and Pille 2010). They focus on enabling the products to be exactly identifiable and also to be autonomous from the beginning of their life. Pille also describes how to cope with related challenges of this engineering process (Pille 2009).

Multi-Agent-System

Even physical objects, which are equipped with Auto-ID technology, have to be made intelligent somehow. By linking the physical object via their unique identifier (RFID) with an agent system, decision-making processes can be set-up. For this reason, an MAS, which is based on JADE, is introduced for enabling the identifiable product to act in a complex network of autonomous objects. The distributed software agents represent the logistics objects and interact in a standard way, which is defined by the Foundation for Intelligent Physical Agents (FIPA) (Gehrke et al. 2006; Foundation for Intelligent Physical Agents 2002). The MAS represents the infrastructure in which decision algorithms can be implemented.

Decision Algorithms

By implementing decision-making algorithms in an MAS environment, physical objects can be enabled to act autonomously in a network. Such an algorithm is based on the "Product Type Corridor". The product moves along this corridor during the manufacturing and assembly process (Windt and Jeken 2009). This allows the Intelligent Product to make decisions online which variant type to choose by considering its degree of assembly. A decision algorithm becomes necessary when the order of demanded products changes during the assembly process. The decision affects also the next possible production steps, which are identified then. Thus, it is required to analyse the all-up situation, which induces the evaluation of every operation alternative (Ludwig 1995). This concept is a precondition for going into decision-making. For this concept a model is used which is able to evaluate multicriterial states. This approach is based on the fuzzy hierarchical aggregation (Rekersbrink et al. 2007). Exemplary criteria are waiting time at potential assembly stations, material in stocks of the stations and current customer orders.

The presented implementation is being developed in the course of the CRC 637, which undertakes basic research in the field of Autonomous Control in logistics. The CRC 637 considers technological innovations and rules like Moore's Law (Moore 1998), so that this research concentrates on the basic issues logistic objects have to be aware of. This means, we assume that necessary technological improvements, like miniaturised processing power, will be available in the near future. This approach allows us to perform research on topics and create results that will be applied when technology is available. Thus, we developed, e.g., decision methods and customised Multi-Agent-Systems and algorithms for decentralised Autonomous Control of logistics objects by using state-of-the-art Internet technologies. We always assume that a bigger framework will be available that allows the interaction of objects and allows our developed methods to be implemented in a big network that goes beyond the used Internet infrastructure. The Internet of Things is deemed to be the complementary part to our research that is working on enabling objects to communicate, while we perform research on how to enhance objects with competencies for acting in this totally networked environment. The presented implementation shows clearly that merging the Internet of Things concept with our research findings can create a positive emergence.

8.7 Challenges by Developing the Internet of Things

Before the Internet of Things can unfold its full potentials in collaborative production environments (e.g., the connectivity of each electronic device), further scientific research is necessary. This section gives an overview of the general challenges as well as necessary developments for a full reliable, secure and all demands satisfying Internet of Things:

Authenticity, Encryption and Integrity of Data
There are already algorithms for the encryption and verification of data in use in the Internet. It will be necessary to check their transferability to the Internet of Things. Especially the encryption of object communication needs further research: symmetric encryption algorithms seem not realistic, due to the necessary exchange of a common decryption key to all objects. However, the usage of the public key cryptography, which does not require an exchange of a secret key, claims comprehensive computing time – potentially more than autonomous objects can offer or than their energy capacities can provide.

Authentication
The Internet of Things will be mainly used for the exchange of object bounded data and instructions. Since the majority of the objects will be non-public objects, significant amounts of sensitive information and execution power will exist. Access to these pieces of information and instruction possibilities need authentication mechanisms, which ensure the determination of the identity and access authorisation of the requesting humans or objects. In collaborative production environments the access authorisation will be very important, due to the enterprise partners, who are still legally independent and who may be still in competition to each other in other markets. The rights for reading and execution need a precise definition so that they can be defined for each single object and resource connected to the Internet of Things. Another option is to design a role model, which can be used for defining the access authorities for groups of users.

Legal Safety for Data Protection

The data exchange within the Internet of Things will be comprehensive and international. Partly these data will consist of sensitive information; their protection is very important for their owners. Existing approaches, which are in use in the Internet, must be checked as to their adaptability to the scenario of an Internet of Things. For that purpose further research about the probably content of data as well as the international cooperation in law is necessary.

Scalability

Billions of objects will communicate across the Internet of Things and will put a strain on its technical infrastructure (COMMISSION OF THE EUROPEAN COMMUNITIES 2009). For a reduction of the data throughput, the Internet of Things needs scalability mechanisms, which enable a data reduction without a loss of important information. Such mechanisms can be provided by clustering methods, which have been developed in the research of Wireless Sensor Networks. However, sensor nodes differ from autonomous objects or intelligent resources (e.g. movement behaviour, energy supply, objectives) so that further research for object clustering in the Internet of Things will be necessary.

Billing and Business Model

The development and operation of the Internet of Things infrastructure cause high costs, which allocation to the beneficiaries have to occur in a comprehensible way, according to the costs-by-cause principle. Additionally, it is necessary to determine the monetary value of individual information, which enables the financial evaluation of the information exchange between objects. This data can be used for the development of billing models. Furthermore, the Internet of Things will offer the chance of a wide range of novel business models, which will provide new services like the innovation of the Internet did (compare to Amazon, eBay or YouTube).

Data Management and Synchronisation

Autonomous objects and intelligent resources are often free in move; they are not bound to a fixed location partly disconnected from the Internet of Things. For that reason they need a representing replication which is permanently connected to the Internet of Things. Another very important question, which has to be answered, is about the storage location of data, which an agent produces during its whole life.
During the offline time the object can make decisions, gather data or change its status; there can be also new instructions or objectives, which are sent to the objects replication. In terms of a previous offline case, the reconnected object requires to synchronise these changes with its replication within the Internet. Moreover, two agents which are responsible for the same object (i.e., within the Internet of Things/on the object itself) need predefinitions about their competences as well as their range of authorisation; this will be necessary in order to enable each agent's autonomy without being connected to each other and to avoid inconsistent decisions.

Human-to-machine Communication

While machines can communicate to each other very quickly via electronic interfaces, humans do not have such interfaces. The communication with humans always requires an access via senses like visual, acoustic or vibration/mechanical signals as well as corresponding input possibilities. The Internet of Things will produce a significant higher degree of human-machine-communication. Especially autonomous objects, which can be offline and which decisions may trigger a human activity in their environment (e.g., if a parcel wants to join a truck load and the employee has to carry it onto the truck), usually are not equipped by a human communication possibility. For those issues developments in wearable technologies as well as in the field of dialogues for reaching a quick, understandable and fault reduced Human-machine-communication will be needed.

Technological Improvements

The main technologies are already available to make the first steps towards the Internet of Things. But there is still a lot of research necessary to achieve a world of real autonomous and agile objects, which represent almost every commodity item in the Internet.

Especially the following improvements are necessary: The energy supply and the energy consumption of the devices, hosting the agents platforms, need to be improved. This could be done for example by developing energy harvesting technologies combined with smaller ICs, consuming less energy as well as energy efficient software algorithms. Moreover, the communication infrastructure needs two main improvements. On the one hand, broad band is needed almost everywhere to handle the increasing network traffic resulting from the high number of intelligent objects. On the other hand, wireless communication technologies are required, which consume less energy and are easily attached to embedded systems to bridge the gap between moving objects and the cable based Internet of Things.

8.8 Conclusion and Outlook

This chapter attends to the benefits of the Internet of Things for a higher degree of autonomy and agility in the processes of collaborative production environments. For that purpose the mutual suitability and synergetic potentials of the concepts of Autonomous Control and Agility in manufacturing were demonstrated. A discussion about the technical implementation and integration of autonomous objects into the Internet of Things showed the theoretical feasibility, but also the necessary improvements. The description of the CRC 637 demonstrator illustrated the potential of a combination of Autonomous Control and a fine-grained informational infrastructure in realising agility on the elemental level of processes and resources.

As mentioned before, the Internet of Things is no unrealisable vision; it will build on the present infrastructure of today's Internet and will include technologies which are currently under development. First steps of evolving the Internet of Things are already in progress, like smartphones, UMTS, LTE, and so on.

Due to the significance of the Internet of Things for globalisation, the research aspects from section 8.7 should be investigated by international research clusters. This will help to evolve common standards and methods within the Internet of Things and will avoid national or continental solos. The research clusters should unify research institutes and companies from those industries which will mainly influence the rising trend of an Internet of Things. Collaborative research in step with actual practice is a promising approach for a wide standardised, accepted and used Internet of Things.

Acknowledgements

This research is partially funded by the German Research Foundation (DFG) as part of the Collaborative Research Centre 637 "Autonomous Cooperating Logistic Processes – A Paradigm Shift and its Limitations" (CRC 637).

References

ACEA (2009) Vehicle Scrapping Schemes in the European Union. http://www.acea.be/images/uploads/files/20090406_Scrapping_schemes.pdf. Accessed 21 May 2010

Bessant J, Francis D, Meredith S, Kaplinsky R, Brown S (2001) Developing manufacturing agility in SMEs. Int J Manuf Technol Manag. doi: 10.1504/IJMTM.2000.001374

Böse F, Windt K (2007) Catalogue of Criteria for Autonomous Control in Logistics. In: Hülsmann M, Windt K (eds) Understanding Autonomous Cooperation & Control in Logistics - The Impact on Management, Information and Communication and Material Flow. Springer, Berlin

CERP-IoT (2009) Internct of Things Strategic Research Roadmap. http://www.grifs-project.eu/data/File/CERP-IoT%20SRA_IoT_v11.pdf. Accessed 21 May 2010

COMMISSION OF THE EUROPEAN COMMUNITIES (2009) Internet of Things - An action plan for Europe. European Commission. http://ec.europa.eu/information_ society/policy/rfid/documents/commiot2009.pdf. Accessed 21 May 2010

EPCglobal (2009) EPCglobal Architecture Framework – Version 1.3. http://www.epcglobalinc.org/standards/architecture. Accessed 21 May 2010

European Union (2007) REGULATION (EC) No 715/2007 OF THE EUROPEAN PARLIAMENT AND OF THE COUNCIL of 20 June 2007. Off J Eur Union. http://eur-lex.europa.eu/LexUriServ/LexUriServ.do?uri=OJ:L:2007:171:0001:0016:EN:PDF. Accessed 21 May 2010

Foundation for Intelligent Physical Agents (2002) FIPA standard status specifications. http://www.fipa.org/repository/standardspecs.html. Accessed 20 June 2010

Fleisch E, Christ O, Dierkes M (2005) Die betriebswirtschaftliche Vision des Internets der Dinge. In: Fleisch E, Mattern E (eds) Das Internet der Dinge. Springer-Verlag, Berlin-Heidelberg

Fleisch E, Thiesse F (2008) Internet der Dinge. In: Kurbel, K., Becker, J., Gronau, N, Sinz, E., Suhl, L (eds), Enzyklopädie der Wirtschaftsinformatik – Online Lexikon.

http://www.enzyklopaedie-der-wirtschaftsinformatik.de/wi-enzyklopaedie/lexikon/technologien-methoden/Rechnernetz/Internet/Internet-der-Dinge/index.html/?searchterm=internet%20der%20dinge. Accessed 31 May 2010
Frese, E (1998): Grundlagen der Organisation : Konzepte – Prinzipien – Strukturen. 7th edn., Gabler, Wiesbaden
Gehrke JD, Behrens C, Jedermann R, Morales Kluge E (2006) The Intelligent Container - Toward Autonomous Logistic Processes. KI 2006 Demo Presentations. http://www.intelligentcontainer.com/fileadmin/template/main/files/pdf/gehrkeDemo.pdf. Accessed 31 May 2010
Goldman SL, Nagel RN, Preiss K (1995) Agile Competitors and Virtual Organisations: Strategies for enriching the customer. Van Nostrand Reinhold, New York
Gunasekaran A (1998) Agile manufacturing: enablers and an implementation framework. Int J Prod Res doi: 10.1080/002075498193291
Hans C, Hribernik K, Thoben K-D (2008) An Approach for the Integration of Data within Complex Logistic Systems. In: Dynamic in Logistics - First International Conference 2007. Springer, Heidelberg
Hejl PM (1990) Self-regulation in social systems. In: Krohn W, Küppers G, Nowotny H (eds) Self-organization: Portrait of scientific revolution. Springer, Berlin
Hightower J, Boriello G (2001) Location Systems for Ubiquitous Computing. Comput. doi: 10.110910.1109/2.940014
Hribernik K, Hans C, Thoben K-D (2009) The Application of the EPCglobal Framework Architecture to Autonomous Control in Logistics. Proceedings of the 2nd International Conference on Dynamics in Logistics, Bremen
Hülsmann M, Windt K (eds) (2007) Understanding Autonomous Cooperation and Control in Logistics. Springer, Berlin et al.
Isidore C (2009) GM bankruptcy: End of an era. http://money.cnn.com/2009/06/01/news/companies/gm_bankruptcy/index.htm. Accessed 21 May 2010
Jedermann R, Lang W (2008) The Benefits of Embedded Intelligence - Tasks and Applications for Ubiquitous Computing in Logistics. In: Floerkemeier C, Langheinrich M, Fleisch E, Mattern F, Sarma SE (eds) The Internet of Things. First International Conference, IOT 2008. Springer, Berlin-Heidelberg
Kärkkäinen M, Holmström J, Främling K, Artto K (2003) Intelligent products: a step towards a more effective project delivery chain. Comput Ind. doi: 10.1016/S0166-3615(02)00116-1
Katayama H, Bennett D (1999) Agility, adaptability and leanness: A comparison of concepts and a study of practice. Int J Prod Res. doi: 10.1016/S0925-5273(98)00129-7
Knirsch P, Timm IJ (1999) Adaptive Multiagent Systems Applied on Temporal Logistics Networks. In: Muffatto M, Pawar KS (eds) Proceedings of the 4th International Symposium on Logistics. Florence
Ludwig B (1995) Methoden zur Modelbildung in der Technikbewertung. CUTEC-Schriftenreihe Nr. 18, Clausthal-Zellerfeld
Mahmoud QH, Yu L (2006) Making Software Agents User-Friendly. Comp. doi: 10.1109/MC.2006.239
Mattern E (2005) Die technische Basis für das Internet der Dinge. In: Fleisch E, Mattern E (eds) Das Internet der Dinge. Springer-Verlag, Berlin-Heidelberg
McFarlane D, Sarma S, Chirn JL, Wong CY, Ashton K (2003) Auto ID systems and intelligent manufacturing control. Eng Appl Artif Intell. doi:10.1016/S0952-1976(03)00077-0
Moore G (1998) Cramming more components onto integrated circuits. P IEEE. doi: /10.1109/JPROC.1998.658762
Morales Kluge E, Pille C (2010) Autonome Steuerung - Intelligente Werkstücke finden selbstgesteuert ihren Weg durch die Produktion. RFID im Blick, Sonderausg Brem 1:44-45
Morales Kluge E, Ganji F; Scholz-Reiter B (2010) Intelligent products - towards autonomous logistic processes - a work in progress paper. PLM 10. 7th International Product Lifecycle Management Conference, Bremen. To appear.

Morgan RE, Strong CA (1998) Market orientation and dimensions of strategic orientation. Eur J Mark 32: 1051 - 1073. doi:10.1108/03090569810243712

Pavlou PA, El Sawy OA (2005) Understanding the 'Black Box' of Dynamic Capabilities 1. ManSci http://agsm.ucr.edu/faculty/papers/pavlou/ms_pavlou_elsawy_rev3%201.pdf. Accessed 21 May 2010

Philipp T, de Beer C, Windt K, Scholz-Reiter B (2007) Evaluation of Autonomous Logistic Processes - Analysis of the Influence of Structural Complexity. In: Hülsmann M, Windt K (eds) Understanding Autonomous Cooperation & Control in Logistics - The Impact on Management, Information and Communication and Material Flow. Springer, Berlin

Pille C (2009) Produktidentifikation, Intralogistik und Plagiatschutz – RFID-Integration in Gussbauteile. In: BDG-Fachtagung Gussteilkennzeichnung. Methoden und Datenmanagement - Praxisberichte. VDG-Akademie, Essen

Porter M (1979) How competitive forces shape strategy. Harv Bus Rev. doi: 10.1225/79208

Prahalad CK, Hamel G (1990) The core competence of the corporation. Harv Bus Rev. doi: 10.1225/6528

Probst GJB (1987) Selbst-Organisation: Ordnungsprozesse in sozialen Systemen aus ganzheitlicher Sicht. Parey, Berlin

Probst GJB (1992) Organisation. Strukturen, Lenkungsinstrumente und Entwicklungsperspektiven. verlag moderne industrie, Landsberg

Pugachev V, Sinitsy, I (2002) Stochastic systems: theory and applications. World Scientific, Singapore

Raju M (2008) Energy Harvesting, Whitepaper, Texas Instruments Inc. http://www.ti.com/corp/docs/landing/cc430/graphics/slyy018_20081031.pdf. Accessed 21 May 2010

Rekersbrink H, Wenning B-L, Scholz-Reiter B (2007) Entscheidungen selbststeuernder logistischer Objekte. Ind Manag 23:25-30. http://www.sfb637.uni-bremen.de/pubdb/repository/SFB637-B1-07-009-IJ.pdf. Accessed 31 May 2010

Stürmann J, Benecke W, Zampolli S, Elmi I, Cardinali GC, Lang W (2005) A micromachined gas chromatographic column to optimize the gas selectivity for a resistive thin film gas sensor. Proceedings of the 13th International Conference on Solid-State Sensors, Actuators and Microsystems. Seoul

Thiesse F, Floerkemeier C, Harrison M, Michahelles F, Rodunes C (2009) Technology, Standards, and Real-World Deployments of the EPC Network. IEEE Internet Comput. doi: 10.1109/MIC.2009.46

Uckelmann D, Isenberg M-A, Teucke M, Halfar H, Scholz-Reiter B (2010) An integrative approach on Autonomous Control and the Internet of Things – Increasing robustness, scalability and agility in logistics networks. In: Ranasinghe D, Sheng Q, Zeadally S (eds) Unique Radio Innovation for the 21st Century: Building Scalable and Global RFID Networks. Springer, Berlin

Ventä O (2007) Intelligent Products and Systems. Technology Theme - Final Report. VTT Publications. http://www.vtt.fi/inf/pdf/publications/2007/P635.pdf. Accessed 31 May 2010

Windt K, Jeken O (2009) Allocation Flexibility – A new Flexibility Type as an Enabler for autonomous control in production Logistics. 42nd CIRP Conference on Manufacturing Systems, Grenoble

Wong CY, McFarlane D, Zaharudin AA, Agarwal V (2002) The Intelligent Product Driven Supply Chain. Proceedings of the IEEE International Conference on Systems Man and Cybernetics, Hammamet

Yusuf YY, Sarhadi M, Gunasekaran A (1999) Agile manufacturing: the drivers, concepts and attributes. Int J Prod Res. doi: 10.1016/S0925-5273(98)00219-9

Zalubowski D (2008): Gas prices put Detroit Big Three in crisis mode - U.S. consumers are moving to hybrids, high mileage models at a fast pace. Associated Press. http://www.msnbc.msn.com/id/24896359/. Accessed 21 May 2010

9 Integrated Billing Solutions in the Internet of Things

Dieter Uckelmann[1], Bernd Scholz-Reiter[2]

[1]LogDynamics Lab, University of Bremen, Germany

[2]University of Bremen, Planning and Control of Production Systems, Germany

Abstract The Internet of Things is one of the most promising technological developments in information technology. It promises huge financial and non-financial benefits across supply chains, in product life cycle and customer relationship applications as well as in smart environments. However, the adoption process of the Internet of Things has been slower than expected. One of the main reasons for this is the missing profitability for each individual stakeholder. Costs and benefits are not equally distributed. Cost benefit sharing models have been proposed to overcome this problem and to enable new areas of application. However, these cost benefit sharing approaches are complex, time consuming, and have failed to achieve broad usage. In this chapter, an alternative concept, suggesting flexible pricing and trading of information, is proposed. On the basis of a beverage supply chain scenario, a prototype installation, based on an open source billing solution and the Electronic Product Code Information Service (EPCIS), is shown as a proof of concept and an introduction to different pricing options. This approach allows a more flexible and scalable solution for cost benefit sharing and may enable new business models for the Internet of Things.

9.1 Introduction

In this chapter there will be a detailed look at costs and benefits in the Internet of Things based on the findings from research on networked RFID. Even though networked RFID covers only a partial aspect of the Internet of Things, there are many similarities and overlaps with it. Numerous studies on costs and benefits of RFID have been published. This is hardly surprising, as one of the problems of RFID adoption has been the difficult calculation of a business case or a positive ROI (Schmitt and Michahelles 2008). Cost benefit analysis has been used as the main tool for economic analysis. According to a study by Seiter et al. (2008), 87% of companies planning to implement and 81% of companies that have already implemented RFID use cost benefit analysis. However, cost benefit analysis of RFID usage is most often based on best guesses (Gille and Strüker 2008, Laubacher et al. 2006).

While RFID and other Auto-ID technologies continue to be major components of the Internet of Things, there are other technologies, such as sensors, actuators, and networked infrastructures that will further add to the ongoing cost discussion. The cost for hardware, software, integration, maintenance, business process reengineering and data analysis are major hurdles in the process of deploying the Internet of Things.

Costs and benefits are not always balanced between all stakeholders. Some Internet of Things related applications may never come true, because some of the stakeholders would need to spend more on technology and integration than can be justified by internal benefits. For RFID adoption across supply chains, cost benefit sharing has been suggested to solve this issue. However, contrary to the widespread usage of cost benefit analysis, cost benefit sharing is not a common instrument (OECD 2007).

There are several problematic aspects in cost benefit analysis and sharing:

- Detailed cost benefit analysis can be time consuming
- It is difficult to identify, measure and analyse all costs and benefits associated with an Internet of Things
- Companies are reluctant to share benefits
- Cost benefit sharing models do not scale, as they are subject to bi-directional negotiations

An alternative solution to cost benefit sharing could be based on selling and buying information that is provided through the Internet of Things. For this, a billing solution is needed to price and bill information. Similar concepts are known from the telecommunications industry, where billing solutions are an integral part of the overall infrastructure, allowing billing of different services, such as voice calls, SMS, Internet access and premium services, across service providers and different countries.

In this chapter there will be a close look at current concepts to evaluate costs and benefits in the Internet of Things. The problems of cost benefit sharing will be discussed and a technical solution to integrate billing software within EPCIS is introduced as one possible approach for pricing and billing of information. This approach does not replace ROI-calculations prior to Internet of Things investments, but it provides a tool to offer billable Internet of Things services and it generates historic data over time that may be used to calculate an ROI for future investments based on real data rather than estimated data.

A prototype installation that has been developed at the LogDynamics Lab in Bremen is used to provide a proof of concept. A scenario from the beverage supply chain will illustrate how physical actions are transformed into EPCIS events that are used to calculate billing orders.

9.2 Cost of RFID and the Internet of Things

There are numerous costs associated with the adoption of the Internet of Things. While the Internet of Things is not synonymous with RFID (even though some publications falsely stimulate this impression), results from cost analysis for RFID can be used as a basis for further calculations. In the following, there will be a short overview of the costs involved for RFID installations. While some of the financial data is based on other publications and cited correspondingly, other data is based on experience from corresponding purchases in the LogDynamics Lab at the University of Bremen between 2006 and 2009.

Agarwal (2001) lists six different costs of RFID deployment for manufacturing firms, including the cost of the tag itself, cost of applying tags to products, cost of purchasing and installing tag readers in factories and/or warehouses, systems integration costs, cost of training and reorganisation, and cost of implementing application solutions. It is not quite clear why Agarwal separates the cost of tags from the application process, while he sees cost for readers and their integration as one subject. Feinbier et al. (2008) list relevant costs for RFID installation in detail, based on experiences in the steel industry. On the basis of both approaches, similar cost structures can be inferred for the Internet of Things (Table 9.1).

Cost level	Cost of tagging (Agarwal 2001)	Cost considerations for RFID (Feinbier et al. 2008)	Cost of Internet of Things adoption
1 Mobile devices	• Cost of the tag itself • Cost of applying tags to products	• Tags	• Cost of mobile technologies, such as data-carriers (e.g., tags), sensors, actuators or smart devices • Cost of applying mobile technologies to things
2 Aggregation devices and software	• Cost of purchasing and installing tag readers in factories and/or warehouses	• Readers • Antenna and cabling • Installation • Tuning • Controllers • Software platform (middleware)	• Cost of purchasing edge devices (e.g., readers, gateways, controllers, accessories) and edgeware for fixed and/or mobile environments • Installation and technical optimisation costs
3 Integration	• Systems integration costs	• Integration (to legacy systems)	• Systems integration costs including new interfaces as well as necessary updates, extensions, or replacements

			of existing systems
4 Training and reorganisation	• Cost of training and reorganisation	• Process (incl. redesign and human elements)	• Training cost • Reorganisation / business reengineering / business model innovation
5 Application	• Cost of implementing application solutions	• -	• Cost of implementing internal application solutions beyond existing applications
6 Networking (technical and organisational)	• -	• -	• Cost for networking in an open environment, including e.g., improved security, fine layered access control, multi-directional communication, product data contracts, service level agreements, standardised syntax and semantics, data conversion, synchronisation, trust concepts
7 Operational	• -	• Maintenance	• Cost for maintenance • Other operational costs for running (e.g., data storage and analysis), extending and improving the system

Table 9.1 Cost Levels for the Internet of Things

The *first cost level* in the Internet of Things includes mobile devices that are linked to physical objects. These can be RFID tags fixed to a product, as well as sensors, actuators (e.g., signal lights, power switches) or smart devices that combine multiple technologies. The price of RFID tags has been an important issue over the last years. User acceptance for tag prices differ in relation to the aggregation level of the product to which they are attached. In a study from 2004, 100 companies were asked what was the highest price they would accept for tags on item and unit level. On item level,a tag price of 0.10 € or less was most often required. On unit level a higher tag price was still reasonable (ten Hompel and Lange 2004). The measured average price for 2008 was 1.13 US-$ per tag, although this average represents High Frequency (HF) as well as Ultra-High Frequency (UHF) tags (IDTechEx 2009). UHF standard smart labels can be bought at a cheaper price, though. The lowest price that was offered to the LogDynamics Lab at the University of Bremen for a standard ISO/IEC 18000/Amd 1 (2006) compliant UHF self-adhesive inlay was 0.08 € in 2009. On-metal UHF tags with a robust housing usually cost in the range of 3 € to 7 €, due to the housing, the adjusted antenna design and the low quantities compared to smart labels. IDTechEx (2009) predicts an average price per tag of 0.22 US-$ by 2014 for both, HF and UHF tags. The discussion on RFID tag costs is mainly focused on passive RFID. For active RFID the cost per tag are considerably higher and will be typically in the range of 15 € to 75 €. While the lower end of the range is mainly defined by the cost for the battery and the housing, the higher end is more determined by the market position of the individual vendors. Usually non-standardised tags and readers have to be bought from the same vendor, thus leading to a long-term tie-up with one company. With the availability of ISO/IEC 18000-7 (2009), providing parameters for active air interface communications at 433 MHz, RFID tags for this frequency range can be bought from different providers. In fact, the Department of Defence (DoD) in the USA – one of the largest customers for active tags – placed its first orders of corresponding tags to Unisys, Savi, Systems and Processes Engineering Corp. (SPEC) and Northrop Grumman. Previously they were tied-up to Savi for sourcing active tags. Savi owns some intellectual property rights that require licensing from Savi to provide ISO/IEC compliant active tags. Nevertheless, the DoD claims that they pay half the price for the Unisys tags, compared to the prices they had to pay for the previous proprietary SAVI tags. Unisys themselves use Identec Solutions and Hi-G-Tek as subcontractors to supply the tags. The active tags need to comply with the DoD military standards, which require safe and reliable operation in helicopters (Swedberg 2009). The corresponding tests are quite expensive and add to the high cost of these tags. Other active tags operate in the range of 860 to 960 MHz, 2.4 GHz or in the Ultra-Wide-Band (UWB) range. These tags sometimes offer additional features, such as location sensing. Considering the prices of active tags and their successful deployment in industry, the isolated price discussion about passive tags seems rather inappropriate. Consequently, the price for the tags should always be compared to the benefit it generates. However, if RFID is compared with other IT-investments, one has to bear in mind the reoccurring costs for tags. When we consider the integration of sensors, actuators and smart devices in the Internet of Things there will be even more expensive ubiquitous mobile technologies that need to be paid for. Therefore, the costs of mobile devices and their installation on things will remain a major topic in the cost discussion for the Internet of Things.

The *second cost level* includes aggregation devices and aggregation software, such as readers, antennas, cabling, controllers and other edge hardware and software as well as the corresponding installation costs. RFID reader kits can be as cheap as 50 € for a HF reader with USB connection, some sample tags and a software that triggers websites or applications[92]. These new offerings will allow RFID to be used in smart home scenarios, and for fun purposes. In the mid-term, they may also put pressure on RFID offerings for industrial purposes. Today, ISO/IEC

[92] A popular example is provided by Violet (www.violet.net).

18000-6c compliant readers with 4 antenna ports can already be bought for under 1,000 US-$ in the USA, while prices in Europe currently are still higher and usually are in the range of 1,300 € to 2,500 €. In some publications (e.g. Feinbier et al. 2008) reader costs are considered to be correlated with functionality. Instead, the price is more related to the company position, the sales strategy of the individual company, and the number of middlemen involved. Corresponding UHF antennas in general are in the price range of 80 € to 300 €. Antenna cables can be considered to cost about 10 € to 30 €. Handheld RFID Personal Data Terminals (PDT) are priced between 1,000 € and 4,000 €. RFID printers start at about 1,000 € and may go up to 30,000 € or more for integrated and automated labelling solutions. Other hardware costs include hardware portal frames to hold the reader and antennas. Some retailers have used large metal housings to shield between dock doors in order to avoid false reads. Newer installations rather use intelligent filtering mechanisms provided by corresponding middleware components. The setup of the gates may require considerable costs for hardware and installation. An RFID site survey will cost about 1.000 € (Feinbier et al. 2008). Feinbier et al. (2008) consider 20,000 € installation cost per read point in a harsh environment, such as the steel industry. This seems rather high for standard dock-door installations, but still illustrates that the cost for installation should not be neglected.

Controllers and middleware are used for managing low-end hardware and abstracting these from the applications. Sometimes the middleware is further divided into solutions interfacing with hardware (edgeware) and the middleware interfacing with applications. In this case, middleware may be considered to be part of the integration level.

The *third cost level* includes all integration costs to legacy systems, middleware and updates of existing system. The cost for the middleware acquisition is further increased by the necessary installation cost. Middleware can be based on freeware, such as the Fosstrak-system[93], or it may also be provided by large integrators, such as IBM[94], software giants, such as Oracle or SAP[95], EDI-specialists, such as Seeburger[96], and RFID-specialists, such as Savi[97] and REVA[98]. In the Internet of Things, middleware does not only link to internal applications, but additionally allows multidirectional communication between companies, end-users and public institutions (see level six).

Additionally, costs for updating applications, such as Enterprise Resource Planning (ERP), Supply Chain Management (SCM), and Product Lifecycle Management (PLM) systems, need to be considered.

---

93 www.fosstrak.org

94 www.ibm.com

95 www.sap.com

96 www.seeburger.com

97 www.savi.com

98 www.revasystems.com

The fourth level includes cost for educating the project team and end-users as well as cost of reorganisation. The necessity of training and education for end-users is quite important, because the Internet of Things requires fundamental knowledge about different technologies, such as Auto-ID and sensors as well as knowledge about real-time data handling and analysis. Additionally, certain aspects of the Internet of Things raise privacy and security concerns of workers and unions, which may lead to a total failure of the project. Training and education help to provide the corresponding skills and to address technology-related fears. The cost of reorganising the business processes result from traditional management tools, such as business process reengineering or newer approaches, like business model innovation. As a result, further infrastructural investments may be required. Ford Cologne (Germany) for example, paved a new roundabout for optimising their car distribution process to vessels, trains, trailers and storage areas, based on RFID and automated access gates (Harley 2008). It can be estimated that the cost for the new roundabout exceeded the cost of the RFID infrastructure. While this example shows an investment in a single process optimisation, new business models may require extensive organisational changes.

The *fifth cost level* includes new internal applications, which are rolled out in a firm to unleash the full potential of the Internet of Things. The costs include standard software, such as PLM or SCM systems, as well as individual software and all associated costs for installation, customisation and training. These applications interface to the Internet of Things and provide tools for data-entry and retrieval, analysis, planning, forecasting and more.

The *sixth cost level* considers the fact that an Internet of Things needs communication and collaboration across enterprise boundaries, non-commercial stakeholders, such as governmental institutions, and end-users. While middleware provides some functionality in the Internet of Things for collaboration and communication, further investments are necessary. Some suppliers, especially in retail, have to consider an investment into an Electronic Data Interchange (EDI-) infrastructure, as EDI represents the current state of the art. Even the EPCglobal network will not replace EDI, as it does not cover issues such as purchasing or forecasting. Software-related costs can start from tens of thousands of Euros and may reach several million Euros in large installations. Others will need to provide Web-interfaces to access and contribute to the Internet of Things.

Negotiations with partners, suppliers and customers about data requirements and service level agreements will be necessary. For machine-to-machine communication, detailed syntax and semantics are required. Finally, trust and security issues need to be addressed in a networked environment.

The *seventh cost level* covers operating costs for maintaining, running, improving and extending the system. The hardware and software need to be maintained and updated regularly. So, an annual amount of 10% to 15% of the hardware and software investment cost should be considered. Electricity costs, to operate the infrastructure, are usually quite low in comparison with the other costs involved. However, as Green IT initiatives are becoming more and more significant, the

Internet of Things is no exception. Above all, the labour involved to provide high-quality product data has to be taken into account. As these costs are difficult to calculate, they are most often omitted from any calculations. Besides keeping the technical infrastructure alive, day to day tasks, such as data storage and analysis as well as overall improvements and upgrades to cope with growth, are adding up to considerable recurring costs.

In an early study from AMR Research (McClenahen 2005), the costs for system integration, changes for supply-chain applications and for data storage and analytics were considered to reach between 8 and 13 million US dollars for a full implementation of RFID for a Consumer Packaged Goods (CPG) manufacturer, shipping 50 million cases per year (see Table 9.2).

Cost category	Assumed cost
Tags and readers	\$5 million to \$10 million
System integration	\$3 million to \$5 million
Changes to existing supply-chain applications	\$3 million to \$5 million
Data storage and analytics	\$2 million to \$5 million
Total	\$13 million to 23 million

Table 9.2 Assumed Cost of Compliance for a Full-fledged RFID System at a CPG Manufacturer (McClenahen 2005)

A study among 137 Wal-Mart suppliers showed though, that the initial average cost was only about 500.000 US-\$ (Incucomm 2004). Hardgrave and Miller (2006) consider that there are three reasons for the deviation between estimated and actual cost. First, they consider that several suppliers have only implemented limited installations. This may change over time, though, if RFID is becoming more ubiquitous. Second, they believe that the RFID cost infrastructure has decreased and continues to do so. However, considering the added costs in an Internet of Things, including multiple different devices (e.g., sensors), it can be expected that the overall cost will be higher than for an isolated RFID deployment. Third, they consider that the deployment costs are lower than expected. But again, this may relate to the limited integration depth of ‘slap and ship’ installations. The Internet of Things requires a deeper integration across company boarders and multiple stakeholders and will therefore add to higher overall cost. Fourth, they consider that the cost of data storage is much less than envisioned by McClenahen. Once more, this may relate to the limited scope of integration on the one hand and missing revenue opportunities on the other. No matter if the final cost will be closer to \$500.000 or to \$23 million – the investments need to be justified by a corresponding ROI.

There are different options to pay for the costs of RFID adoption. These differ between implementation and operation. In a study from Bensel and Fürstenberg

(2009), more than 100 end-user companies have been asked which payment options they prefer for implementation and operation. For implementation there was a clear preference towards a target agreement-based payment scheme. Variable payment options based on number of tags, data volume, process times or pay-per-read were not well accepted (see Table 9.3).

	Transponder volume	Data volume	Process times	Pay per read	Work package	Target agreement	Fixed monthly payment	Single payment
Always true for me (weighting factor 2)								
Implementation	12	0	0	0	19	26	2	12
Operation	17	0	2	2	5	22	7	7
Usually true for me (weighting factor 1)								
Implementation	12	3	5	2	21	30	5	12
Operation	10	12	7	5	12	17	12	5
Neutral (weighting factor 0)								
Implementation	9	7	15	10	21	14	17	15
Operation	19	12	12	12	28	21	19	21
Usually not true for me (weighting factor -1)								
Implementation	19	26	17	20	14	5	14	7
Operation	14	14	14	16	7	7	9	12
Not at all true for me (weighting factor -2)								
Implementation	48	64	63	68	25	25	62	54
Operation	40	62	65	65	48	33	53	55
Weighted results / average								
Implementation	-0.79	-1.51	-1.38	-1.54	-0.05	0.27	-1.29	-0.79
Operation	-0.50	-1.26	-1.33	-1.37	-0.81	-0.12	-0.89	-1.03

Table 9.3 Preferred Payment Options for Implementation and Operation (based on Bensel and Fürstenberg 2009)

One of the reasons for this could be the missing technical infrastructure to measure and bill the corresponding usage. For operation, a usage-based accounting did receive higher acceptance levels. While pricing based on target agreements still was preferred, a pricing scheme based on transponder volume, followed as second preference.

It may be assumed that one of the reasons for the reluctance to use usage-based pricing schemes, based on pay per read, process times or data volume, may be once more the lack of an integrated technical billing solution.

9.3 Benefits of RFID and the Internet of Things

There have been numerous analyses to identify and structure benefits of RFID in supply chains. While the benefits are named in relation to RFID adoption, the corresponding IT infrastructure, including e.g. the EPCglobal Network, is most often implied. Baars et al. (2008) have identified four different approaches towards systemisation of RFID benefits:

- Collecting and grouping – benefits are collected and grouped. Examples for these types of studies are Agarwal (2001) and Li and Visich (2006)
- Layer of impact – benefits are structured to impact layers such as short term and long term automation, informational and transformational benefits, proven or potential (Bovenschulte et al. 2007, Hardgrave et al. 2008)
- Locus of impact – these studies highlight who benefits, thus it automatically considers benefits to multiple stakeholders (Wong et al. 2002, Hardgrave et al. 2008, Tajima 2007)
- Indicator system – established evaluation systems, such as Balanced Scorecards, are used to structure RFID benefits (Schuster et al. 2007, Scholz-Reiter et al. 2007)

Sometimes combinations of these structures are used (e.g. Hardgrave et al. 2008). For this chapter it will be important to understand who benefits (locus of impact) from RFID and the Internet of Things usage on an inter-organisational or even end-user level. The following list is based on Wong et al. (2002), Hardgrave et al. (2008), and Tajima (2007), but additionally includes benefits to society. Service and infrastructure providers are not included, as they benefit only indirectly, for example through sales, services and new business opportunities, rather than directly from accessing the Internet of Things.

Collective benefits can be achieved by all of these stakeholders. These include:

- *Reduced product shrinkage:* reduction of loss of goods through misplacement, spoilage, and theft
- *Improved information sharing:* product related data may be exchanged to benefit multiple stakeholders, problems resulting from converting paper-based information to digital information are avoided and manual data-entry is drastically reduced
- *Compensatory benefits:* benefits provided through other stakeholders, including for example cost benefit sharing, funded research, bonus payments, vouchers, information (e.g. sales data)

Companies in general may benefit from:

- *Increased inventory, shipping and data accuracy:* e.g., differences between real stock numbers and assumed stock, based on false data[99]
- *Subsequent fault reduction:* inaccurate and incomplete visibility may lead to false decisions and can be avoided through the Internet of Things[100]
- *Faster exception management (agility):* capability of responding to unplanned events in a timely manner before critical problems escalate
- *Asset management:* better asset utilisation may lead to an opportunity to reduce asset inventory, reduced asset shrinkage, better shipment consolidation, reduced energy consumption and improved reverse logistics
- *Product rotation:* methods of inventory control, such as First In, First Out (FIFO) can be used more accurately to ensure efficient stock rotation e.g. in time sales for perishable goods (Hardgrave et al. 2008)

Manufacturers and suppliers benefit mainly from:

- *Production tracking:* tracking of raw material, work-in-progress inventory, assembly status tracking and finished products
- *Quality control:* ensured quality control in production
- *Supply / production continuity:* enabled through improved material tracking
- *Compliance:* e.g., in case of mandates issued for example by large retailers (Aberdeen 2007) or legislators and regulators

Distributor and logistics provider as well as internal distribution and logistics departments benefit from:

- *Material handling:* time (labour) savings for loading / unloading of trucks, administrative overhead at the goods receipt[101], cross-docking, customs clearance,

---

[99] In a survey among 141 companies, 70% estimated a deviation between real and IT-data of up to 10%. 13% of the companies even estimated a higher inaccuracy of 10% to 30% (Gille and Strüker 2007).

[100] As an example Wal-Mart reduced unnecessary manual orders, due to inaccurate stock visibility by 10% (Hardgrave et al. 2008).

delivery lead times and reduced delays, faster inventory, goods receiving, loading and unloading as well as reduced human errors through Auto-ID

- *Space utilisation:* achieved through reduced buffers and reduction of product storage incompatibilities (e.g., placement of hazardous goods[102]), based on better data accuracy through RFID usage

Retailer benefits include:

- *Customer service:* RFID can be used to simplify checkouts and payments as well as for promotion management (Thiesse and Condea 2009)
- *Lower inventory:* reduced stockouts and smaller buffer stocks, due to improved inventory data
- *Reduced stockouts:* substantially reduced stockouts can be achieved through RFID if movements to the shop floor can be tracked[103]
- *Promotion execution:* RFID and the Internet of Things may be used to obtain better visibility for timely placements of promotional items[104]
- *After sales services:* in after-sales service, RFID may be used for warranty issues, repair and goods authentication

Benefits for consumers are:

- *Personal access to product specific information:* e.g., to be able to access the product history of a car, based on a vehicle identification number
- *Active participation opportunity:* e.g., through beta testing, product ratings, field reports, applications and more
- *Interaction with other stakeholders:* e.g., automatic updates and repairs, dynamic safety warnings, product recalls, public applications
- *Home automation and leisure applications:* e.g., room monitoring, smart devices, intelligent toys

Benefits to society include:

- *Consumer protection / safety:* e.g., food and health safety, environmental monitoring
- *Security:* e.g., to avoid terrorist attacks, customs support
- *Trade facilitation:* comparable with the introduction of UN/EDIFACT in 1988

---

[101] Times for loading and unloading of trucks can be reduced up to 13%, administrative overhead may be reduced up to 70% and time savings at the goods receipt may be as high as 90%, if bulk reading can be applied (Grote 2006).

[102] A solution approach for incompatible products has been researched in the OPAK project (Schnatmeyer 2007).

[103] Wal-Mart has achieved up to 30% reduction in out-of-stocks by using RFID-tagged cases to improve shelf-stocking processes (Hardgrave et al. 2006). Other companies report 10% to 50% reduction on out-of-stocks resulting in a gain of 7.5 to nearly 25 sales basis points (Laubacher et al. 2006).

[104] Procter & Gamble estimates an average of 20% increased sales by timely placements (Collins 2006).

- *Infrastructure optimisation:* e.g., roads, public transportation

These benefits are based on technologies in the Internet of Things. Some (e.g., quality control in production) may not require an overall Internet of Things implementation, but the Internet of Things will improve these individual tasks by sharing information in networked environments. The list of benefits mentioned above shows quite clearly that numerous stakeholders may benefit from an Internet of Things, but unfortunately not to the same extent. Additionally, several of these benefits cannot be achieved alone, but only in collaboration with other stakeholders.

The measurability of the benefits should be considered. While measurable benefits most often refer to monetary aspects, there are as well qualitative benefits that can be measured, such as customer satisfaction. Measurability may be subjective to individual projects, for example time measurements are not allowed in some companies.

9.4 Cost Benefit Sharing

Costs and benefits of the Internet of Things that have been explained in detail in the last paragraphs are not evenly distributed between the stakeholders. Cost benefit sharing models may be used as a tool to balance these asymetries. Cost benefit sharing in combination with RFID has been researched by several authors (Riha 2009, Hirthammer and Riha 2005, Bensel et al. 2008, Wildemann et al. 2007). Sharing benefits and investments in multi-tiered situations is seen as a core requirement for wide-scale deployment of RFID (Schuster et al. 2007). Hirthammer and Riha (2005) define cost benefit sharing as:

> "A systematic and system-oriented incentive system that motivates companies in a network to participate in joint projects that do not benefit them directly. … A Joint Project is a cooperative effort to improve the processes or resource allocation in the network. It involves at least two parties in the network."

This rather limited definition with a focus on providing an incentive to otherwise non-profiting companies is extended by Riha (2009):

> "Cost benefit sharing (CBS) is a method to accomplish process changing projects in networks. It is based on a stakeholder oriented total cost analysis of all packages of measures in a project. Based on the achieved transparency of positive and negative effects a win-win situation is provided through reallocation strategies for all stakeholders. Therefore an incentive to a network-wide optimisation is given."

In this definition a cost and benefit transparency between the stakeholders is suggested to achieve a win-win situation. Unfortunately, this level of transparency is quite often not wanted by companies.

The structural requirements for cost benefit sharing can be quite complex and cost intensive. Hirthammer and Riha (2005) even suggest having different institutions on a structural level, including a board of company representatives, a mediator, and a company independent controller. According to Hirthammer and Riha (2005), the cost benefit sharing process loop can be structured in several subtasks:

1. Detailed process analysis in the network through auditing
2. Enquiry of weak points through benchmarking
3. Development of corresponding actions to solve or lessen the effect of the weak points based on overall strategies and goals
4. Cost benefit sharing
    a. Calculation of costs
    b. Evaluation of benefits
        i. Calculate monetary benefits
        ii. Calculate qualitative benefits
        iii. Evaluate total benefit
        iv. Calculate share of benefit
    c. Distribution of costs
5. Implementation of actions proposed in step 3
6. Controlling
7. Feedback loop to adjust the system to external dynamics

While tools have been developed to calculate costs as well as benefits, it becomes apparent, why cost benefit sharing approaches have failed to gain wider acceptance. The effort involved to install and maintain such a system exceeds the advantages, in most cases.

One of the fundamental mistakes in the usual cost benefit sharing models is to look for a 'fair' scheme to level cost and benefit, rather than to look for a model that accepts market forces. Hirthammer and Riha (2005) suggest using a mediator to settle disputes, which does not seem appropriate for highly-dynamic information sharing processes.

An IT infrastructure that supports a self-regulating approach, based on supply and demand of information and assisting free competition, may be more promising.

9.5 A Technical Framework for Integrating Billing Capabilities into the EPCglobal Network

As discussed in chapter 1, a possible solution to overcome the problems of cost benefit sharing in the Internet of Things may be based on an integration of a billing solution into the EPCglobal Network. In a prototype test scenario that has been set up at the LogDynamics Lab in Bremen, two open source products have been chosen for implementation. The well-known Fosstrak[105] EPCIS software has been integrated with jBilling[106], an open source billing solution that is mainly being used in telecommunication companies. The jBilling system has been chosen for the following three reasons. Firstly, it does not require an upfront investment in software. Secondly, it is open source and, therefore, allows modification to the software. And thirdly, it aligns well with the technologies used in Fosstrak and therefore may allow a tighter integration.

Both products use Tomcat as Web-server, but there are two different relational databases in use – Hypersonic for jBilling and MySQL for Fosstrak. jBilling can run on MySQL, so that Hypersonic could be eliminated in a further integration effort[107]. To combine the two different systems, there are two initial requirements:

1. There should be an integrated login procedure
2. Selected EPCIS events should be translated to jBilling purchase orders

Any charge to a customer corresponds to a purchase order[108] in jBilling. These include subscriptions, single purchases, taxes, and interest.

Figure 9.1 shows the overall billing process between Fosstrak and jBilling. The accounting process may be triggered by an event, such as a pallet with an RFID tag passing a dock-door (1a, 1b).

[105] www.fosstrak.com

[106] www.jbilling.com

[107] A first trial of using MySQL for jBilling has produced several error messages.

[108] For brevity, purchase orders are referred to as orders in jBilling.

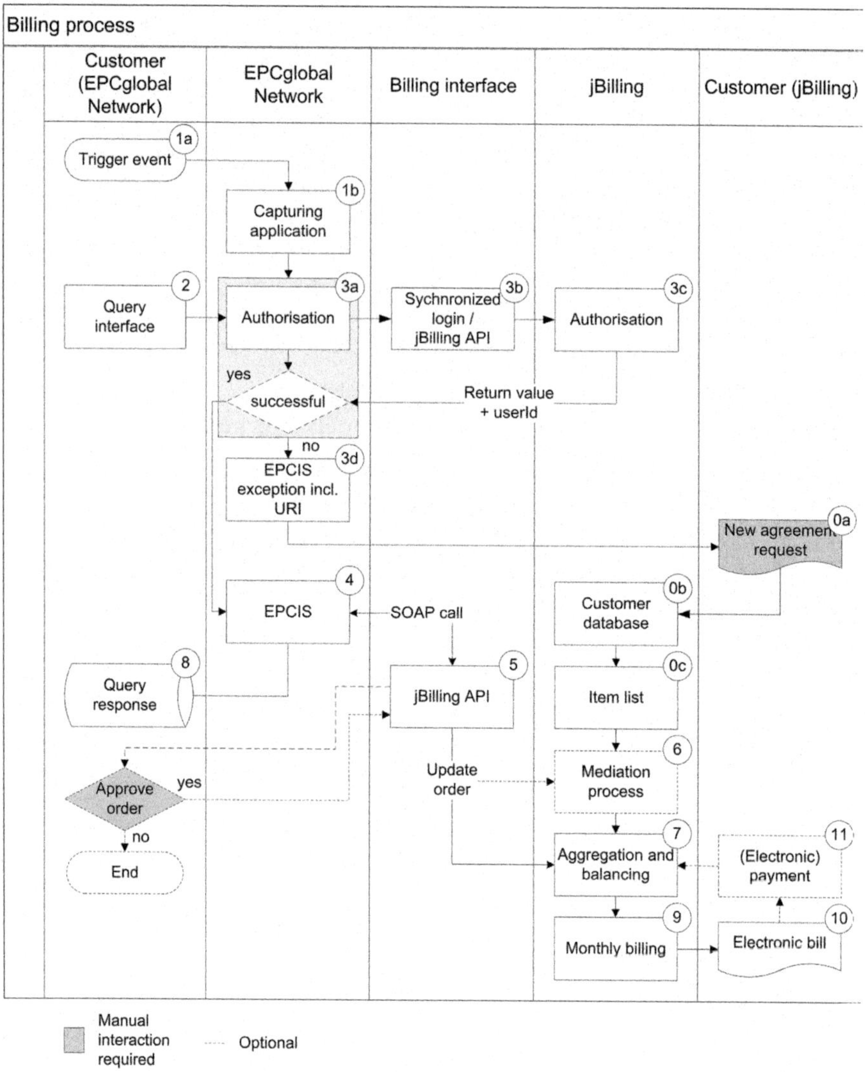

Fig. 9.1 The Billing Process between Fosstrak and jBilling

This event may already start a billing process if we consider for example deposit fees for returnable transport items. Other billing activities may be started through a query for payable information (2). As part of the Fosstrak authentication process (3a), the access rights, including the availability of a billing account (3c), are checked via the jBilling Application Programming Interface (API). For this purpose, a combined login process has been implemented as an option in the Fosstrak EPCIS query interface (Figure 9.2) at the LogDynamics Lab.

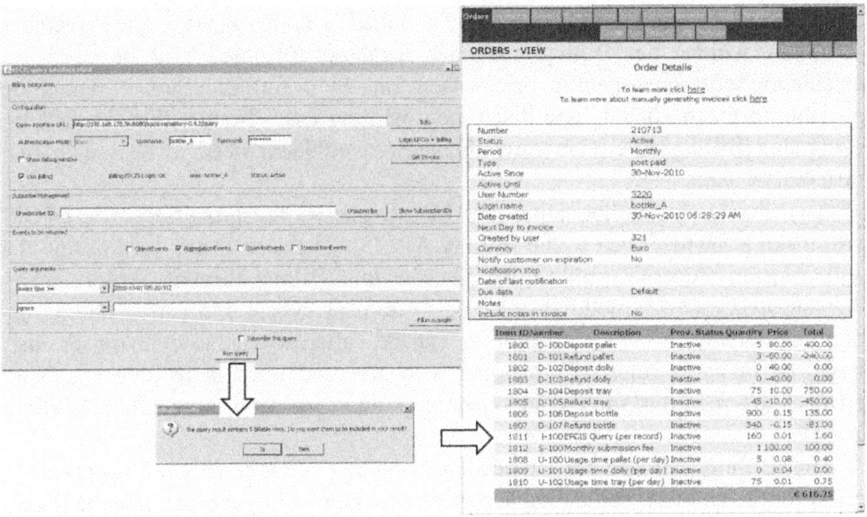

Fig. 9.2 Integrated Login Procedure and Workflow between Fosstrak (EPCIS Query Interface) and jBilling

Currently, only basic authentication is enabled. More sophisticated security functions could be supported in a future version. For the prototype installation we use the same login and password data for both systems. If the input data is null or missing, jBilling generates an API exception (jBilling 2010). Otherwise jBilling returns different integer values as described in Table 9.4.

Integer value	Description
0	The user was successfully authenticated and his current status allows him entrance to the system.
1	Invalid credentials. The user name or password are not correct.
2	Locked account: The user name or password are incorrect, and this is the last allowed login attempt. From now on, the account is locked.
3	Password expired: The credentials are valid, but the password is expired. The user needs to change it before logging in.

Table 9.4 Return Values for the Authorisation Process from jBilling (jBilling 2010)

An integer value for the user ID will be used further on to link purchases (orders) to a specific account. If no valid contract in jBilling can be found, *JbillingAPIException* could be converted into an EPCIS exception, containing a Uniform Resource Identifier (URI) that links to a new agreement request (3d). The agreement may contain pricing information, financial details, such as preferred payment service, and payment options (e.g. monthly). For further usage in the Internet of Things it would be favourable, if individual service level agreements and information quality details could be included or linked as well. The agreement is stored within the jBilling customer database (0b) and will be used for calculating customer-specific prices later on. In a further effort it would be possible to create, update and delete new jBilling users from within the EPCIS, using the jBilling API. Consequently, users would not need to deal with two different systems.

After successful authorisation, the EPCIS queries are processed. The EPCIS will make a SOAP[109] call to the jBilling API (5). The *userID* provided during the authorisation process is used to link an order to a jBilling account. The *createOrder* and *updateOrder* methods are used to transfer events into corresponding orders. Optionally, a mediation process can be called to enable dynamic pricing, based on business rules. If prices per item are predefined in jBilling and if no changes are required, the mediation process does not need to be called (jBilling 2010).

The jBilling API updates the account balance (7). Optionally, an approval request for the end-user can be implemented. An approval by the user may be necessary, for example. if the information purchase is not covered by a flat-fee subscription. Finally the query response is delivered and the account balance is updated by jBilling. Usually, monthly billing will be used to invoice the aggregated values in business scenarios (9). In order to avoid problems resulting from analogue to digital media conversions and cost intensive manual labour, electronic bills (10) and electronic payment (11) will be preferred. The login screen to jBilling and the EPCIS (Figure 9.2) also offers an opportunity to retrieve last invoice values. Additionally, an invoice is sent via e-mail or traditional postal services to a defined recipient.

A Usage Scenario within the Beverage Industry

The described integration of a billing solution offers flexible usage for multiple industries and applications. To illustrate the prototype installation, a scenario from the beverage industry has been used. There may be different events that need to be processed for the billing system. Querying information is just an example. Usage-based fees and deposits for Returnable Transport Items (RTI), or initial costs for infrastructure could also be handled through the billing system. Any event where customers are using measurable services may be communicated to the billing system. The billing mediation process is able to differentiate the different events and to calculate individual prices, based on business rules.

---

[109] SOAP is a standard WEB services protocol for exchanging structured information in distributed environments.

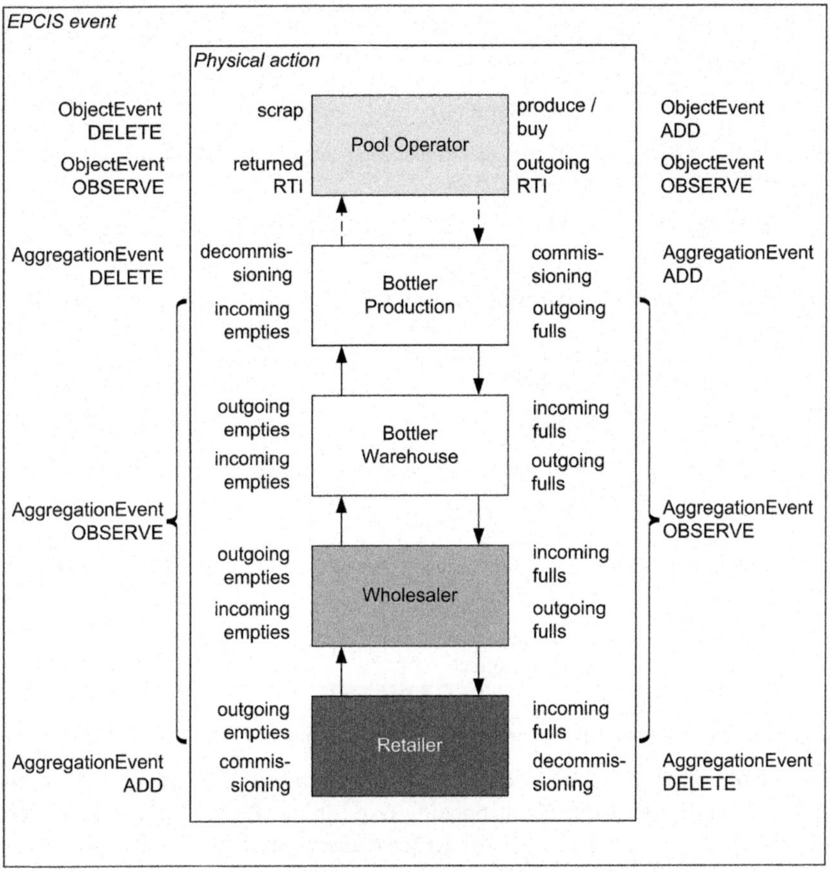

Fig. 9.3 A Simplified Supply Chain Scenario in the Beverage Industry

In a simplified scenario in the beverage supply chain, the EPCglobal Network may be used to track the flow of goods between pool operator, bottler, wholesaler and retailer (Figure 9.3). The pool operator provides RTIs (e.g., pallets, dollies, trays) to the bottler, who fills the pallets, stores and ships them to the wholesaler. At the end of production the different RTIs are aggregated to one pallet. The wholesaler delivers the pallet to the retailer and in return collects pallets with empty bottles. These are returned to the bottler to be refilled or to the pool operator in case of over-capacities or repair requests. The integrated billing times may include beverage prices, usage-based pricing and deposits for RTIs, as well as initial, monthly and usage-based fees for information access. Table 9.5 lists the different cost types, the corresponding EPCIS events and the associated price structure.

Cost type	Event / calculation	Price structure
Beverage price	e.g. BottlerOutgoingGoods (AggregationEvent, OBSERVE)	Per pallet
Deposit (pallet, dolly, tray, bottle)	e.g. BottlerEndOfProduction (AggregationEvent, ADD)	Fixed price, per event
Deposit refund (pallet, dolly, tray, bottle)	e.g. BottlerEndOfProduction (AggregationEvent, DELETE)	Fixed price, per event
Usage-based fee (pallet, dolly, tray)	e.g. RetailerOutgoingGoods (AggregationEvent, OBSERVE) - RetailerIncommingGoods (AggregationEvent, OBSERVE)	Usage-based, per day
Initial fee (optional)	Account opening	Fixed price, only once
Monthly IT infrastructure rent or lease (optional, e.g. for readers)	Initial contract, contract period (e.g. 12 month)	Percentage of purchasing cost, per month
Monthly information access	Initial contract, contract period (e.g. 12 month)	Flat fee, per month
Premium query	Queries that are not covered through monthly contract	Usage-based, per event

Table 9.5 List of Different Options for an EPCIS-based Pricing in a Beverage Scenario

The table shows different pricing schemes for products (beverage), RTI (usage-based fees and deposit), account opening (e.g., initial fee for new stakeholders), infrastructure rent or lease (e.g., for RFID readers), monthly information access, including standard queries, covered by a subscription, and premium services that require extra payments. It is quite obvious that this is just an example of using a billing system in combination with applications in the Internet of Things, such as the EPCIS. Nonetheless, it illustrates the flexibility that can be achieved for pricing beyond 'physical' product pricing. The actual pricing scheme will depend on the individual business model.

Instead of an internal billing solution, billing service providers in the Internet of Things could offer their services. Unfortunately, these services usually require a minimum fee (e.g., 0.15 €) per transaction, which is much too high for low-value queries. A company offering information services through the EPCglobal Network, could have millions of billable low-value events. However, there is no need for a micro-payment system, as these events may be consolidated in a periodic (e.g., monthly) bill. If the proposed integration of billing and the Internet of Things proves to be beneficial, billing service providers may change the pricing models to participate in this market. A further advantage of an internal billing solution is a higher level of flexibility in dynamic pricing and a tighter integration possibility with internal applications. However, the effort for installing and maintaining an internal business solution should not be underestimated.

9.6 Discussion and Outlook

In this chapter, costs and benefits of the Internet of Things have been presented and the concept of cost benefit sharing has been evaluated. A technical solution has been provided for integrating billing into the future Internet of Things. Therefore, a synchronisation of material, information and financial flows has been achieved. The concept has been validated by developing a prototype that combines an open source billing solution with an open source implementation of the EPCglobal EPCIS standard. A beverage scenario has been used to illustrate the technical prototype.

The overall goal of the prototype integration of a billing system with EPCIS was to provide a means for charging for information access, thus enabling a free trade of information within an Internet of Things, based on market forces. It may be used as a simple alternative to timely and costly cost benefit sharing agreements. Another important effect will be to collect historical data about the value of information over time, based on real values rather than on guesses for future ROI calculations.

A phased approach will probably be necessary to validate the acceptance and applicability of the concept. Firstly, there may be a trial for internal purposes, for example. to split IT infrastructure costs between different departments. Secondly, limited networks, such as closed loop RTI-applications may adopt billing as described in the beverage scenario above. Thirdly, an open billing opportunity in a ubiquitous Internet of Things would not only solve current problems, such as a missing ROI in a lot of calculations, but it would also enable new business models.

Acknowledgments

Our thanks go to Mark Harrison and Jeanette Mansfeld for their constructive criticism and several interesting discussions about billing in an Internet of Things.

References

Aberdeen (2007) Winning RFID Strategies for 2008. http://www.barco.cz/data/download/rfid/Winning_RFID_Strategies_for_2008_Aberdeen.pdf. Accessed 1 November 2009

Agarwal V (2001) Assessing the benefits of Auto-ID technology in the consumer goods industry. http://www.autoidlabs.org/uploads/media/CAM-WH-003.pdf. Accessed 11 November 2009

Baars H, Sun X, Strüker J, Gille, D (2008) Profiling Benefits of RFID Applicationshttp://aisel.aisnet.org/cgi/viewcontent.cgi?article=1262&context=amcis2008. Accessed 4 November 2009

Bensel P, Fürstenberg F (2009) Partnerintegration im Rahmen von RFID-Projekten. In: F Straube (ed), RFID in der Logistik – Empfehlungen für eine erfolgreiche Einführung. Universitätsverlag der Technischen Universität Berlin, Germany

Bensel P, Günther O, Tribowski C, Vogler S (2008) Cost-Benefit Sharing in Cross-Company RFID Applications: A Case Study Approach. Proceedings of the International Conference on Information Systems (ICIS 2008) Paris, France

Bovenschulte M, Gabriel P, Gaßner K, Seidel U (2007) RFID: Prospectives for Germany. http://www.bmwi.de/BMWi/Redaktion/PDF/Publikationen/rfid-prospectives-for-germany,property=pdf,bereich=bmwi,sprache=de,rwb=true.pdf. Accessed 7 November 2009

Collins J (2006) P&G Finds RFID 'Sweet Spot'. RFID Journal. http://www.rfidjournal.com/article/articleview/2312/1/1/. Accessed 25 May 2010

Feinbier L, Schittko L, Gallais G (2008). The benefits of RFID for slab- and coil-logistics. http://www.accenture.com/NR/rdonlyres/20C9A517-C0E4-40D8-81A9-60FAC76CC735/0/Accenture_Metals_The_Benefits_of_RFID.pdf. Accessed 16 November 2009

Gille D, Strücker J (2008) Into the Unkonwn – Measuring the Business Performance of RFID Applications. In: Golden W, Acton T, Conboy K, van der Heijden H, Tuunainen, V (eds.) 16th European Conference on Information Systems (ECIS 2008). Galway, Ireland

Grote W (2006) RFID – eine Technologie mit hohem Nutzenpotenzial. In: Nagel K, Knoblauch J (eds) Praktische Unternehmensführung, 62. Subsequent delivery. OLZOG, Munich, Germany

Hardgrave BC, Miller R (2006) The Myths and Realities of RFID. Int J Global Logist Supply Chain Manag 1:1-16. doi:10.1.1.113.5565

Hardgrave B, Riemenschneider CK, Armstrong DJ (2008) Making the Business Case for RFID. In: Kreowski HJ, Scholz-Reiter B, Haasis HD (eds) Dynamics in Logistics. doi:10.1007/978-3-540-76862-3_2

Harley S (2008). Shipper's eFreight vision: RFID technology in vehicle logistics at Cologne vehicle operations plant. http://www.euro-case.org/documents/HARLEY.pdf. Accessed 16 November 2009,

Hirthammer K, Riha I (2005) Framework for cost-benefits sharing in logistics networks. http://publica.fraunhofer.de/documents/N-35547.html. Accessed 7 November 2009

IDTechEx (2009) RFID market forecasts 2009-2019. http://www.idtechex.com/research/articles/rfid_market_forecasts_2009_2019_00001377.asp. Accessed 1 November 2009

Incucomm (2004) Wal-Mart's RFID Deployment – How is it Going? http://www.incucomm.com/releases/Wal-Mart%20Jan%202005%20Status%20-%20Executive%20Summary.PDF. Accessed 7 December 2010

ISO/IEC 18000/Amd 1 (2006) Information Technology -- Radio Frequency Identification for Item Management - Part 6: Parameters for Air Interface Communications at 860 MHz to 960 MHz, Extension with Type C and Update of Types A and B

ISO/IEC/18000-7 (2009) Information technology -- Radio frequency identification for item management -- Part 7: Parameters for active air interface communications at 433 MHz

jBilling (2010) The Open Source Enterprise Billing System – Integration Guide. http://www.jbilling.com/files/documentation/integration_guide.pdf. Accessed 20 May 2010

Laubacher R, Kothari S, Malone TW, Subirana B (2006) What is RFID worth to your company? Measuring performance at the activity level. The MIT Center for Digital Business. ebusiness.mit.edu/research/papers/223%20Laubacher_%20APBM.pdf. Accessed 15 November 2009

Li S, Visich JK (2006) Radio Frequency Identification: Supply Chain Impact and Implementation Challenges. Int J Integr Supply Manag 2:407-424. doi:10.1504/IJISM.2006.009643

McClenahen JS (2005) Wal-Mart's big gamble. IndustryWeek. http://www.industryweek.com/articles/wal-marts_big_gamble_10055.aspx. Accessed 15 November 2009

OECD (2007) Radio Frequency Identification (RFID) Implementation in Germany: Challenges and Benefits. doi:10.1787/230687544816

Riha I (2009) Entwicklung einer Methode für cost benefit sharing in Logistiknetzwerken. https://eldorado.tu-dortmund.de/handle/2003/26103. Accessed 15 November 2009

Schmitt P, Michahelles F (2008) Economic Impact of RFID Report. http://www.bridge-project.eu/data/File/BRIDGE_WP13_Economic_impact_RFID.pdf. Accessed 15 May 2010

Schnatmeyer M (2007) RFID-basierte Nachverfolgung logistischer Einheiten in der Kreislaufwirtschaft. Dissertation. University of Bremen

Scholz-Reiter B, Gorldt C, Hinrichs U, Tervo JT, Lewandowski M (2007) RFID – Einsatzmöglichkeiten und Potentiale in logistischen Prozessen. Mobile Research Center Bremen. http://www.mrc-bremen.de/fileadmin/user_upload/mrcMobileResearchCenter/RFID.pdf. Accessed 24 November 2009

Schuster EW, Allen SJ, Brock DL (2007) Global RFID: The Value of the EPCglobal Network for Supply Chain Management. Springer, Berlin, Germany

Seiter M, Urban U, Rosentritt C (2008) Wirtschaftlicher Einsatz von RFID - Ergebnisse einer empirischen Studie in Deutschland. http://www.ipri-institute.com/wissen_verbreiten/research_paper.htm. Accessed 7 November 2009

Swedberg C (2009) DOD Tests, Buys New ISO 18000-7 Tags From Four Companies. RFID Journal. http://www.rfidjournal.com/article/print/5317. Accessed 16 November 2009

Tajima M (2007) Strategic Value of RFID in Supply Chain Management. J Purch and Supply Manag 13:261-273. doi:10.1016/j.pursup.2007.11.001

ten Hompel M, Lange V (2004) RFID 2004: Logistiktrends für Industrie und Handel. Praxiswissen Service UG, Dortmund, Germany

Thiesse F, Condea C (2009) RFID data sharing in supply chains: What is the value of the EPC network? Int J of Electron Bus 7:21-43. doi: 10.1504/IJEB.2009.023607

Wildemann H, Wahl P, Boeck B (2007) NutzLog – Vorteilsausgleich-Nutzenverteilung. http://www.forlog.de/pdf/ForLog_ZB07.pdf. Accessed 7 November 2009

Wong CY, McFarlane D, Ahmad Zaharudin A, Agarwal V (2002) The intelligent product driven supply chain. IEEE International Conference on Man and Cybernetics Systems 2002. http://ieeexplore.ieee.org/stamp/stamp.jsp?tp=&arnumber=1173319. Accessed 20 May 2010

10 Business Models for the Internet of Things

Eva Bucherer[1], Dieter Uckelmann[2]

[1]SAP Research, St. Gallen, Switzerland

[2]LogDynamics Lab, University of Bremen

Abstract The emerging Internet of Things provides a networked infrastructure that enables incremental business transformation as well as radical business changes. So far, the full potential of possible business opportunities has not been leveraged. Within this chapter we propose the concept of business models and business model innovation as a means to align "technological development and economic value creation" (Chesbrough and Rosenbloom, 2002) in the Internet of Things. A central point of this paper is the value and revenue creation in the Internet of Things. We consider information to be the main source for value proposition. To investigate resulting impacts, we draw on the "laws of information" proposed by Moody and Walsh (2002) and deduct specifics for the Internet of Things. Building on this, we describe four exemplary business model scenarios. These are visualised using the business model framework by Osterwalder and Pigneur (2009). This framework, the fundamental rules of value creation through information in the Internet of Things and the provided examples may serve as a tool-set for practitioners to analyse and change their business models when implementing the Internet of Things.

10.1 Introduction

The Internet has significantly changed the way products and services are marketed and distributed and thus let to a series of new types of business models. Similarly, the Internet of Things provides – yet mostly unleashed – potential for business transformations. This will be reinforced by its popularisation through progress in miniaturisation of technical components and falling costs.

The Internet of Things links uniquely identifiable things to their virtual representations in the Internet. Current applications in the Internet of Things generally focus on the optimisation of existing processes and associated cost reductions within companies and along value chains. Product Life Cycle Management, Customer Relationship Management, and Supply Chain Management are typical application scenarios. New application scenarios, sometimes referred to as smart technologies and smart services, are more focused on revenue generation (Fleisch et al. 2005). This chapter builds on findings from e-commerce and traditional business models to derive a new business model understanding for the Internet of Things. Envisioned scenarios, including Product as a Service (PaaS), enhanced end-user consumer involvement through the integration of social platforms, as well as right-time business analysis and decision making, demand an economical rethinking. These changes will have major influence on how companies are involved in the Internet of Things. The cost-centric approach therefore has to be replaced by a value-focused perspective. In the long term, a financial or non-financial pay-off that exceeds the efforts of information provisioning is needed to provide sustainable business models (cf. chapter 1).

This article provides a foundation for discussing the Internet of Things from an economic perspective, based on the business model concept. We will demonstrate that technical innovations in the Internet of Things do have economical and business implications. Moreover, they hold the potential of changing existing or creating new business models. The implications will be illustrated by the use of exemplary cases.

With advancements in the area of mass participation, openness, scalability and security, the personal involvement grows and clear boundaries between business and consumer use are vanishing. Social platforms to share experience and personalised insights will be integrated with business-centric applications. Mash-ups and end-user programming will enable people to contribute to the Internet of Things with data, presentation, and functionality. The success of these changes becomes more and more dependent on "valid" business models rather than on burning venture capital.

The structure of the chapter is as follows. Section 10.2 gives a short overview of the state of the art in business models and business model innovation. To create a common understanding, a framework describing the components included in a business model will be introduced. In section 10.3 we examine the value creation in the Internet of Things. We will have a closer look at the differences between information and product flows that need to be considered for new business models. The economics of information, such as information providers and information flows will be assessed. Potential products and services will be evaluated. Based on the previous findings and considerations, section 10.4 gives exemplary business model scenarios for the Internet of Things. It will be depicted how the configuration of business models can help companies to monetise on the Internet of Things. Finally, section 10.5 summarises the findings and gives an outlook on future research.

10.2 Business Models and Business Model Innovation

The term "business model" has been predominantly coined in practice during the last decades of the 20th century. Only gradually it has been adopted and researched by the scientific world. Thus, the business model can be seen as a "fairly recent concept" (Morris et al. 2006). "Business model innovation creates new or reinvents existing business models. Both terms are described in more detail in the following.

10.2.1 Business Models

For a long time research on firms focused on industry (Porter 1980) and resources (Barney et al. 2001, Wernerfelt 1984). The business model has to be seen as the replacement or complement of the traditional unit of analysis, as a result of the altered surrounding conditions (Amit and Zott 2001, Venkatraman and Henderson 1998). Already in 1998, Sampler called for a redefinition of the traditional value chain. The changed competitive environment, influenced by dramatic technological progress, entailed a series of new types of businesses. Today's business condition is determined by technological progress, service orientation, the digitalisation of products as well as increasing relevance of cooperation and ecosystems of different companies, which blur the boundaries of the individual enterprise. The unit of analysis must therefore be holistic and comprehend various different aspects. A business model can add to the competitiveness of a firm by offering a logical and consistent approach to the (innovative) design and execution of the business. Its increasing popularity with the emergence of electronic commerce and particularly during the dot.com phase can be explained by shortcomings in existing frameworks and theories to address all aspects of the novel possibilities defying conventional ways of doing business (Chesbrough and Rosenbloom 2002). However, the ideas and principles which underlie the concept are not new. Aspects characterising the business model can already be found in Drucker (1954)[1] and in concepts of strategic management (see e.g. Hedman and Kalling 2003, Morris et al. 2005).

Every business activity can be reduced to its core elements, which in the simplest case comprise the value proposition, distribution channels and the customers of the company, explaining how a company produces and sells a good or service. Accordingly, each business is implicitly based on a business model, even though it is not always explicitly presented.

Although the expression "business model" is frequently used both in research and practice, a common definition is missing (Morris et al. 2005). One of the most cited definitions of the term can be found in Timmers (1998). He defines a business model as "an architecture of the products, services and information flows [...]". This includes the involved actors and roles as well as the potential value created for all participants and the source of revenue.

---

1 What is our business? Who is the customer? What is value to the customer? What will our business be? (p.51ff)

Considering existing definitions and the presented characteristic features of business models, we define the business model as an abstraction of the complexity of a company by reducing it to its core elements and their interrelations. It facilitates the analysis and the description of business activities. Besides, the business model is gaining in importance as a starting point for business innovation and transformation. It can serve as means to align "technology development and economic value creation" (Chesbrough and Rosenbloom, 2002). In relation to the Internet of Things we see the business model as a major element to unite its technical developments with its economical business perspective.

According to Afuah and Tucci (2000), "a business model can be conceptualised as a system that is made up of components, linkages between the components, and dynamics". Components refer to the elements to be addressed by a business model. Just like the definitions of the term "business model" the proposed components vary largely between different authors.

In the following, we will base our work on the framework by Osterwalder and Pigneur (2009), which is referred to as the "business model canvas". The applicability of the model is proven by its use in practice, but it has also been referenced by a number of publications (e.g., Chesbrough 2009).

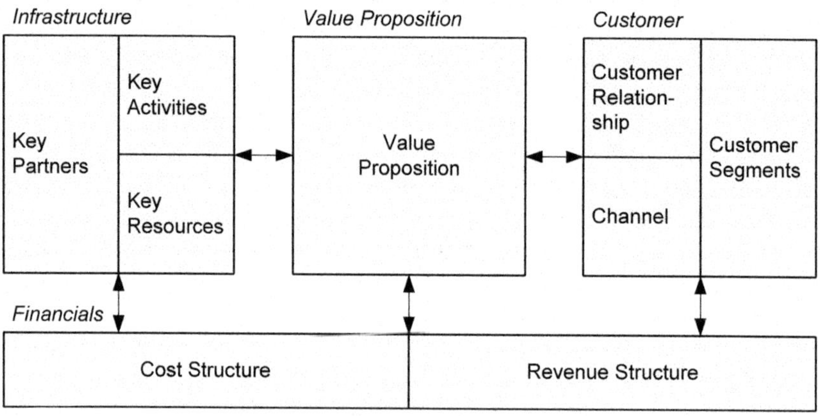

Fig. 10.1 Business Model Framework (Adapted from Osterwalder and Pigneur 2009)

The business model framework depicted in Figure 10.1 includes four main perspectives of the business model, namely the value proposition, the customer, financials and the infrastructure. The components are not stand-alone but mutually influence each other.

The *value proposition* specifies what is actually delivered to the customer. This goes beyond the product or service offered. It describes which customer needs are satisfied and details what other quantitative (e.g., price or speed of service) and qualitative aspects (e.g., brand, design, cost/risk reduction) contribute to the offered value. In the Internet of Things we consider raw data about physical objects as well as any aggregated or processed information a core component of the value proposition.

The *customer perspective* includes the *customer segments* addressed by the company, such as related channels and customer relationships. The customer segments define the different groups of people that are served. Different types of customer segments can be distinguished: mass market vs. niche market, segmented vs. diversified or multisided platforms. Multisided platforms will exist, if two or more interdependent customer segments are served by the company (e.g. credit card companies). The company can reach its customers, respectively customer segments through different *channels*. These can be direct or indirect and owned by the company itself or by partners. Channels can be aligned to the different phases of the lifecycle, such as creating awareness for the value proposition, evaluation of the value proposition through the customer, purchase, delivery and after sales. *Customer relationships* are often determined by the channels used. Relationships can range from very loose (self-service, automated services) to highly engaged (personal assistance, communities, co-creation).

The *financial perspective* comprises the costs as well as the revenues. The *revenue structure* depicts the sources and ways of revenue generation. Here, too, different types of revenue streams can be distinguished: asset sale, usage fee, subscription fee, lending / renting / leasing, licensing, brokerage fee, and advertising (see section 10.3.2). The *cost structure* describes the most important costs (variable and fixed) inherent to the business model. The business model can be rather value or cost driven (cost leadership vs. differentiation strategy). Companies can use economies of scale or economies of scope to create a successful business model.

Key partners, *key activities* and *key resources* can be referred to as the *infrastructure components*. The *key resources* are the assets required to make the business model work. Key resources can be physical, intellectual, human or financial. The *key activities* describe the most important actions to be performed by the company in order to create, offer and market the value proposition. These can be producing, problem solving or developing and maintaining a platform, respectively network. *Key partners* are the network of suppliers and collaboration partners (strategic alliances, outsourcing partners, co-creation) the business model depends on.

10.2.2 Business Model Innovation

Business model innovations are becoming increasingly critical in practice. In a study conducted by IBM (2008) 98% of the CEOs interviewed stated that their company would undertake extensive (69%) or moderate (29%) business model innovation within the following three years. In order to stay competitive in times of change, companies have to adapt and innovate in every dimension. Mere product and process innovations are seen as insufficient (e.g., Chesbrough 2007). The new business conditions require companies to change their whole way of doing business.

External factors, such as technological innovations, increased competition, and market changes as well as legal or regulatory changes are seen as the dominant triggers of business model innovation (IBM 2008, Linder and Cantrell 2000). Through business model innovation companies can differentiate from competitors and establish a competitive advantage. By pursuing an opportunity driven approach, companies can benefit from the first-mover advantage.

> "When external changes undermine a model, it typically cannot be recalibrated, a new model must be constructed" (Morris et al. 2005).

However, once the existing model is undermined, it can already be too late to change course. We therefore suggest a forward looking approach, where business model innovation is used proactively to capture new market shares or enter new markets.

Business model innovation can help to align innovation activities within the company (Venkatraman and Henderson 2008):

> "Innovations have been piecemeal and disconnected across different functions and locations without overarching logic for corporate-wide innovations. Best practices exist for localised, incremental innovations, but there is a clear lack of management frameworks for business model innovations that create new rules of competition."

A general deficit in business model innovation literature seems the discrimination of product or service innovation from business model innovation. The specifics of business model innovation need to be researched and pointed out in more detail, as for example done by Venkatraman and Henderson (2008):

> "[...] we need to innovate more holistically – namely: the entire business model (which encompasses customer value proposition, operating model, management processes, and roles and responsibilities of multiple partners with shared incentives and decision rights)."

In line with the definition of innovation by Hauschildt (1997), we see business model innovation as a process resulting in a qualitatively new business model, which differs distinctively from the previous. A deliberate change of one or more key elements of the business model, respectively their interrelations, has to take place. The resulting business model can range from an incremental improvement to a radical new way of doing business.

Some of the most successful companies that have used a distinctively new business approach based on the Internet are shown in Table 10.1.

Company	Traditional business	Initial business model innovation	Further developments
Amazon[2]	Book trade	Online shopping	Shopping portal

[2] http://www.amazon.com

		Automated distribution model Collaborative filtering	Digitalisation (mp3, books) Terminals (Kindle) Mobile payments Amazon web services (incl. billing)
eBay[3]	Classifieds Flea markets Auctions	Online auctions	Shopping portal Payment services (PayPal)
Google[4]	Yellow pages	Hypertext web search Prioritised advertisements	Terminals (Android) Video (You Tube) Maps (Google Maps) Web based software (e.g. Google Docs) Digitalised books Payment services (Checkout)
Apple iTunes[5]	Music shops	Music digitalisation Terminals (iPod, iPhone, iPad) Applications (apps)	Videos, Newspapers

Table 10.1 Traditional Business vs. Business Model Innovation

Their success builds on a technological innovation (the Internet) and on services that replaced some traditional businesses, such as online shopping or online auctions. When physical goods are shipped, a fast and agile logistic service provides an advantage over traditional concepts. The growing digitalisation of music, books and videos allows instant delivery. Another key to success is based on well accepted billing systems, such as PayPal or Checkout. These have lead to the increased usage of Amazon and eBay as shopping portals. Lately, there is a clear move towards mobility to allow ubiquitous access to digital content. Google (Android), Amazon (Kindle) and Apple's iPod, iPad and iPhone are some of the examples for further integration of mobility platforms and web based services.

It can be expected that new business models based on the Internet of Things will change and replace some of the traditional business approaches in a similar manner.

[3] http://www.ebay.com

[4] http://www.google.com

[5] http://www.apple.com/itunes

10.3 Value Creation in the Internet of Things

A typical business transaction today is defined by a physical product, information stream, and money stream (Alt and Zbornik 2002). It should be noted though, that business transactions may be focused on services instead of physical product transactions, as well. However, in the Internet of Things, there always is a link to a physical product. The product stream includes order processing from procurement via storage and production to distribution of products to the customer. The information stream includes processes, such as order processing, supply chain and product life cycle data sharing.

The Internet of Things may be seen as an approach to align these different streams. It provides a higher level of visibility and control mechanisms. Moreover, in the Internet of Things, information itself may become a major source for value creation and thus the value proposition. This includes information only made possible through Internet of Things technologies as well as the association of existing information to physical products.

Traditionally, the money stream is exclusively dependent on the product stream prices. A separate price for the information is not defined. Instead, information is most often expected to be free of charge. It is obvious that the costs of information are hidden in the product price. However, the reluctance to pay for information may change over time. In B2C-markets, the willingness to pay for digital goods has increased to 88%, according to a survey with more than 15.000 participating consumers (Krüger et al. 2008). Even though digital goods (e.g. software, tickets, travel, songs, and videos) and information are not synonymous, it is still obvious that there is a change in society to accept the Internet as a business transaction platform. In addition to direct information payments, alternative revenue streams should be considered. Approaches, such as advertising or the less well known idea of freemium have untapped potential, even for B2B relationships. Freemium – a word derived from the terms "free" and "premium" – refers to the offer of free basic services and the revenue creation through paid premium services (see Anderson 2009).

10.3.1 Laws of Information

Even though information is recognised as an asset on its own right, quantitative measurements are difficult to achieve. It consumes a growing number of organisational resources for data capturing, storage, processing and maintenance. While hardware and sometimes software may be capitalised, the value of information in general is not financially recognised in the balance sheets. Information may be considered a product that is produced out of raw data through hard- and software utilisation. The cost of information is mainly not related to hard- and software, but to the people that feed the information systems with data. Their salary is usually hidden in the budgets of the corresponding departments. Therefore, a way of measuring the value of information is required (Moody and Walsh 2002).

Moody and Walsh (2002) define seven "laws of information", explaining the specifics of information compared to other (physical) assets. From these "laws" we can deduct approaches to the value creation in the Internet of Things. These "laws of information" provide opportunities for new business and pricing models for the Internet of Things:

First Law of Information: Information is (Infinitely) Shareable and Can Be Shared with Others Without a Loss of Value

The Internet of Things eases the sharing of product related information and allows information distribution to all participating stakeholders. The information provided through the Internet of Things can be monetised through paid access to the provided information. A win-win situation is achieved, when the *cumulated* amounts of accessing information exceeds the efforts of information provisioning. Therefore, the individual amount of accessing information may decrease with the number of information consumers.

Second Law of Information: The Value of Information Increases with Use and It Does Not Provide Any Value, If It Is Not Used at All

The mayor cost factors are related to data collection, storage and maintenance, while marginal costs of using are considerably small. The Internet of Things eases and consequently increases the distribution and usage of information. However, people have to be aware of the existence of information. Discovery services can be used as an "information asset register", as requested by Moody and Walsh (2002). Additionally, decision-makers have to be capable of interpreting and using the information in a beneficial way. The Internet of Things therefore needs integration to existing and proven business applications as well as new tools that visualise and analyse information and assist in decision making processes. If a pay-per-use model for information access can be applied, it will be possible to charge the users per information request, thus leveraging the second law of information to its full extend.

Third Law of Information: Information Is Perishable and It Depreciates Over Time

The Internet of Things provides real-time information and thus provides high value information. However, one of the beneficial applications in the Internet of Things is focused on life cycle information access. Therefore, historical information about a product may keep or even increase its value over time. Pay-per-use pricing models for information with decreasing or increasing prices over time would correspond to the time-dependency concerning value of information.

Fourth Law of Information: The Value of Information Increases with Accuracy

However, "*100% accurate is rarely required in a business context*" (Moody and Walsh 2002). The Internet of Things provides a fine grained view of the real world and therefore enables "high resolution management". Automatic identification helps to avoid mistakes from manual data entry, but the corresponding product information needs to provide a high level of accuracy as well. In Electronic Data Interchange (EDI) product data contracts are a common instrument to agree on data quality standards. Pricing models can be based on service level agreements and reoccurring assessments of information accuracy compliance.

Fifth Law of Information: The Value of Information Increases When Combined with Other Information
For example the identification number of an electronic component may have little value, if it is not combined with its firmware release number or its service history. In this respect standardisation of small percentage of *identifiers* and *coding schemes* can lead to high benefits in information integration (Moody and Walsh 2002). By its nature, the Internet of Things links different sources of information to specific objects (things). This provides new business opportunities for third party data aggregators and information service providers. Data sharing between different information providers is favourable in order to increase the value of aggregated data. End-user participation and co-creation further add to the overall value of information in the Internet of Things. Freemium models offer the ability to provide basic information for free, while access to enriched or aggregated information would require a premium account.

Sixth Law of Information: More Information Is Not Necessarily Better
While the value of information increases to a certain level if more information is supplied, it decreases, when more information than can be processed is provided (information overload). The linkage of things and related information binds information to a specific object and therefore eases information consumption in the Internet of Things. Filtering, personalisation, customised information feeds, and pre-processing can help to further reduce the information overload and to tailor the information to specific user requirements. A business opportunity exists for monetising customised or pre-processed information, such as alert messages.

Seventh Law of Information: Information Is Not Depletable.
Information instead is rather self-generating as summarising, combining or analysing information leads to more information. All possible sources of information generation and data processing that provide value to the Internet of Things should be considered, including for example sensors, users, software agents, and business intelligence software. Co-creation models, where for example access to information is free, if this information is further enriched through data analysis, may provide a win-win business situation in this context. Data-mining will enable further business opportunities for companies with access to multiple data sources.

Other opportunities can be achieved through reinvention of classical business models (e.g., PaaS), based on better information capabilities provided through the

Internet of Things. In these cases benefits are not directly generated through the value of information. Instead, the Internet of Things rather acts as an enabling technology. The following major consequences for business models result from the new possibilities offered.

10.3.2 Revenue Generation in the Internet of Things

As stated above, information may become the main source of value creation and thus a major part of the value proposition in the Internet of Things. More and especially more detailed information is made available. Information can be directly associated to things or products and instances of products. The usage, status, and location of things become traceable. This allows for new value proposition scenarios, such as the provision of additional product-related data to the consumer (e.g., carbon footprint) or the exact billing of products or services based on the actual use (e.g., rental car, returnable transport items).
The following requirements constitute the specifics for the value proposition:

- **Providing the *right information* ...**
  - Linked through a unique identifier to a physical product
- **... in the *right granularity* ...**
  - High information granularity, providing a new dimension of clarity and insight
- **... and the *right condition* ...**
  - High information accuracy
  - Aggregation of information from various sources, such as tags, sensors or embedded systems
  - Correlation, integration, and further analysis of information in a way that allows new insights to be derived
  - Defined syntax and semantics
- **... at the *right time* ...**
  - Timeliness of information
  - Access to real time information as well as to historical data for business analysis
  - Real-time analytics and business intelligence for high resolution management
  - Intelligent real-time decision-making capability based on real-time physical events
- **... *anywhere* in the network ...**

  - Online access and possibly offline usage
  - Mobile access

- **... at an *appropriate* price.**
  - Price transparency
  - Low premium for billing service, the price should be paid for the information rather than the infrastructure

New value propositions require a rethinking of financial aspects. Historically cost discussions have dominated the Internet of Things. Costs for tags, sensors, actuators, readers, soft- and hardware can be calculated quite well. An ROI, instead, has been more difficult to find, as only small parts of the overall financial benefits could be raised within an enterprise.

Therefore, revenue generation should play a more important role in the Internet of Things, to generate new money streams. Pricing of information as well as other benefits or bonuses provide the basis to compensate for the provided infrastructure and information generation. Usage based pricing will require usage data acquisition, including metering and collection of data. Subscription fees are an easier alternative to usage based billing or may be combined as known from offerings in the telecommunications industry. Information brokers may be included in the framework through brokerage fees. Advertising is another source of income but requires manual interaction with the Internet of Things and does not provide a valid business model in machine-to-machine (M2M) scenarios.

Considering that mechanisms to measure, collect and bill information may be integrated in a future architecture of the Internet of Things (see chapter 9), the separated billing capability for physical products and information, and thus a decoupling of information and product prices, will enable new business models.

Whereas the exchange of physical products spans along the value chain and usually ends with the delivery to the consumer, the exchange of information in the Internet of Things goes beyond and may include different actors. In order to fully understand the information exchange on the Internet of Things the information flows and actors involved have to be considered. Figure10.2 depicts information providers in the Internet of Things and information flows between them.

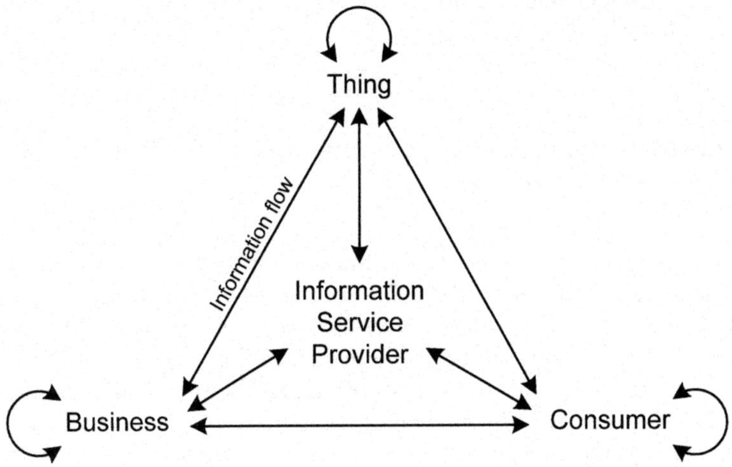

Fig. 10.2 Information Providers and Information Flows in the Internet of Things

The actors, respectively information sources identified, include things, consumers, businesses and a special form of business, the service provider or information service provider. They can be depicted as a triangle of information exchange. Information flows can be direct, such as for example thing-to-thing, business-to-consumer or consumer-to–thing, or indirect, such as from thing-to-business through an information provider or from business-to-business through a thing. Things include products that communicate their ID and status through sensors as well as data processing units and actuators. Additional information is provided by businesses or consumers. This covers information from information systems (e.g., ERP systems) or manually entered data (e.g., product ratings). Information service providers aggregate information from different sources. Additionally, they may combine and enrich data to add value.

In case of thing-to-thing (including M2M) relations it has to be kept in mind that there are companies or consumers owning these things. But still the distribution channels for information will require different interfaces than used in classical B2B and B2C scenarios.

The resulting customer relationships may be structured according to the information flow, including unidirectional, bi-directional and multi-directional information flows. While the Internet of Things is designed to support multi-directional information flows, there are still only few applications that utilise its full potential. Additionally, self-servicing and automation play an important role in the customer relationship.

The question that has to be asked here is how to create a win-win situation for all stakeholders involved in the information exchange? The consideration of different business model scenarios might help in answering this question and helps to understand how new possibilities can be commercialised through businesses or information service providers.

10.4 Exemplary Business Model Scenarios for the Internet of Things

Based on the previous considerations and findings, different exemplary business model scenarios are developed within this section. The field of application for Internet of Things technology is much wider then we have seen so far. The control of processes and the quality of goods in manufacturing, logistics, service and maintenance are still valid applications. Moreover, new areas of applications have to be considered. End-user integration through data provision and end-user programming as well as the implementation of autonomous services will take the Internet of Things to the next level, where the Internet of Things is more than a pure B2B infrastructure.

The following exemplary scenarios will include the use of Internet of Things technology to support the offer of *PaaS*, the role of *information service providers* in the Internet of Things, the *integration of end-users* and opportunities through *right-time business analysis and decision making.* With the help of the business model framework, it will be depicted how the configuration of business models can help companies to monetise on the Internet of Things.

10.4.1 Scenario 1: Product as a Service (PaaS)

The shift from providing products to providing services is a major trend in business model innovation. Not only software companies provide SaaS instead of selling software licenses, but more and more manufacturers follow this trend. As a reaction to increasing competition through low-cost manufacturers, Hilti[6], an international manufacturer and supplier of professional construction tools, launched what they call “Fleet Management”. The customer is no longer required to own a tool. Instead, Hilti offers its customers access to a range of tools on a contract basis and monthly fee, including additional services, such as repairs. Customers benefit from lower upfront investments, no cost repairs, flexible inventories, less downtime and up to date tools (Johnson et al. 2008). In a further step the pricing schemes can be based on service performance. Popular examples are Power by the Hour (PbH) or Performance Based Logistics (PBL) (Kim et al. 2007). Consequently, measureable performance values are needed to provide a reliable calculation fundament.

6 http://www.hilti.com

Problem Statement

Today, the shift to PaaS is often hindered by missing means of performance measuring and billing as well as unsuitable pricing models. Current implementations are only isolated instead of integrated offerings.

The Internet of Things as Enabler

The Internet of Things offers a range of possibilities to support such PaaS scenarios. Sensors allow for the tracking of a product and the location of its current position. In addition, the usage times of a product can be exactly documented as well as the condition under which a product was used (e.g. the speed at which a car has been driven). Sensors also enable a company to monitor the condition of the product or parts and tools and thus support maintenance and repairs. Through an open Internet of Things infrastructure, different offerings can be combined.

Possible Scenario

A scenario that utilises the Internet of Things can be envisioned in the sector of car rental (a similar scenario is currently implemented by Daimler under the name of Car2Go[7]). Up to now, usually time-based fees depending on the class of car plus gasoline are charged for. In a future scenario the pricing could be based on the exact usage of the car, with the calculation based on actual emissions as well as on (engine) speed, acceleration, transport weight, streets used or any other measurable value. This would motivate an environmentally friendly usage of rental cars if a direct feedback channel to the driver, such as a current meter for cost per distance, accumulated costs, and current as well as average emissions is provided. All services such as refuelling, insurance and possibly toll payments may be included in the usage fee. Third party provider can remotely monitor the condition of the cars through the Internet of Things and can react to emergency signals emitted by the cars. Finally, when returning the car, it is not necessary to bring it to a local car rental station. Instead, it can be parked at third-party service stations (e.g., gas stations) for cleaning and visual inspection as the real location and technical status is always known through the Internet of Things..In a longer term, even visual inspection can be automated through a corresponding drive-through video gate.

[7] http://www.car2go.com/

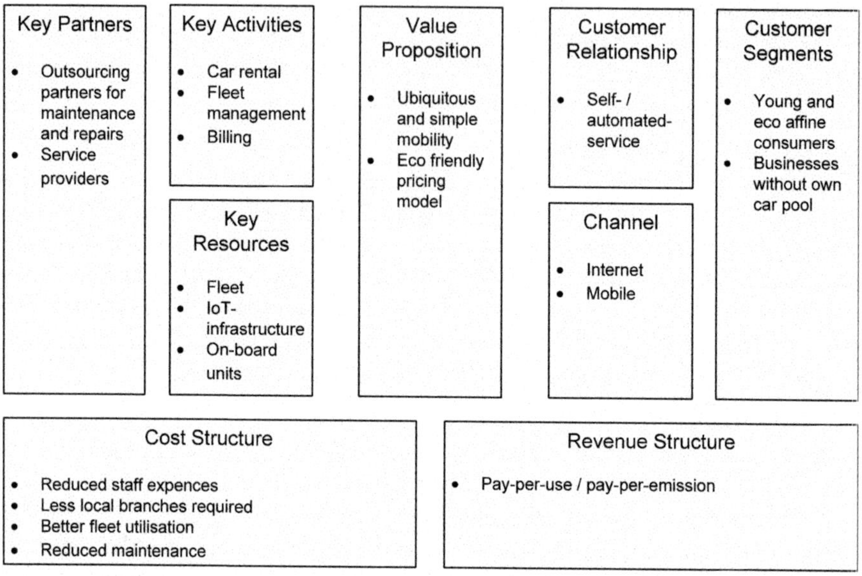

Fig. 10.3 Business Model for a Car Rental Scenario in the Internet of Things

Compared to the traditional rental car business model, the proposed business model results in viewer fixed costs through the omission of local subsidiaries and a decreased need in staff as the car rental process is implemented as an automated self-service. New costs originate in particular in the Internet of Things infrastructure employed. The monitoring of the cars' conditions through an outsourcing partner allows for timely repairs, less downtime, and reduced maintenance costs of cars.

10.4.2 Scenario 2: Information Service Providers

If information can be measured and billed, new business opportunities for information service providers will be enabled. IT departments can become profit centres instead of cost centres. Data centres can provide storage and processing capabilities for Internet of Things-related data. Additionally, information service providers can aggregate and process information from different sources, thus providing a higher value of information.

Information service providers in the Internet of Things may revolutionise market research, as sample sizes are increased, costs of information collection are reduced and real-time analytics provide instant feedback. A potential application scenario for information service providers is anti-counterfeiting. The problem of anti-counterfeiting is prevalent in the consumer goods market. Brand items, such as apparel and accessories or even worse drugs or spare parts are copied and sold as original products. This results in economic damage and can have severe impacts on the consumer side.

Problem Statement

Hitherto, the definite and non-manipulable identification of product instances is most often impossible. Product identification is mostly restricted to the product category. The EPCglobal Network allows identification and tracking of products along the value chain. However, setting and maintaining the infrastructure is still costly and incentives for sharing product data are missing.

The Internet of Things as Enabler

The Internet of Things supports this scenario through the association of information to a product instance. In addition, it allows easing sharing information across different parties, especially if billing capabilities are added as a core functionality.

Envisioned Scenario

To fight the problem of counterfeiting, the following service could be offered to a manufacturer by an independent information service provider. The information service is aimed at the verification of the originality of a certain product in order to detect counterfeits. In our case, the information service provider has specialised on the verification of spare parts in the machinery and equipment industry as well as the automotive industry. He thus collaborates with a series of manufacturers and their business partners, supplying them with the needed information. The consumer – the buyer of the spare part or a service partner installing it – can submit a request to the information provider through the Internet of Things. Another important customer segment is customs. The verification of the product can be based on the serial number. The information service provider could query its information systems where information from different sources is aggregated and find out whether the serial number is valid, whether the spare part had been already used by another client and which route the spare part had been taken through the value chain. Two different pricing models are offered for this service: pay-per-use or subscription for customers who want to use the service more often. A similar scenario is currently implemented by "Original1"[8], a joint venture of SAP, Nokia, Giesecke and Devrient. This can be achieved by for example. the EPCglobal Network, but major problems in this context are the cost for the corresponding infrastructure and the missing incentive for sharing data. Both issues can be tackled by integrating billing and balancing capabilities. The service provider needs to acquire the information at a price (or non-financial benefit) that is worthwhile to the information provider and needs to offer his service at a price to the information requesting party that is exceeded by the potential benefits of the aggregated and processed information.

---

[8] http://www.original1.net/

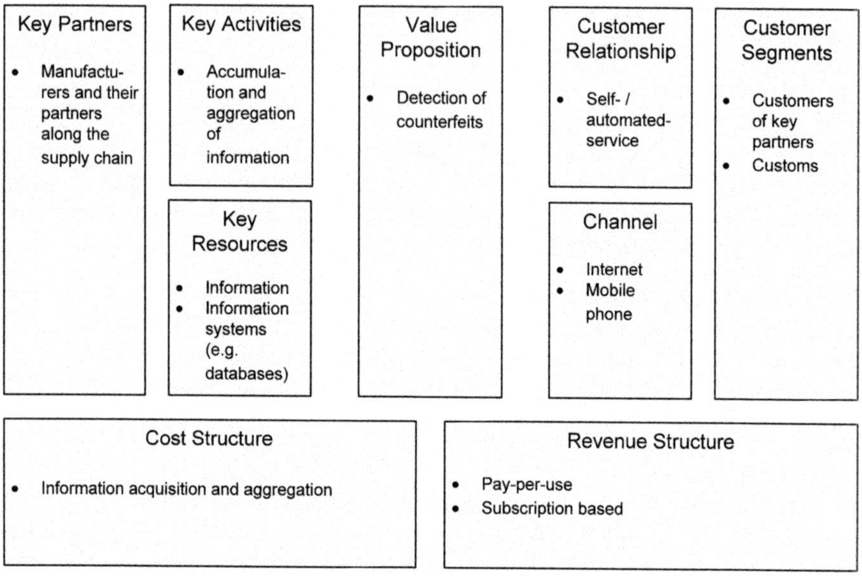

Fig. 10.4 Business Model for Anti-counterfeiting Based on the Internet of Things

The business model of the investigated case of a service provider on the Internet of Things does not differ much from a traditional service provider business model. However, the value proposition which was only made possible through unique identification and billing in the Internet of Things differs significantly. Most important cost factors are the acquisition and aggregation of information (data) and the purchase and maintenance of needed information systems.

10.4.3 Scenario 3: End-userInvolvement

The Internet of Things provides a new level of consumer integration into co-creation processes. While "living labs" have been used to integrate limited user groups into product and service development at a certain stage in the product life cycle, the Internet of Things will link all consumers across the life cycle of a product. Companies that will know how to utilise this huge potential will be in the lead for new business models in B2C scenarios.

The motivation to participate in co-creation can be motivated through financial and non-financial benefits. Again, an integrated billing solution in the Internet of Things would allow a seamless bi-directional flow between businesses and consumers. Currently, vouchers, e-coupons, lotteries and free products are used in

lack of an integrated billing system. Available offerings include Stickybits[9] and my2cents[10]. Other services include a payment scheme for product reviews, that is based on positive review ratings. Ciao[11] is offering their users a small financial benefit as low as 0.5 pence every time their product rating is positively reviewed (Ciao 2009). There may be other, non-financial benefits, such as personalised products. Sometimes end-users are motivated only because the Internet (of Things) provides a platform for their self-expression. In any case a high level of security and privacy as well as the freedom of choice to participate are mandatory.

In B2B scenarios, mandating is a common instrument to motivate participation. Sometimes mandates include financial penalties in case of non-compliance.

Problem Statement

To date, there are only few interconnections between information collection, buying and product rating. Amazon.com is one of the exceptions where consumers look to obtain information, buy and rate their products. Still, a direct identification link to the product is missing. Different firmware released on electronic equipment, for example, may lead to different ratings and cannot be distinguished without unique identification.

The Internet of Things as Enabler

Through the Internet of Things information can be related to specific product instances. In addition, the local access to automatic unique identification increases particularly through the integration of Near Field Communication (NFC) and bar code reader software into mobile camera phones. Another important innovation is the use of image or sound recognition.

Envisioned Scenario

Through the use of a mobile phone, the end-user is enabled to supply and retrieve product related information at the point of sale – in this scenario a large supermarket chain. Both actions are supported by the use of RFID-chips or barcodes. The supermarket supplies the customer with information from internal systems, such as ingredients of a product or the price history. In addition, information related to the product instance, such as its carbon foot print can be retrieved. The user can enter additional information for a product, such as a rating, using the mobile phone or an internet connection at home. In return the supermarket may reward end-users with special bonuses.

To make this service more individual, the user can create a profile with his / her preferences and needs. This allows the supermarket chain to inform the user about current promotions, suggest new products or warn the user in case of food intolerance. The information entered by the customer can be made available to other cus-

[9] www.stickybits.com

[10] www.my2cents.ca

[11] www.ciao.de

tomers but it can also be used for internal analysis. By supplying information, the customer can earn bonus points that can be redeemed for discounts.

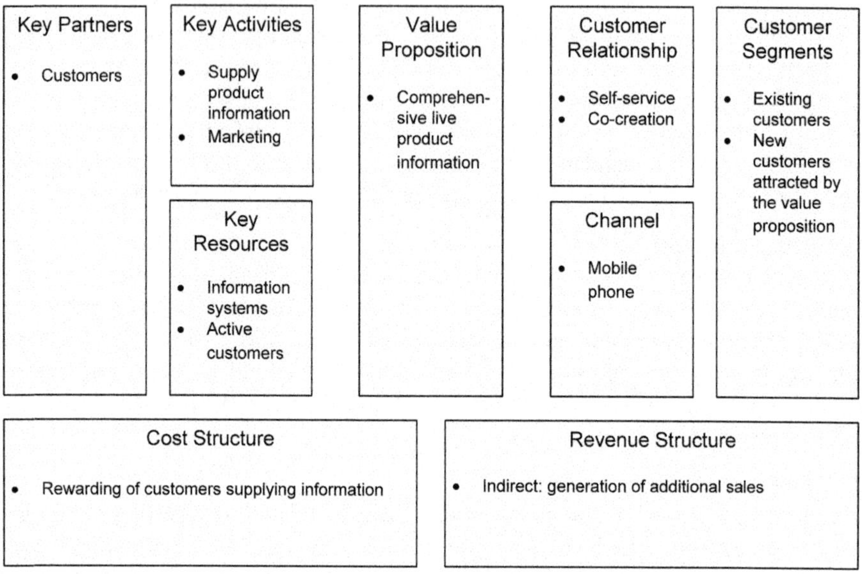

Fig. 10.5 Business Model for End-user Involvement in a Supermarket Scenario

The business model described above is part of a higher-level business model of the supermarket. The particularity of this business model is the generation of revenue through the indirect increase of sales. Successful end-user participation always requires the offer of an incentive and the usefulness of end-user participation for business. Low-quality product providers may not have an interest at all in product ratings. Only highly competitive companies will be interested to distinguish there offerings from their competitors'. Vice versa, expenditures for financial or non-financial benefits of end-users have to be justified by an increase on the revenue side. The question remains if the consumer is willing to pay for additional information directly or indirectly through product buys. From experience we know that consumers are willing to pay more for organic food and for products that comply to quality standards. With the Internet of Things consumers can instantly drill down on related information rather than being limited to rely on simple and sometimes meaningless "compliance labels".

10.4.4 Scenario 4: Right-time Business Analysis and Decision making

In production engineering, real-time usually refers to M2M-systems that record events and responds within milliseconds. In logistics, the time frame is not as well-defined, yet seconds, minutes, or even hours are sometimes still considered real-time, compared to longer traditional processes, such as transportation that causes information gaps of days or weeks. Real-time is often used as a qualitative rather than a quantitative value to differentiate timely from out-of-date information distribution thus allowing acting instead of reacting. Therefore, it is more appropriate to use *right-time business analysis and decision making*. The amount of time between a business event and a decision is influenced by time periods, including data capturing latency, analysis latency, and decision latency (Hackathorn 2004). Real-time business analysis capability remains a core requirement of each enterprise, as it provides the basis for agile management strategies. In the Internet of Things perishable goods represent an interesting research topic for real-time business analysis, especially during long transportation processes that may lead to drastic quality changes. Depending on the current status of the goods and the calculated best-before-date, different management strategies may be applied. As part of the *Collaborative Research Centre 637 "Autonomous cooperating logistic processes – a paradigm shift and its limitations"*, scenarios about intelligent trucks and intelligent containers have been evaluated, based on RFID, sensor integration, communication infrastructures and decentralised decision making through software agents (Jedermann et al. 2007). While these scenarios where based on autonomous strategies and are not directly linked to the Internet of Things, a further integration of both concepts would enable a higher level of agility in logistic processes (Uckelmann et al. 2010).

Problem Statement
Today, right-time business analysis and decision making is mostly restricted to internal processes or bi-directional business relations. For perishable goods manual spot tests and visual inspections are common, but these cannot provide real-time monitoring or proactive strategies.

The Internet of Things as Enabler
The Internet of Things provides real-time access and analysis opportunities across supply-chains or product lifecycles. Data analysis can be provided in proximity to things (smart objects), at the business premises or anywhere in the Internet of Things. Agile management strategies are enabled based on real-time availability and analysis of data.

Envisioned Scenario
The envisioned scenario is based on an intelligent truck that combines different technologies and applications to increase the value of information (*5. law of information*) and to a boost utilisation of the Internet of Things infrastructure. The

truck communicates data to the Internet of Things and receives responses in real-time. While some easy tasks, such as navigation and dynamic routing, can be achieved without the Internet of Things, more complex tasks, such as tracking and condition monitoring, would largely benefit from it.

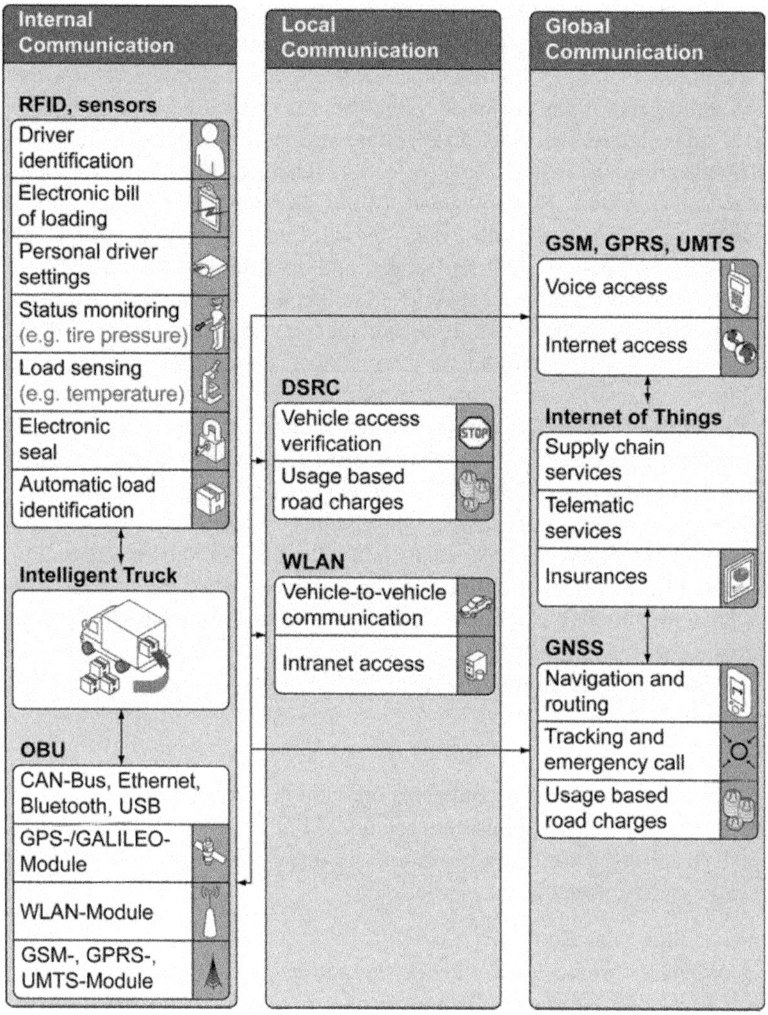

Fig. 10.6 The Intelligent Truck Scenario as an Example for Real-time Analysis and Decision making

The business model depends largely on monetising the benefit from information that depreciates in value over time (*3. law of information*). The goal is to achieve an optimum between proactive (agile) acting and cost of infrastructure required. Therefore, the best response time to a business event is not necessarily the fastest possible response time.

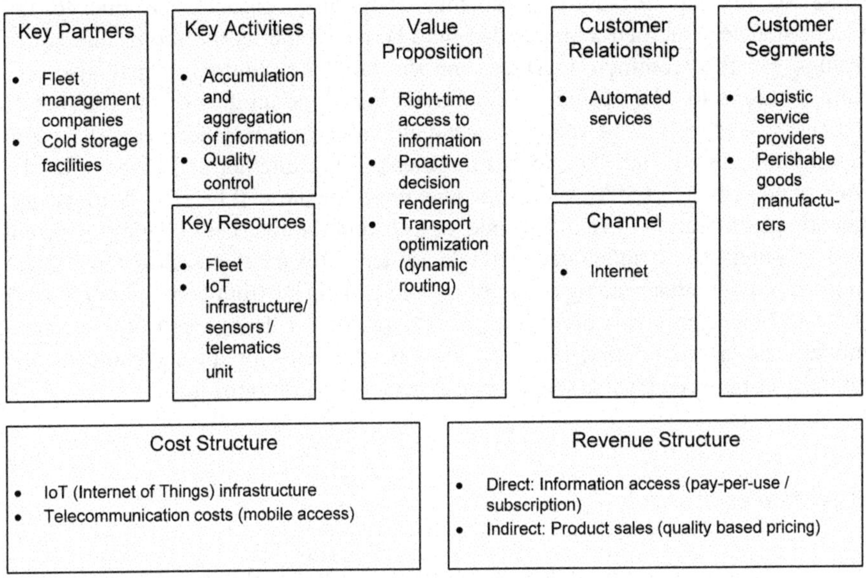

Fig. 10.7 Business Model for the Intelligent Truck

10.5 Conclusion

Moody and Walsh (2002) claim that *"of all the corporate resources (people, finances, assets, information), information is probably the least well managed"*. While the Internet of Things currently helps to overcome some of the more technical problems, such as finding the right information and providing anywhere and anytime access, the business perspective of information as an asset in its own right remains an open issue.

This chapter provided an overview concerning business models and business model innovation and their relation to the Internet of Things. The value of information in relation to its specific "laws" has been explained. Additionally, the value proposition of the Internet of Things and possible effects on existing or new business models have been investigated in detail. The business model concept helps to gain a holistic overview and may serve as a means to identify new opportunities for business model innovation. Based on the given scenarios, we derived that business models can be an important driver for the Internet of Things, to motivate companies to invest, reach new markets and generate new revenues.

While we have investigated the role of business model innovation in the Internet of Things, we have left out user-acceptance for new business models. Numerous approaches, such as the “Intelligent Refrigerator”, have failed up to now, because of missing end-users acceptance. This may partially be due to the inconsistencies and media brakes that may be overcome by the future Internet of Things. For this, common interfaces and standards are required. It may also take time for users to adapt to new technologies and opportunities of the Internet of Things. Just as mobile Internet access and e-commerce have taken years to be successful and are still far from their full potential, the Internet of Things will take time to be actively used by end-users. One pre-requisite will be an easy to use automatic identification device that links objects and the Internet of Things. Current mobile phones with integrated barcode reader software that utilise the camera module provide poor reading capability. Even NFC is far from being ubiquitous. If these technical problems are solved in the Internet of Things and new business models can be found that provide win-win situations for all stakeholders, the boundaries between businesses and consumers will be diminishing.

References

Afuah A, Tucci C (2000) Internet business models and strategies: Text and case. McGraw-Hill Higher Education, New York

Alt R, Zbornik S (2002) Integrierte Geschäftsabwicklung mit Electronic Bill Presentment and Payment. In: Weinhardt C, Holtmann C (eds) E-Commerce: Netze, Märkte, Technologien, Proceedings zur Teilkonferenz der Multikonferenz Wirtschaftsinformatik 2002. Physica, Heidelberg

Amit R, Zott C (2000) Value drivers of e-commerce business models. INSEAD Working Paper. Fontainebleau, France

Anderson C (2009) Free: The Futurc of a Radical Price. Hyperion, New York

Baatz E (1996) Will your business model float?. WebMaster Magazine 10

Barney J, Wright M, Ketchen DJ (2001) The resource-based view of the firm: Ten years after 1991. J of Manag 27:625-641. doi:10.1177/014920630102700601

Chesbrough H (2009) Business Model Innovation: Opportunities and Barriers. Long Range Planning.

Chesbrough H (2007) Business model innovation: it’s not just about technology anymore. Strategy & Leadership 35:12-17. doi:10.1108/10878570710833714

Chesbrough H, Rosenbloom RS (2002) The role of the business model in capturing value from innovation: evidence from Xerox Corporation’s technology spin-off companies. Ind Corp Chang 11:529-555

Ciao (2009) Earning money – Earn money by writing reviews. http://www.ciao.co.uk/faq.php/Id/2/Idx/5/Idy/1. Accessed 5 Mai 2010

Drucker PF (1954) The Practice of Management. HarperCollins, New York

Fleisch E, Christ O, Dierkes M (2005) Die betriebswirtschaftliche Vision des Internets der Dinge. In: Fleisch E, Mattern F (eds.) Das Internet der Dinge. Springer, Berlin, Heidelberg

Hackathorn R (2004) Real-Time to Real-Value. Information Management Magazine. http://www.information-management.com/issues/20040101/7913-1.html. Accessed 10 May 2010

Hauschildt J (1997) Innovationsmanagement. 2nd edn. Vahlen, München

Hedman J, Kalling T (2003) The business model concept: theoretical underpinnings and empirical illustrations. Eur J Inf Syst 12:49-59. doi:10.1057/palgrave.ejis.3000446

IBM (2008) Global CEO Study – The Enterprise of the Future. http://www.ibm.com/ibm/ideasfromibm/us/ceo/20080505/resources/IFI_05052008.pdf. Accessed 10 May 2010

Jedermann R, Behrens C, Laur R, Lang W (2007) Intelligent Containers and Sensor Networks Approaches to apply Autonomous Cooperation on Systems with limited Resources. Springer, Berlin, Heidelberg

Johnson MW, Christensen CM, Kagermann H (2008) Reinventing Your Business Model. Harv Bus Rev 68:50-59

Kim SH, Cohen M, Netessine S (2007) Performance Contracting in After-Sales Service Supply Chains. Manag Sci 53:1843–1858. doi:10.1287/mnsc.1070.0741

Krüger M, Leibold K, Smasal D (2008) IZV9 – Internet Zahlungssysteme aus der Sicht der Verbraucher. http://www.iww.uni-karlsruhe.de/reddot/download/izv9_Endbericht_v2.pdf. Accessed 24 November 2009

Linder J, Cantrell S (2000) Changing business models: surveying the landscape. Working Paper, Accenture Institute for Strategic Change. http://www.accenture.com/NR/rdonlyres/0DE8F2BE-5522-414C-8E1B-E19CF86D6CBC/0/Surveying_the_Landscape_WP.pdf. Accessed 1 May 2010

Moody D, Walsh P (2002) Measuring the value of information: an asset valuation approach. In: Morgan B, Nolan C (eds), Guidelines for Implementing Data Resource Management. 4th edn. DAMA International Press, Seatle USA

Morris M, Schindehutte M, Allen J (2005) The entrepreneur's business model: toward a unified perspective. J Bus Res 58:726-735. doi:10.1016/j.jbusres.2003.11.001

Morris M, Schindehutte M, Richardson J, Allen J (2006) Is the business model a useful strategic concept? Conceptual, theoretical, and empirical insights. J Small Bus Strateg 17:27-50

Osterwalder A, Pigneur Y (2009) Business Model Generation. A Handbook for Visionaries, Game Changers, and Challengers. OSF

Porter ME (1980) Competitive strategy techniques for analysing industries and competitors. Free Press, New York

Sampler J (1998) Redefining industry structure for the information age. Strateg Manag J 19:343-355. doi:10.1002/(SICI)1097-0266(199804)19:4<343::AID-SMJ975>3.0.CO;2-G

Timmers P (1998) Business models for electronic markets. J Electron Market 8:3-8. doi:10.1080/10196789800000016

Uckelmann D, Isenberg MA, Teucke M, Halfar H, Scholz-Reiter B (2010) An integrative approach on Autonomous Control and the Internet of Things. In: Ranasinghe DC, Sheng QZ, Zeadally S (eds) Unique Radio Innovation for the 21st Century: Building Scalable and Global RFID Networks. Springer, Berlin

Venkatraman N, Henderson J (1998) Real strategies for virtual organizing. Sloan Management Review 40(1): 33-48

Venkatraman N, Henderson JC (2008): Four vectors of business model innovation: value capture in a network era. From Strategy to Execution: Turning Accelerated Global Change Into Opportunity. Springer, Berlin

Wernerfelt B (1984) A resource-based view of the firm. Strateg Manag J 5:171-180. doi:10.1002/smj.4250050207

11 The DiY Smart Experiences Project

A European Endeavour Removing Barriers for User-generated Internet of Things Applications

Marc Roelands[1], Johan Plomp[2], Diego Casado Mansilla[3], Juan R. Velasco[3], Ismail Salhi[4], Gyu Myoung Lee[5], Noel Crespi[5], Filipe Vinci dos Santos[6], Julien Vachaudez[6], Frédéric Bettens[6], Joel Hanqc[6], Carlos Valderrama[6], Nilo Menezes[7], Alexandre Girardi[7], Xavier Ricco[7], Mario Lopez-Ramos[8], Nicolas Dumont[8], Iván Corredor[9], Miguel S. Familiar[9], José F. Martínez[9], Vicente Hernández[9], Dries De Roeck[10], Christof van Nimwegen[10], Leire Bastida[11], Marisa Escalante[11], Juncal Alonso[11], Quentin Reul[12], Yan Tang[12], Robert Meersman[12]

1 Alcatel-Lucent Bell Labs, Antwerp, Belgium

2 VTT, Helsinki, Finland

3 UAH, Madrid, Spain

4 ENSIE, France

5 Institut Telecom SudParis, Paris, France

6 UMONS, Mons, Belgium

7 Multitel asbl, Mons, Belgium

8 Thales, Paris, France

9 UPM, Madrid, Spain

10 CUO, Leuven, Belgium

11 ESI, Bilbao, Spain

12 STARLab, Brussels, Belgium

Abstract In this chapter we discuss the wide range of challenges in user-generated Internet of Things applications, as being worked on among the large consortium of the *DiY Smart Experiences* (DiYSE) project (DiYSE, ITEA2 08005). The chapter starts with a discussion on the context of 'DiY' as a phenomenon to be leveraged, and eco-awareness as an example application area. The main body of the chapter is devoted to the technical outline of the DiYSE architecture, starting at the lower Internet of Things layers of sensors, actuators and middleware, over the role of semantics in device and service interoperability, up to requirements for the service framework and the application creation process. Furthermore, the chapter adds considerations concerning tangible interaction in the smart space, assumed in DiYSE both for the context of experiencing as well as shaping the user experience. With the chapter, we thus take a holistic view, sampling the range from lower-layer technical implications of enabling DiY creation in the Internet of Things, up to the human-level aspects of creative communities as well as tangible interaction.

11.1 Drivers, Motives and Persona in the DiY Society

With the 'DiY society' (Von Hippel 2005) a world is imagined where anybody could become a creator of objects. With the DiYSE project taking the Do-it-Yourself (DiY) phenomenon as a starting point, we discuss its broader context in this section.

At first sight, the idea of creating objects might seem like nothing new. People have been creating things from the very start of civilisation, dating back to the prehistoric ages where people created very basic tools out of materials at their disposal. Ever since, the process of creating things has evolved and has become more complex, as the world and society itself became more complex (Sterling 2005). If we make a time warp to today's modern world, we see that the introduction of technology into our lives is at least one of the aspects that have influenced the way we create, use and perceive objects. Computerised systems are nowadays allowing us to create very complex products that not everyone is capable of creating from scratch anymore. In order to incorporate a computerised, electronic system into an object a certain amount of expertise is needed for programming the system or to integrate the various hardware and software elements.

So, a major challenge to make the DiY society possible is to make people more capable of creating meaningful objects again in the context of today's object complexity, beyond the intended use as driven and orchestrated by solution vendors, opening up e.g. the physical and electronic customisation possibilities. In an optimal *utopian* scenario, this means that the creation of technological and purely physical 'analogue' products should be a seamless activity, allowing people to create things that enhance their lives in a pervasive world. The way this process of creation is done by someone is inherently linked to characteristics such as personal background, intention, expertise and motivation.

Of course, all this is to be seen in the context of a complete next generation 'manufacturing' ecosystem, of which the viability depends on finding a sustainable balance, a multi-sided 'win-win', between the various involved actors. This will eventually determine the economical, next to the evident societal impact.

11.1.1 Evolution of DiY

Recently, Do-it-Yourself as a phenomenon has, again, started to take the central stage in research and development. To understand why this is the case, several things can be learned from the history of DiY and can be projected onto how the phenomenon could be perceived in the future.

11.1.1.1 The Past

Looking at the past, DiY can be seen as a variety of activities. Obviously people have been making objects themselves since the prehistoric ages, but looking at the evolution in history regarding the creation of objects one can observe several elements that had significant impact in how we approach DiY today. A good illustration of this is the way objects were created in the Middle Ages, as at that time people started having a *marketplace* to buy and sell things and there were established communities to share and learn new skills (Sennet 2008). In the Middle Ages, the creation of things was mostly done by *skilled craftsmen* who grouped together in *guilds*. In order for someone to 'learn' how to create something one had to go through a learning process in which a 'master' taught his skills to one or more apprentices.

Since then, we have evolved into a society where skills knowledge is more distributed among people and is less confined to one person or group. When nowadays the term DiY is coined, often the first associations made are about shops selling home improvement materials and people refurbishing their houses themselves.

11.1.1.2 The Possible Future

With the advent of computers and in particular of the Internet, the notion of DiY has taken new dimensions. First of all, it is now a lot easier to share and talk about DiY activities of all kinds through dedicated online community platforms (Dormer 1997). Secondly, more and more people are creating their own electronics, both hardware and software. Both these facts result in an increasing *accessibility* of technology, making tools available for people to enhance the quality of their lives on other levels than the purely functional.

11.1.2 Why Do People Build Things Themselves?

A central question that still remains is why people would at a certain point decide to build something themselves. There are at least two ways to approach this question. On the one hand it can be seen as part of a *motivational psychology*, where a person does something based on intrinsic motivation. This means that the motivation comes from the person himself wanting to solve a problem in his own life for example, possibly but not necessarily with cost savings in mind[121]. On the other hand, DiY can be interpreted on various levels depending on the background of the person or people involved in the DiY activity, the so-called types of 'people logic'. For instance, a person customising his bought shoes is on a different level than a person who is creating shoes from scratch.

11.1.3 People Motivation as Driver

A major driver behind the reason that people at some point decide to create something themselves instead of buying a ready-made solution from a shop is the relationship they create with the thing they created themselves. In the context of interior decoration, Elizabeth Shove describes this type of motivation as follows: "The house objectifies the vision the occupants have of themselves in the eyes of others and as such it becomes an entity and process to live up, give time to, and to show off. What is important are end results, not the actual physical involvement in the tasks and projects of 'doing it yourself'" (Shove et al. 2007). Doing something yourself allows people to identify and relate to objects on a much deeper level than merely the functional. Von Hippel (2005) also states that "A thing is not merely a material object, but a frozen techno-social relationship", which points out that the relation between a person and an object is something quite delicate. This emotional link between a person and an object is what defines the meaning a person gives to something. It is this process of giving meaning that is highly stimulated through DiY activities.

11.1.4 People Logics, Distinguishing Motivation Levels

With regard to the previously mentioned motivational aspects, it should be nuanced that it does not work the same way for all people. The concept of DiY can be approached and understood by various people on different levels, depending on their personal background, personal skill or experience. To understand and comprehend these levels better, it should be made clear that what really matters in a

---

121 Note that, while one would expect DiY activities to *save* costs to the one performing it, as he is the one investing time, effort and creativity, the modern DiY in many cases rather is motivated by feelings of 'ownership', 'passion for creating' and other psychological motives as discussed here. So DiY often implies a willingness to pay which is higher as compared to buying off-the-shelf products solving the same problem. Both *cost saving* and *spending* have a place in the DiY societal phenomenon.

DiY activity is the *mindset of a person*. Depending on the way people think about a subject, they will interpret it as being something, DiY or not. We here use the concept of 'people logics' introduced by Mogensen to illustrate this (Mogensen 2004):

- **Industrial logic:** This way of thinking is mostly straightforward, no-nonsense. In order for people of this kind to have a drive for DiY, a very small action would be needed. For example, mounting a device on the wall may give such a person a feeling of satisfaction.
- **Dream society logic:** In the dream society, people do things in order to show themselves to the outside world. Thinking about DiY from such perspective, a deep customising of a product could suffice to trigger the feeling of 'I did this myself'. This could be, for example, choosing the colour and materials of a pair of shoes.
- **Creative man logic:** The creative man wants to create things from scratch by himself based on his own personal needs. Starting from this point of view, this person could follow an *instructable* to create his own windmill to provide power to his house, as an example.

The logics presented here may need to be extended to cover every possible aspect of DiY, but the main point is that approaching people based on their mindset may proof to be the key to getting the masses to engage in (Internet of Things) DiY activities.

Next to the DiY mindset of people, as the main and basic driver, we mentioned before that this is to be seen in the context of overall *ecosystem dynamics*, where *economical constraints* are into play. DiY for the practitioner in fact can mean a cost saving or can be rather a higher spending for the same problem solved, depending on the degree and level of motivation as discussed. This last case is an obvious opening to business opportunities, as in fact leveraged for years already in the creative and *hobby crafting* sector as well as by vendors of high-end *modular* systems in various domains.

But with the evolution to cheaper and more accessible electronics, and the potential to easily connect wirelessly and ubiquitously to the internet, fuelling the Internet of Things as a grassroots economic platform, DiY may also become a *game changer*, forcing many product and solution vendors to reconsider opening up to their products to customisation and interconnectivity as a quality, rather than pursuing 'locking in' consumers into a single-vendor buying track. In fact, this is what the *Institute for the Future* (IFTF) predicts in their map on *'The Future of Making'* (IFTF 2008). As with other market evolutions, commercial actors anticipating taking a strategic role in such a new ecosystem – a 'Web 2.0 of the Internet of Things' – may develop a clear first-mover advantage, comparable to what happened with the *Apple iPhone App Store*.

One particular theme, driven by economical but also broader societal choices, is *eco-awareness*. The following sections elaborate on this as an example area of DiY Internet of Things activities.

11.1.5 Eco-awareness, an Example Application Theme in DiYSE

One example area where new DiY user-generated applications could have a large socio-economic impact is the theme of *eco-awareness*, including but not limited to energy-efficient infrastructure. One scenario cluster in the DiYSE project considers leveraging user-generated pollution data and possibly also safety-related data in the city for community-building of mass-consumable applications, supporting this societal awareness. Another set of applications considered is about energy-efficient comfortable living, with energy consumption monitoring and control using smart objects. In this section, we discuss the requirements for applying the DiY concept to this example area.

11.1.5.1 Energy Consumption in a DiY Internet of Things

With the emergence of the Internet of Things, everything is becoming connected, and so, networks have evolved from primarily a source of information to the most important platform for many types of applications, involving all kinds of devices and objects. Likewise, connected communities of people using the ‘connecting to anything’ capability of the Internet of Things are also expected to grow more and more. Therefore, the need is emerging for solutions for interdisciplinary fusion services that combine Information Technology (IT) with other technologies.

Among several applications for interdisciplinary fusion services in relation to ecological themes, aspects such as *energy harvesting* and *low power consumption* are also quite important elements for Internet of Things smart experiences to become a reality. Current technology seems inadequate for the emerging low-power processing requirements. The development of new and more efficient and compact energy storage like fuel cells, printed/polymer batteries, etc; as well as energy generation devices, coupling energy transmission methods or energy harvesting using energy conversion, will be pivotal for implementing autonomous wireless smart systems.

In the DiYSE project we address the challenge of eco-awareness for energy efficiency by making it more tangible to people, and more ‘DiY’ in people’s mindset by introducing network-connected smart objects in the setting. Taking advantage of this paradigm, one can indeed imagine that *consumers start monitoring* their energy consumption and thus better understand how their habits relate to their energy consumption. This not only provides a more fine-grained picture of the energy consumption in houses, buildings and vehicles to the *energy suppliers*, but ultimately, with DiY involvement, it also provides citizens with impactful participation means, *sharing good practices* and energy saving 'tricks' using self-made hardware or software enhancements, fuelling a collective green society mindset. Also, energy suppliers would be able to interact with their customer households in a less 'black-and-white' fashion, for example activating appliances that consume much energy, such as washing machines or laundry dryers, at times when energy can be produced and provided in ways that is environmentally friendly and well-priced. With service creation technology around connected smart objects, as researched in DiYSE, a lot more could be offered, like tracking energy consumption peaks, providing consumer notifications on appliances still running possibly inadvertently, or other, more complex applications for which personalisation is an essential factor in mass market acceptance.

11.1.5.2 DiY Engagement in Eco-aware Applications

In the *home environment*, the big paradigm shift could come when every smart object knows the interoperable protocols, removing the need for the dedicated systems developed independently today. Even beyond that, without putting any pre-established 'high-end' solution in place for which – a priori – cross-system interoperability standards would have been established, building intrinsically more *open* systems – in a DiY Internet of Things fashion – would encourage inhabitants to participate effectively in an ecological engagement. For example, maintaining a comfortable temperature and heating of water are the most energy consuming tasks in a typical house, with a dramatic potential for energy conservation and as a consequence a potentially significant positive impact on the environment. With the family engaging in more elaborate 'self-configuration' as a DiY activity, the house could become so fine-tuned, that the comfort of each of its inhabitants is simultaneously maximised through learning the individual preference profiles, while keeping energy consumption within desired limits. In the *vehicle environment*, smart objects in the car will be able to manage better the energy needed. Optimal route planning will reduce the distance driven, and better control systems for the car itself will make the ride more energy efficient, all combined contributing to reduced emissions and less pollution. Here also, awareness among citizens can be amplified by giving them the means to customise the experience, and even contribute data and measurements to related electronic communities.

Among several applications envisioned in DiYSE, Figure 11.1 shows an applications overview on the theme of eco-awareness and energy efficiency. For the home and building environment, objects such as energy saving controllable sockets, smart metering, and home automation controllers are used for energy management. In vehicles, devices taking part in the navigation control, and devices for safety, can be used for energy saving.

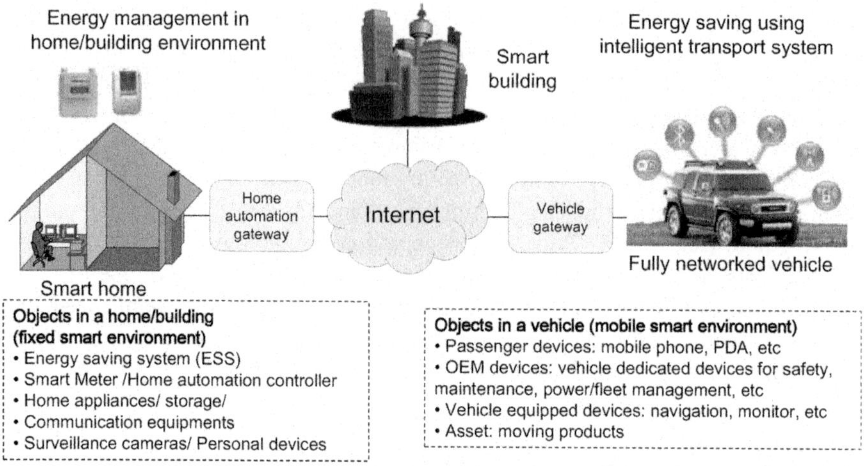

Fig. 11.1 Applications Using Eco-awareness for Energy Efficiency

The smart home and smart building in fact may cover a wide range of services, applications, equipment, networks and systems that act together in delivering the *'intelligent' environment* for domains such as security and control, communications, leisure and comfort, environment integration and accessibility. Particularly smart building entails a suite of technologies used to make the design, construction and operation of buildings more efficiently, applicable to both existing and newly built properties (GeSI and The Climate Group 2008). Example systems are building management systems (BMS) that run heating and cooling systems according to occupants' needs, or software that switches off all computers and displays after everyone has gone home. BMS data can be used to identify additional opportunities for efficiency improvements.

So, various concepts and approaches are possible in optimising the energy efficiency of buildings and homes, leveraging the intelligent building control in an attractive cost-benefit ratio. A few example applications from this view are:

- *intelligent/automated light control*, allowing users to lighten their homes before entering, both for safety and to create a welcoming environment, or to mimic activity while away – even reconfiguring activities remotely when away from home / office;
- *auto-regulation of heating based on non-occupancy detection*, maximising energy savings, while remote temperature controls still allow for adjustments; and
- *media/entertainment control*, integrated more comfortably with home activity compared to stand-alone entertainment systems, and remotely accessible.

In these examples, the home or office becoming a 'smart space' with DiY Internet of Things capabilities would allow much easier 'programming by doing' configuration of the otherwise (semi-)professional home automation configuration, and allow for originally unforeseen improvements from community 'wisdom of the crowd'.

Automotive transport, as the other area of applications mentioned, represents one of the main sources of green house gas emissions, but with the generalised availability of ultra-high-speed broadband access and the ubiquitous provision of next generation mobile telecom services, many tasks and movements could be coordinated much better, for minimising power consumption. While the main focus of applying ICT to transport through the development of *Intelligent Transport Systems* (ITS) is *safety*, the efficiency management of transport systems through ITS can also reduce the *environmental impact* of transportation. Example applications for this area currently considered by industry (ITU-T 2008) are:

- *enhanced navigation or vehicle dispatch*, considering alternative routes, possibly proactively, reducing journey time and energy consumption;
- *parking guidance systems*, additionally reducing engine time;
- *road pricing schemes*, such as the congestion charge applied in London, encouraging use of public transport during congestion periods.

Furthermore, *vehicles* can serve as mobile environmental pollution *sensors*, and electrical vehicles can play an important role as *energy storage*, production or consumption elements in smart energy grids.

Here again, the engagement of people can be dramatically improved, by providing the flexibility of high (DiY) personalisation, and even having them actively contribute to the roll-out of the mobile and fixed pollution sensing infrastructure by sharing personal data and devices. A car vendor, for example, opening up the car information system to community driven sensor (and other) applications could have a unique selling point in enhanced in-car navigation, with this maybe even becoming a must-have feature when the market evolves.

11.1.5.3 Requirements for Enabling DiY in Eco-awareness Applications

In light of the range of possible eco-awareness applications, a number of further requirements on capabilities at various levels need to be considered, in order for such DiY, sometimes crowd-sourced value to become possible, for example with respect to:

- *easy installation and integration of everyday objects* in our home environment, locally as well as remotely, for monitoring and control purposes, in particular for intelligent energy consumption monitoring and control functions;
- *public IP network data connectivity and access*, directly or indirectly, to all objects involved in the service or application;
- *device* virtualisation, installation and *provisioning*;

- *meaningful and permanent, globally unique identification of objects*, allowing for the linking of devices into associated functions, and the unambiguous derivation of meaningful information from raw sensor data;
- *search and discovery* of suitable, available appliances according to properties and capabilities;
- *web-based* information processing, notification and *visualisation*;
- *personalisation* of associated services, considering *context information,* such as in personal home energy profiles and scenes;
- support for a *creation workflow* entailing activities involving all the above;
- *secure access* limitation or sharing of personal or household data across Internet; and
- *identity-based user management.*

Such groups of requirements may be extrapolated to get to *generic* requirements as need to be covered for *any* creation architecture on top of the Internet of Things as is aimed at in the DiYSE project. Further in this chapter, specific solution parts of DiYSE are discussed, addressing these requirements as relevant for diverse use cases.

11.1.5.4 Technologies and Standards Relevant for DiY Eco-awareness

While energy efficiency in buildings clearly would benefit environmental sustainability, there is still a technological barrier to DiY creation by the masses on this theme, or, as a start, just even for involvement of all related non-IT professional parties. Therefore, a way must be found to *disseminate* and promote technological *good practices* for energy efficient buildings – a typical ingredient of a DiY community phenomenon – and to increase the accessibility for non-technical experts to a range of available technologies.

As was introduced already in the eco-awareness examples in previous sections, this could be highly accelerated by the availability of properly standardised, general and domain-specifically profiled, *interoperable protocols* for smart objects and associated applications, in order to keep users agnostic from the underlying technology.

A number of existing technologies and standards should therefore be taken into account, in a first step to enabling DiY creation on top of the Internet of Things:

- *Web technologies*, including information processing, notification and visualisation, but also so-called *mash-up* technologies, should be easily usable in combination with the Internet of Things, as a basic communication platform like needed for example in energy efficiency in building automation and smart homes. In fact, the leveraging of this technology has led to early definitions of a *Web of Things*, in which a REST-based convention is taken as a first step to realise *physical mash-ups* (Guinard et al. 2009).

- *Internet of Things and device application programming interface (API)* workgroups in standards bodies like Internet Engineering/Research Task Force (IETF[122]/IRTF[123]), International Telecommunication Union - Telecommunication Standardisation Sector (ITU-T)[124] and World Wide Web Consortium (W3C)[125], play a pivotal role.
- *Service deployment, service life cycle management and device management* standards alliances like the Open Service Gateway initiative (OSGi) Alliance[126], the Universal Plug-and-Play (UPnP) Forum[127], and the Digital Living Network Alliance (DLNA) for home devices configuration and functional abstraction.
- Beyond this, the Internet of Things Research Cluster (IERC)[128] is making an effort to concertise standardisation and interoperability activities among the many European projects working around the Internet of Things. In this context, subject of debate is, for example, the *unique identifiers for objects*, which, while mainly stemming from early Internet of Things applications in logistics and supply chain management, are also required when extending the eco-awareness theme into the DiY realm. Related especially to naming and addressing for the Internet of Things also is the work of the European Telecommunications Standards Institute (ETSI) Technical Committee on Machine-to-Machine Communication (TC M2M).
- *Optimisation techniques from Cloud Computing* could particularly be applied and combined with the notion of Internet of Things in the context of smart energy grids, as power needs to be 'routed' according to the distributed fluctuations in energy capacity and consumption needs, requiring bidirectional, real time information exchange among customers and energy management operations.

11.2 Sensor-actuator Technologies and Middleware as a Basis for a DiY Service Creation Framework

For applying the freedom of creativity of Web 2.0 to the Internet of Things, as aimed at in the DiYSE project, it is essential that non-expert users are enabled to easily search for public devices or share their own, privately bought or DiY-built

---

[122] http://www.ietf.org/

[123] http://www.ietf.org/

[124] http://www.itu.int/ITU-T/index.html

[125] http://www.w3.org/

[126] http://www.osgi.org/

[127] http://www.upnp.org/

[128] http://internet-of-things-research.eu/

devices, and as such personalise the physical environment by combining and 'mashing-up' device functions, regardless of whether the system has prior knowledge about the devices or not. A large, heterogeneous set of device types needs to be considered, with devices 'speaking' a wide range of 'languages', having varying specific constraints in terms of mobility, battery, computation, etc., and serving different usages, the same device even having different purposes in different contexts for different users, all in a constantly evolving manner.

Traditional computing approaches are not intended to cope with such complexity. Therefore, this section explores how DiY service creation environments, as envisioned in the DiYSE project, can deal with *plug-and-play* connectivity of heterogeneous device types, how the function of appearing devices can be understood, and how the data generated by these devices can be interpreted.

11.2.1 Device Integration

In the following subsections we introduce the notion of enhanced device drivers, as a means of first-level abstraction for heterogeneous device types, and the DiYSE Gateway, serving as a proxy for resource-constrained devices. Finally, we discuss ways to identify and address the discovered devices.

11.2.1.1 A first Level of Abstraction Addressing Device Heterogeneity

As a DiY creation system needs to support legacy devices, and cannot assume that future devices will respect any specific standard, the only viable solution is to make the system accept *any* kind of device interface and *describe* it using a common 'meta-language' understandable by a machine.

A similar abstraction mechanism is commonly used for peripherals in every computer operating system, and is known as a *device driver*. It contains only the programming interfaces required for the system to communicate with the device, while hiding specific implementation differences within one device class. However, the device driver does not contain any information about the different ways of using the device.

As an example, a user may want to control his motorised pan-tilt-zoom camera using a *WiiMote* controller and gestures. This interaction may seem conceptually straightforward for a human being, but technically it is unfeasible for a non-expert user unless the specific software exists. It may seem obvious to a human that both devices could 'talk' about pointing a given direction, but machines need additional knowledge to achieve it. The information about the *meaning* of actions such as *'get pointed direction'* or *'turn to direction'*, required for their automatic mapping, needs to be provided by a human in every case because there is no computer algorithm enabling to find a logical relationship between those.

One solution as investigated in the DiYSE project is to *embed* the knowledge about the capabilities of a device in an *enhanced* driver, so that it is understandable by machines without low-level programming intervention. In the example, the *WiiMote* driver would be augmented with information such as *'can control direction'*, whereas the camera would have *'can have its direction controlled'* as a property exposed by the enhanced driver. Semantic reasoning mechanisms, as discussed later in this chapter, would use such conceptual information to assist the user in describing device interactions that make sense conceptually and are at the same time technically feasible and well-described, so that the desired device interaction is executable without the need to develop dedicated software. For instance, in a most basic scenario not even considering the higher layer semantic reasoning capabilities of an application creation environment above it, a straightforward request to the system to link the *WiiMote* and the camera can already default to the automatic realisation of the intuitively expected interaction of controlling the orientation of the camera by means of the *WiiMote*. So, for such basic scenarios, behaviour creation can be as simple as defining a *Lego* crane control, without such control being predesigned as a fixed function, only relying on the basic semantic annotations obtained from the enhanced device driver.

With this approach, the problem of complexity due to the heterogeneity of devices is solved at a low-level stage. Even at this basic, not further enhanced level, non-expert users will not experience a barrier of low-level technical details or compatibility issues anymore.

Beyond this, in the context of DiY creation of Internet of Things applications, as an important potential enabling element for the DiY Internet of Things ecosystem, web communities are envisioned to emerge in which experts can publish and enrich enhanced drivers, so that the spectrum of possible applications constantly broadens, including newly supported devices as well as new ways of applying existing device features.

11.2.1.2 Achieving Device Data Connectivity for Resource-constrained Devices

Despite the progress in leveraging the IPv6 protocol for connecting smart objects into the cloud, in the foreseeable future many, in one or multiple aspects *resource-constrained* devices will remain supporting only dedicated – but nevertheless often standardised – protocols specifically designed for the resource-constrained nature of the devices. Examples are *Wireless Sensor Network* nodes, or *Zigbee* or *Bluetooth* peripherals, for which hardware cost, related to memory and processing power, but especially also energy consumption are important factors.

As such resource-constrained devices are also considered *key* in the DiYSE context, the project considers an *intermediate gateway function* for exposing also these devices in a uniform way to the overall framework and make them IP-addressable, to connect them into local or global IP networks for data retrieval, control, and device management. Such a *DiYSE gateway*, as we named this function, requires a flexible abstraction layer hiding the underlying network technology heterogeneity, while supporting fast and seamless device deployment. Also, this layer should relay unique identification of the devices, for transparent interoperability and remote device querying, control and monitoring. Figure 11.2 shows the main modules of a DiYSE Gateway, distinguishing:

- a discovery module for devices being plugged in,
- means to install and execute enhanced drivers, and
- the bookkeeping of connected devices both for keeping track of local execution and for southbound device exposure.

For devices that directly connect into the cloud via IP, equivalent functions for proper exposure to the middleware may be provided also, according to a notion that we could call a *cloud DiYSE gateway*.

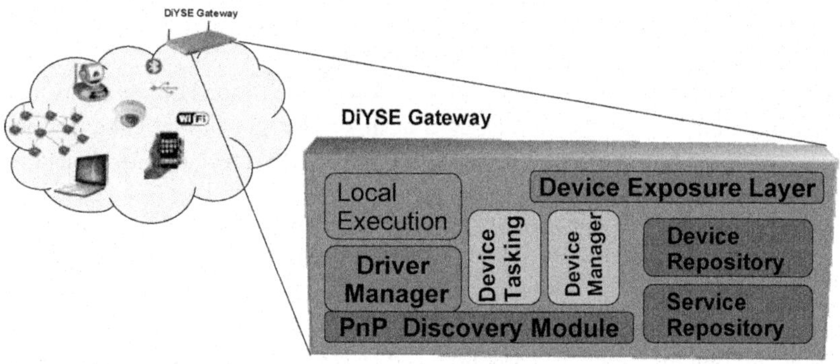

Fig. 11.2 Main Modules of a *DiYSE Gateway* Function

In the next subsection we discuss some further aspects of the installation, registration, and integration of these resource-constrained devices, or *nodes* as we call them.

11.2.1.3 From Hardware to Device Description

As previously described, we consider the use of *enhanced drivers* to cope with the high heterogeneity of *nodes*. Typically for each new device, the device manufacturer, or an expert developer, can make available appropriate driver software, which eventually gets automatically installed on local DiYSE gateways, thus providing a description of device capabilities and an interactions-set that users, or other devices, may leverage.

For this automatic driver installation, a DiYSE gateway triggers a *driver-lookup* operation among the different repositories by using uniquely characterising meta-data extracted from the new node, like type of device, vendor, hardware MAC address, etc.. On top of the base set of handling functions that devices from the same hardware family may have, specific additional handlers may be required for managing additional functionalities, like for instance a special night-capture function that only a specific type of pan-tilt-zoom camera may have.

A further essential requirement for the integration of large numbers of devices in the cloud is the capacity to individually *identify and address* them. Several solutions for that can be considered, such as the use of a generic syntax like *Uniform Resource Identifier* (URI, RFC 3986[129]) as a permanent and unique identifier included in the device description, or, alternatively, the association of an *IPv6 address* with every node, or still, the use of *application level identifiers* on top of the network addresses, like logical peer-to-peer (P2P) identifiers or Dynamic DNS. As many nodes may however not be able to store or compute their identifiers, such operation often needs to be performed on the connecting DiYSE gateway.

In the DiYSE architecture, also a device discovery service is foreseen, providing high-level descriptions of the capabilities and the services that the installed nodes can provide. While more and more devices today are discoverable via direct embedded support for communication protocols like UPnP[130], DPWS[131] or DLNA[132], most of today's devices in the surroundings remain to be incompatible or not equipped with such self-description mechanisms. Thus, DiYSE gateways use the enhanced drivers to map the node functionality and attributes into a *common description language* like DPWS for remote description of connected nodes, and offer the functionality to expose devices and services and send events beyond the local network domain boundaries across the internet.

11.2.2 Middleware Technologies Needed for a DiY Internet of Things

A middleware, being a software infrastructure that ties together hardware, operating systems, network stacks, and applications, should provide a runtime environment supporting functions such as multi-application coordination, standardised system services (e.g. data aggregation, control and management policies), and mechanisms for adaptive, efficient resource handling. As such, middleware support is essential for interworking with so-called *Reduced Functionality Devices* (RFDs), such as DiYSE *nodes,* which are by definition resource-constrained devices, and which moreover are using one out of a heterogeneous range of commu-

---

[129] http://www.ietf.org/rfc/rfc3986.txt

[130] http://www.upnp.org/

[131] http://docs.oasis-open.org/ws-dd/ns/dpws/2009/01

[132] http://www.dlna.org/home

nication standards, including *IEEE 802.15.4*, *ZigBee*, *Z-Wave*, *Bluetooth*, and *6LoWPAN*.

In the middleware that exposes RFDs as a set of generic services for applications, we distinguish in the DiYSE project:

- *Low Level Services*, containing the vital, intensively used functions always needed for interaction with the hardware, like *Real-Time Management*, *Communication and Context Discovery Management*,
- *High Level Services*, typically less critical, providing a first level of application support, more adaptable to different scenarios, with functions such as *Query*, *In-node Service Configuration*, and *Command*,
- *Cross-Layer Services*, providing mixed high and low level functions, such as *Reasoning*, *Portable Code Execution Environment*, and *Security*, and
- *Control Services*, providing the middleware's core functions of component deployment and lifecycle management as well as inter-component communication through event communication, by means of entities called *Software Component Container* and *Eventing Service Manager*.

So, one of the interesting research challenges as investigated in DiYSE with respect to middleware for RFDs is to *translate the service-oriented computing (SOC) paradigm to wireless sensor and actuator networks*. The SOC approach is promising for easy assembly and deployment of interoperable, platform and operating system independent services in such networks, but should also fulfil typical additional requirements for smart environments, such as lightweight business logic optimised for low computational overhead and low battery consumption.

In order to evaluate the performance in low-resources devices in the DiYSE framework in terms of processor power, memory size, bandwidth and battery lifetime, the RFDs-based approach has been implemented in a Wireless Sensor Network. As illustrated in Figure 11.3, in order to provide advanced sensor services to the envisioned end user applications, service composition and management tasks are performed in specialised nodes in the sensor network. To this end, sensor nodes with *Broker*, *Orchestrator* and *Trunk Manager* roles has been defined.

Broker nodes represent the interface between the Wireless Sensor Network and external networks. They receive semantic descriptions of the simple services provided by each sensor using lightweight notation languages, such as *Service Mapping Description* (SMD), on one side and receive service requests from external networks on the other side. *Orchestrator* nodes are responsible for implementing a virtual sensor service paradigm. They perform the composition of the offered simple services into potentially complex and sophisticated composed services, using the semantic descriptions of those primitives. The service control plane is implemented in the *Trunk Managers*, which perform tasks associated with service state supervision, such as self-configuration, self-adaptation and self-recovery, in order to increase service availability and network resilience.

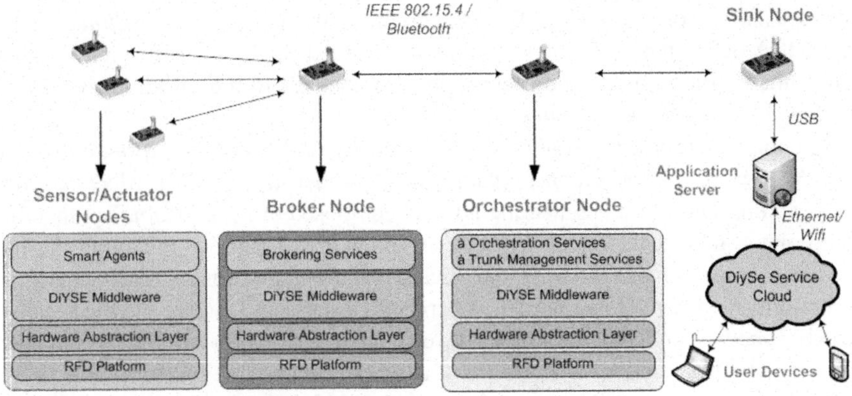

Fig. 11.3 Network Architecture and Middleware for WSANs in DiYSE

11.3 Semantic Interoperability as a Requirement for DiY Creation

As discussed in the previous sections, the interoperability among devices supporting different lower layer communication protocols is a crucial requirement that must be fulfilled to enable the ad-hoc mixing and matching of devices and sensor nodes in DiY applications, as is aimed at with the DiYSE architecture. As *ontologies* have been proposed as a solution not only to provide semantics for the data to be exchanged, but also for describing the devices themselves, they indeed provide the necessary means for compositions of devices in applications – by web-based composition of service-level exposure of the devices, or even beyond that, by locative, on-the-spot creation of applications in the smart space.

In this section, we first introduce the concepts concerning ontologies in computing science, and then describe through examples how the ontologies are envisioned to achieve semantic interoperability in an Internet of Things creation environment as researched in DiYSE.

11.3.1 Ontology

Ontology, as a discipline, is the branch of philosophy that is concerned with the nature of things that exist in the universe (Smith 2003). More specifically, it is the science that aims to provide an *exhaustive* and *definitive* classification of things based on their similarities and differences. By *exhaustive*, we mean that it provides an explanation for everything that is ongoing in the universe. By *definitive*, we mean that every type of things should be included in the classification. In this sense, the ontology discipline tries to provide answers to the question: *What are the features common to all things?*

In computing science, an *ontology* is commonly defined as: *"a formal, explicit specification of a shared conceptualization"* (Studer et al. 1998). More specifically, an ontology is an engineering artefact composed of *(i)* a *vocabulary* specific to a domain of discourse, and (ii) a set of explicit assumptions regarding the intended *meaning* of the terms in the vocabulary for that domain. This set of assumptions is generally expressed in terms of unary and binary predicates, by which *concepts* and the *relations* between them are expressed. In its simplest form, an ontology defines a hierarchy of concepts related by their taxonomical relationships, whereas in more complex cases, additional relationships can be expressed between concepts to constrain their intended meaning. As an ontology captures knowledge about a domain, this needs to happen *by consensus* among a group of people, in order to reach a common agreement on its conceptualisation.

Different languages have been developed over the last two decades to represent ontologies. For instance, *Simple HTML Ontology Extension* (SHOE) (Luke et al. 1997) was developed to annotate web pages with semantics, whereas the *Ontology Exchange Language* (XOL) (Karp et al. 1999) was primarily developed to exchange ontologies in the bioinformatics domain. Since the World Wide Web consortium has published several recommendations to express ontological content on the Web. For example, the *Resource Description Framework* (RDF) (Miller and Manola 2004) allows users to describe the relation between different web resources, while the *Web Ontology Language* (OWL) (van Harmelenand and McGuinness 2004) extends on the RDF vocabulary to provide precise meaning through formal semantics.

11.3.2 Ontology Engineering Methodologies

Over the last two decades, several ontology engineering methodologies have been developed. Gruninger and Fox (1995) propose a method inspired by knowledge-based development using first order logic, starting by identifying a number of motivating scenarios from which a number of natural language competency questions are extracted. These questions subsequently lead to the identification and formalisation of the terminology and axioms that constitute the ontology. Finally, the ontology is evaluated by proving that the original questions can be answered. In this way, competency questions are used to determine the scope and adequacy of the ontology. Uschold and King (1995) propose a method for building ontologies based on their experience with the Enterprise Ontology for system interoperation, while *Methontology* (Fernandez et al. 1997) builds on the main activities of software development and knowledge engineering methods, proposing an ontology development life cycle based on evolving prototypes. *CommonKADS* (Schreiber et al. 1999) is a methodology for knowledge engineering in general, which is used to design and analyse knowledge-intensive, structured systems. A knowledge engineer such as a risk analyst in an enterprise, or a knowledge engineer in an ontological domain, can use it to detect the knowledge expansion, e.g. the opportunities based on the available knowledge resource, and the knowledge acquisition bottleneck.

Alternatively, the *Developing Ontology Grounded Methods and Applications* framework (DOGMA) is a formal ontology engineering framework inspired by various scientific disciplines, such as database semantics and natural language processing (De Leenheer et al. 2007). Although DOGMA partly draws on the best practice of the other methodologies, it differs from these approaches by providing a strict separation between the lexical representation of concepts and their relationships and the semantic constraints. This separation results in higher reuse possibilities and design scalability, and eases ontology engineering, as the complexity is divided and agreement can be more easily reached. Furthermore, the definition of terms in a natural language and the grouping of terms have been incorporated. By grounding knowledge in natural language, domain experts and knowledge engineers can use ordinary language constructs to communicate and capture knowledge. Therefore, domain experts do not have to tackle or learn to think in new paradigms, e.g. without the need to express their knowledge in RDF or OWL. Indeed, the complexity of just capturing knowledge is difficult enough already. Based on this approach, end users are able to represent the domain of discourse in terms they understand. Once the elicitation process is finished, and the ontology is formalised, the DOGMA tools can output the information to the requested paradigms. For example, simple linguistic structures like *lexons* can be transformed into *RDF triples,* which results in data (i.e. *facts*) being available as part of the *Linked Open Data* (LOD) project[133].

[133] http://linkeddata.org/

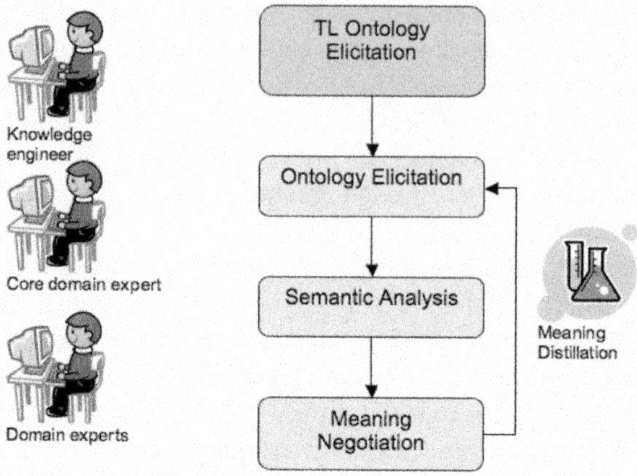

Fig. 11.4 DOGMA-MESS Iterative Process.

The *Meaning Evolution Support System* (DOGMA-MESS) is STARLab's methodology and tool to support community-driven ontology engineering (de Moor et al. 2006). The benefit of DOGMA-MESS is that it allows the domain experts themselves to capture meaning, while relevant commonalities and differences are identified, so that each iteration in the process results in a useable and accepted ontology. Hence, it provides an efficient, community-grounded methodology to address the issues of relevance. Figure 11.4 illustrates how a domain ontology is created through the interaction among three different types of stakeholders, namely the *domain expert*, the *core domain expert* and the *knowledge engineer*. The *domain expert* is a professional within the domain of discourse, while the *core domain expert* has a deep understanding of the domain across different organisations. The *knowledge engineer*, who has excellent expertise in representing and analysing formal semantics, is responsible to assist the domain experts and core domain experts in the processes of ontology creation, validation and evolution.

11.3.3 Application of Ontology Engineering in the Internet of Things

In this section we describe three ontology-based services that would enable three different areas of interoperability as needed for DiY application creation in the Internet of Things.

11.3.3.1 Knowledge Integration and Sharing

As the Web has changed from a mere repository of documents to a highly distributed platform where new types of resources can be discovered and even easily shared, one can extrapolate the Web as an Internet of Things making everyday objects addressable via IPv6 (Sundmaeker et al. 2010), as well as an *Internet of Services* making services easy to implement, consume, and trade.

However, the diversity of this increasing volume of data, services, and devices implies that it is impossible for them to work together, as many are designed independently, with particular, different application domains in mind. Making knowledge transparent to users and services thus requires the development of a *formal* and *precise* vocabulary that (i) defines concepts shared by a community and (ii) can be processed by machines.

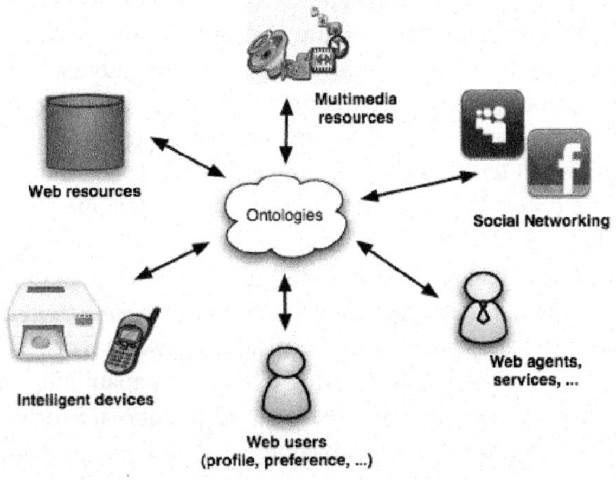

Fig. 11.5 Ontology-based Knowledge Integration and Sharing

Figure 11.5 shows how a semantic layer can be used to facilitate knowledge integration and sharing on the Web. For example, the annotation of devices with concepts from the FIPA *Device Ontology Specification*[134] would enable users to retrieve devices, like smartphones, based on their capabilities. Similarly, Eid et al. (2007) have developed an ontology to discover sensor data like GPS or temperature, consisting of three components:

- the *Suggested Upper Merged Ontology* (SUMO) (Niles and Pease 2001) is an upper level ontology developed by the IEEE to promote interoperability, information search and retrieval, automatic inference, and extendibility;
- the *Sensor Hierarchy Ontology* (SHO) includes models for data acquisition units and data processing and transmitting units; whereas

[134] http://www.fipa.org/specs/fipa00091/PC00091A.html

- the *Sensor Data Ontology* (SDO) describes the context of a sensor with regard to spatial and/or temporal observations.

Alternatively, the *Web Service Modelling Ontology* (WSMO)[135] provides a conceptualisation for the core elements of Web Services. For example, the web service component provides a vocabulary to describe the capabilities, interfaces, and internal working of a Web Service. In turn, this allows the discovery of services, their invocation (context based parameterisation and data transformation), and their mediation (e.g., the composition and orchestration of services).

11.3.3.2 Ontology-based Search

The DiYSE project aims to develop a framework to enable communities to create and exchange applications (i.e. software components) for ubiquitous computing and ambient intelligence, leveraging the Internet of Things. In practice, these applications are likely to come from a number of repositories and in a variety of formats. For example, software components could be indexed or tagged, based on the type of devices used to run the software. However, different repositories are likely to use different terminologies for the indexing.

This is where *ontology-based search* comes in as a solution, as in that approach the indexes are expressed in terms of an ontology which translates and hides the different repositories that have committed to it. The advantage of this method is that users can retrieve information based on *unambiguous terms*, thus enabling interoperability when *interpreting both queries and replies*. For example, a community may define their own ontology to define the capabilities of their software components. This ontology would then be used to annotate software components, thus enabling interoperability. The advantage is that a user may search for existing solutions *according to the community's vocabulary*. If the DiY user finds existing solutions, then he can either reuse the solution directly or extend the solution to solve other problems. Otherwise, the DiY user may submit his own annotated solution, which can then be retrieved by other members of the community.

In DiYSE, the ontology-based search needs to serve two types of users. *Technical users* are likely to use technical terms to define the capabilities of a hardware or software component, whereas *non-technical users* will use other non-technical terms that are meaningful to them. As a result, semantic 'translation' methodologies are researched that resolve the differences between technical and non-technical terminologies in order to get to a true *DiY ecosystem* on top of the Internet of Things.

---

[135] http://www.wsmo.org/

11.3.3.3 Context-aware Computing

A further challenge, which is typical for mobile distributed computing in general but becomes even more explicit in a sensor-rich environment, is to exploit the *dynamical changes* in the environment in applications, by means of those applications having the capability to *adapt to the context* in which they are running. Therefore, *context-aware computing* (Schilit et al. 1994) focuses on gathering information about *users*, like status, location, preferences and profile, next to *environment factors* such as lighting conditions, noise level, network connectivity, nearby things and even social aspects. This information is then used to adjust the behaviour of an application to suit user needs and preferences.

As is done in the DiYSE framework, context management can be supported semantically by two core elements, namely the *contextual ontology* and the *context model*. The *contextual ontology* provides a conceptualisation of the characteristics of *real world objects*, while the *context model* provides access to the *contextual knowledge*. For example, the *iHAP* ontology (Machuca et al. 2005) provides a vocabulary to represent (i) spatial description, (ii) actor description, (iii) context features description, (iv) service description, and (v) device description in smart environments like vehicles, homes or public buildings. So, based on for example this ontology, agents and users are able to interoperate to provide context-aware services in the dynamically changing smart environments.

In short summary of this section, we have discussed three areas in which ontology engineering methodologies are important for the Internet of Things, and for DiY creation on top of it in particular, as researched within the scope of the DiYSE project, namely *knowledge integration and sharing*, *ontology-based search*, and *context-aware computing*. Essentially, ontologies are a means to the agreements made among a community and are intrinsically community-based, and so form an enabling step needed for effective sharing and creation activities among DiY communities. Even when a software agent 'commits' to such ontology by using the same vocabulary in a consistent manner, it shares the same knowledge as the agents designed by others in the same community. So, this kind of shareability, as originating from the community, also enables the agents to seamlessly interoperate with each other. In other words, the fundamental principle of ontology engineering is 'autonomy' (Meersman 2010), granting many engineering advantages to the application builders and professionals, up to occasional DiY users.

Furthermore, many other generally applicable ontology engineering techniques can interestingly be leveraged in the Internet of Things, like for instance around *modelling* (Spyns et al. 2002; Baglioni et al. 2008), *querying* (Loiseau et al. 2006), *reasoning* (Baglioni et al. 2008), *annotating* (Kim and Park 2005) and *matching* (Tang et al. 2010).

11.4 The DiYSE Service Framework

On top of the network of connected sensor and actuator hardware, uniformly abstracted via sensor abstraction middleware and semantic annotation, the DiYSE service framework provides a *number of service-level functions*, in turn needed to support the application creation layer above it, in which professional developers up to non-technical end users can shape the smart space by collaboratively creating and deploying Internet of Things applications. Next to the service functionality for *composition, deployment and execution*, this in particular entails also functionality to *adapt and personalise* applications, as well as the *creation* thereof, to context of use, respectively creation.

Figure 11.6 positions the DiYSE service framework in the high-level overview of the overall DiYSE architecture. Mediating between all the identified actors and application areas versus the underlying Internet of Things technologies, three main functional areas for the framework can be distinguished. We discuss each of them in the next sections.

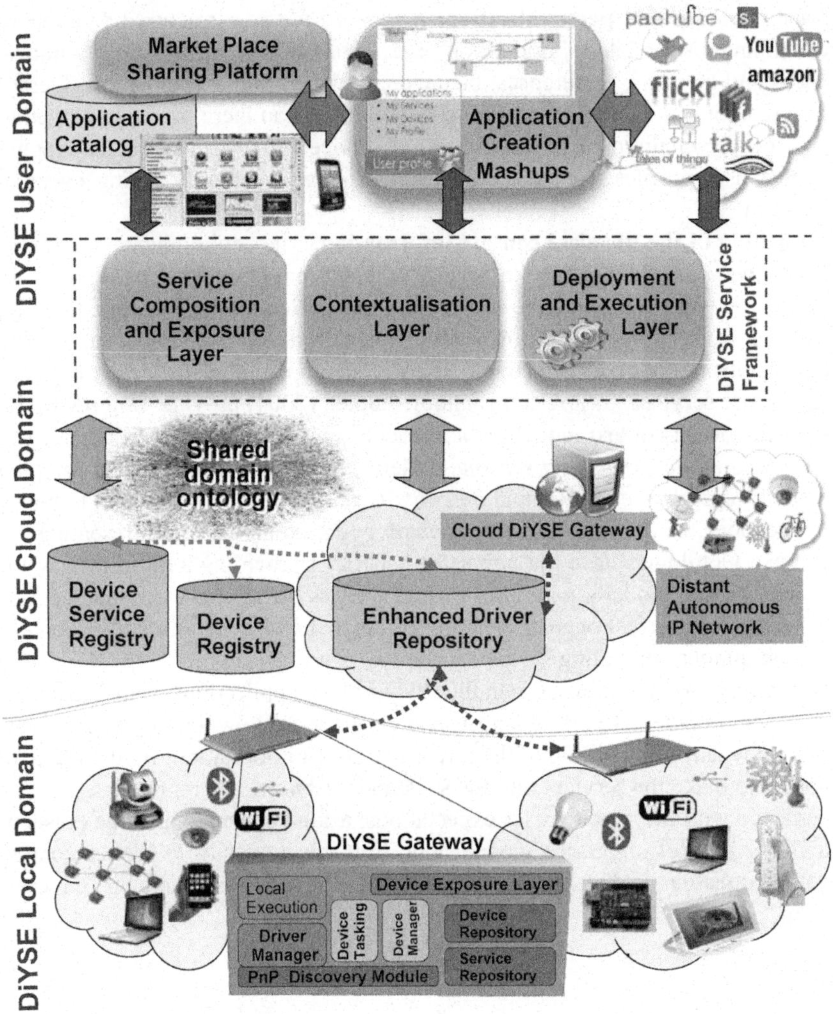

Fig. 11.6 Position of the Service Framework in the DiYSE Overall Architecture View

11.4.1 Contextualisation Layer

Under the *Contextualisation Layer* for DiYSE we group the components for contextualisation and personalisation, serving the application and application creation environment on top of it. In particular, we distinguish three, highly interrelated functions in it:

- **User profiling and personalisation:** A *user profile* is a structured data record containing user-related information like identifiers, characteristics, abilities, needs and interests, preferences, behavioural history and extrapolations thereof for predicting and anticipating future behaviour. It can therefore be exploited to provide personalised, user-context-aware service recommendation, leveraging related user profiles from the crowd, and context-awareness during eventual service use.
- **Modelling of the physical context information:** The environmental context is also a very relevant feature in service oriented environments, particularly in 'smart' environments, where services are expected to behave intelligently, learning from and anticipate on what happens in the surroundings. In general, the establishment of an effective *context model* is essential for designing context-aware services. Strang and Linnhoff-Popien (2004) provide a survey of the most important context modelling approaches for pervasive computing, such as *key-value* models, *mark-up scheme* models, *graphical* models, *object oriented* models, *logic-based* models and *ontology-based* models. As discussed in previous sections, DiYSE has selected an *ontology-based* model for representing the context. Ontologies have the important benefit of providing a uniform way for specifying the model's core *concepts* as well as an arbitrary amount of *sub-concepts* and *facts*, altogether enabling contextual *knowledge sharing* and reuse in a ubiquitous computing system.
- **Reasoning:** Another key issue in the study of DiY applications is the *reasoning* about environmental context and user information, allowing for deduction of *new* knowledge in addition of the directly detected information. As the ultimate goal is to make the services and the surrounding *smart*, i.e., more closely in accordance with the specific user expectations, a fundamental challenge exists in deriving *correct and stable* conclusions from the typically *imperfect* context data acquisition in the highly dynamic and heterogeneous ambient environment.

11.4.2 Service Composition and Exposition Layer

The *Service Composition and Exposition Layer* in DiYSE groups the functions that enable the upper user-facing tools to list and access the different available services and service parts as provided by any actor of the DiY community, i.e., third parties and professionals as well as any users.

It comprises the following functions:

- **Service exposure:** This function provides a unified access to the services and components made available by different levels of users, professionals and third parties (Blum et al. 2008), which is essential for the envisioned DiYSE creation process. It thus enables the different types of users to discover, compose and publish at a properly abstracted service-level. Besides that, functionality such as instantiation and the related exception handling, authentication and authorisation, layered functional exposure, configuration and service user interface representation in the DiYSE creation process is envisioned.

- **Semantic engine:** The semantic engine function provides the service exposure function with the abstractions to semantically mediate interaction of devices, services and actors, according to the methods discussed in the sections on *semantic interoperability*, section 11.3, leveraging a set of *shared ontology repositories* for that purpose.
- **Orchestrator-compositor:** As a key part in the DiYSE creation process, the dynamic composition and orchestration of hybrid and composite services is needed, leveraging the semantic engine as well as the Contextualisation and Personalisation Layer, and closely interacting with the service exposure function, registering also newly composed applications (ESI 2008).

11.4.3 Execution Layer

The main objective of the DiYSE *Execution Layer* is to execute the composed, distributed applications in a dynamic, context-aware manner.

One of the main challenges in this layer is to establish a mechanism for managing dependencies on all the context data at runtime, ranging from user profiles and user context up to the various aspects of the environment's context, sensor data streams as well as events of new devices appearing or disappearing in the environment. Moreover, there is a *tight relation to the devices in the environment* because of the tangible interaction envisioned in DiYSE, as discussed in later sections, requiring device-level mediation mechanisms at lower layers, as previously discussed.

Solutions that have been proposed for execution at the end of a creation process include the use of *software variability* for defining those parts in a workflow that may vary at runtime (Bastida et al. 2008). Before the workflow is executed, its variable parts are instantiated according to the relevant contextual parameters.

A further aspect to be considered for the Execution Layer is the potential semantic binding of *running* component instances, and the *dynamic* adaptation of these bindings when the environment changes.

11.4.4 DiYSE Application Creation and Deployment

As the main objective of the DiYSE project is to eventually enable non-technical users to create their own Internet of Things applications based on available devices and service parts, leveraging the environmental and user context, we have stipulated the overall phases in the DiYSE creation process as follows:

- **Installation of sensors, devices or actuators:** The main challenge in this phase is the dynamic and correct registration of all required device information in a device registry, a driver registry and the service registry, while providing the user with a highly intuitive procedure, ideally not requiring any 'unnatural', i.e. seemingly unneeded interventions. Support at the hardware and network level for this in DiYSE was discussed in section 11.2.
- **User design of the application:** This phase is where the user creates, configures or composes a (partially) new application, taking into account the capabilities (expected to be) available in the environment, according to his/her own profile and context. The challenge here clearly is to provide the user with the right kind of tools at the *right level of abstraction*, according to his/her expertise level. Also, validation or simulation of device interactions and device data as part of the design is an important aspect. This is partly related to the discussions in section 11.3 on *semantics*, but is also closely combined with the *interaction* as discussed in section 11.5.
- **Production of the application runtime code:** After designing the new application, its runtime code can be factored, among other processes using variability techniques and semantic mapping, as related to what was previously under section 11.4.
- **Deployment of the application:** The run time application code is eventually deployed in a consecutive phase, possibly in a distributed or mobile manner, according to dynamic device and network resource conditions and other context factors, as applicable at that moment in time, and adapting to environmental context changes.
- **Execution of the application:** After deployment, the regular execution life cycle phases comes in action, effectively starting or stopping dependable subservices for the application, for the assumed context and user reach.

Finally, users start interacting with the newly created application.

11.5 Interactions, Using and Creating in Smart Spaces

The smart space consists of interactive components, sensors and actuators and allows for very versatile interaction with the services given shape by the tangible, distributed interface. One could say that the smart space ideally forms an ecology, in which people seamlessly interact with the environment to achieve specific goals, in particular also the creation goal which the previously discussed service framework is aiming to support at the software level.

11.5.1 Service Interaction and Environment Configuration

The interaction will be related to *using* the services provided by or through the environment, or to *configuring* the environment itself. The latter task receives special attention in the DiYSE project, as we want to enable the DiY end users to perform this task in most cases. This requires the interaction to be very intuitive and 'programming'-like solutions are out of the question. Configuring the environment, or defining the 'intelligence' of the environment, consists for instance of:

- associating input *events*, like a button press, a sensor reading change, or GUI widgets, to *actions* in the environment, like motors, valves or other actuators, or application settings, via a set of *behaviour rules*,
- defining dependencies on *context* information, like presence detectors, time of day, or temperature,
- personalising services, like *look and feel* adaptation to the user's identity, taking into account preferences for content, adapting to use patterns, switching to the preferred input or output modality, or
- creating *mash-ups* of existing controls, defining a "macro" of control operations specifying a personal *remote* interaction to services.

In this project we envision the use of *physical browsing* techniques to help selecting the physical target objects for use in the configuration by means of touching or pointing actions. The project is also researching the use of templates, wizards or *'define by doing'* approaches to simplify the configuration. The *'define by doing'* approach requires the environment to be completely observable. The user will define complex functionality by creating the specified circumstances and performing a series of actions that *show* the system what is expected as application behaviour in those circumstances.

11.5.2 Ecological Design Approach

Refining beyond the essentially different phases as discussed in the previous sections, the design of a smart space ecology does not happen in one step, but consists of the design of the components (i.e., devices, sensors, actuators, single device applications) for use in the environment, *enabling design*, and the design of the functional environment, *local design*. While the *enabling design* will typically be performed by professionals as part of their product development, *local design* can be performed also by the end user, tailoring his environment and combining the functionality of the enabled products, as exemplified by the configuration task in the previous paragraph. We coin this to be the *Ecological Approach to Smart Environments* (EASE). Note that the approach emphasises the importance of *in-*

volving the users at all phases of the design (Keinonen 2007; Norros and Salo 2009).

11.5.3 Architectural Support and Modelling for Interaction

To get the various interactive components in the environment to work together to provide such a context-aware personalised interactive experience is not a trivial task. Also from the user perspective, the interaction capabilities of the components must be described properly and be advertised, dependencies on context information must be specified, and, as indicated before in the discussions on context, the user's preferences and abilities must be taken into account.

The architecture as described in the previous sections provides a good base for this interactive environment. Interactive component capabilities can be described using the same ontology-based semantic methods when applying a suitable interaction ontology. Interactive events can be channelled through the available interoperability solutions. Context information is provided at a suitable level by the brokers in the system.

The interactive applications themselves need to be modelled in such a way, that their interactions are easy to map to the components available in the environment. This requires new solutions. Most user interfaces are designed for a specific platform, even a specific device, and cannot be transferred to other platforms, let alone a set of interactive components. Remote interfaces, using *HTML Forms* for example, have partly solved this problem, as the platform rendering the user interface may be different from the one running the application. This approach has also allowed for scaling the user interface to devices with various viewports. The method is sometimes referred to as a *multi-channel* approach. It works well with '*window, icon, menu, pointing device*' (WIMP) devices and has successfully been mapped to mobile devices as well. Mapping for multi-modal interaction or distributing the interaction over a multi-device solution requires more versatile modelling of the user interface. A starting point for this kind of modelling can be found in the *Abstract UI* solutions found in UsiXML (Limbourg et al. 2005), Teresa (Paternò 1999) and the like. But unlike those modelling solutions, the DiYSE system must moreover be able to resolve the mapping issues *at run time*, in the *changing* environment. This is one of the central themes of research in the project.

11.5.4 Example Personalised Interaction Method: Smart Companion Devices

11.5.4.1 Multimodal Mood Detection in Smart Companions

In contrast to the multi-channel, spatially distributed user interaction discussed as a critically needed paradigm leveraging the possibly 'thin' nature of sensors and actuators as an intuitive, natural user (creation) interface in the Internet of Things, another asymptote of rich, intuitive user interaction is the one of a single- or multi-object, 'thick' *smart object* paradigm, offering and embodying a human-like counterpart for the user-creator. In the DiYSE project, this is seen as an advancement beyond classical multimedia interfaces, which ads up to the 'things' available in the smart space for use and DiY creation.

In particular, for *smart companions*, being robotic pets whose appearance and behaviour are tailored to human interaction, a comfortable user experience requires the establishment of a meaningful robot behaviour illusion. This can be achieved employing a variety of techniques, aimed at the recognition of auditory and visual cues, such as speech/speaker and face/gesture recognition, of which the most advanced variant is the *multimodal* approach, combining voice, image and gesture recognition to derive context. So, context data available through the smart companion device can be exposed in the DiYSE environment for use by other services and applications and vice versa, forming a rich connection to the interacting user.

In the current state of the art, smart companions lack the ability to detect what is arguably the most important factor present in normal human interaction: the *mood of the speaker*. While speech and non-verbal analysis methods can be extended to detect mood or emotions, also such *affect recognition* can be further enhanced by associating image analysis to it for face and gesture recognition. These affect detection techniques have been the object of extensive research, but the associated computational cost has generally kept their application restricted to relatively powerful computing platforms, restricting the interaction illusion to being it 'via' the computer.

The aim of the smart companion work in DiYSE is therefore entailing two steps: (i) to research and implement affect detection algorithms on a PC platform, and (ii) to migrate them to an embedded platform present in a state-of-the-art smart companion robot. A standard back-end software interface is also foreseen for tying into the DiYSE context-awareness functions, integrating user mood as well as adapted companion behaviour as enrichments of the DiYSE smart user interaction.

11.5.4.2 Embedded Systems for Autonomous Smart Companions

So, as indicated, the enhancements proposed for the smart companion require a sufficiently compact, computationally powerful, and relatively low cost hardware platform. In fact, while serving a different purpose, such hardware requirements are of a similar nature as those needed for DiYSE Gateways needed to connect the 'thin' sensor and actuator nodes. Indeed, fortunately, nowadays available typical DiY electronics boards could be selected[136] as appropriate for this purpose too, namely *Beagle Board*[137] and *Gumstix Overo*[138].

11.5.4.3 Affect Recognition in DiYSE

By not taking into account the affective state of the user, the traditional *Human-Computer Interaction* systems are often perceived as cold and unnatural when compared to human-to-human communications. In the past decade, advances have been achieved toward the collection of large databases of affective displays, as well as toward the analysis of human behaviours by means of *audio-based, video-based,* and *audiovisual* methods[139] (Zeng et al. 2009).

A prerequisite in designing automatic affect recognition systems is the availability of *databases containing labelled data* of human affective expressions. Since manual labelling of emotional expressions is time consuming, subjective, error prone, and expensive, many databases consist of 'artificially' acted emotions, but also recordings of real, spontaneous affective behaviour were collected from human interviews, phone conversations, meetings, computer-based dialogue sys-

---

[136] The most popular embedded systems are built around the *ARM* architecture. Lately, Intel has introduced the *Atom* processor to cater for the same range of applications. These considerations narrowed our choices to (i) *Texas Instruments OMAP* based single-board computers and (ii) *Intel Atom* based system, presenting a power consumption versus code portability trade-off. From that, *OMAP* (v3), supporting *WindowsCE, Symbian, Android* and *Linux*, was eventually selected, leading to Beagle-Board and Gumstix Overo as the preferred platforms for the smart companion.

[137] http://beagleboard.org/

[138] http://www.gumstix.net/Overo/

[139] Most *audio-based* systems are trained and tested on acted speech in order to recognize prototypical emotions. Beside the selection of the classifier, another issue concerns the optimal feature set among linguistic and paralinguistic descriptors, as well as the reliable extraction of these cues (e.g. pitch-related prosodic features). *Vision-based* affect recognition studies mainly focus on facial expression analysis by means of pattern recognition approaches. The best choice for designing automatic recognizers seems to be the combined use of both geometric and appearance features. However, an important challenge remains the robustness to arbitrary head movement, occlusions, and scene complexity. Finally, while the vast majority of the *audiovisual-based* systems implement a decision-level fusion strategy and some other studies focus on the feature-level fusion approach to recognize coarse affective states (e.g., positive, negative, or neutral), the model-level fusion methods have the advantage of making use of the correlation between audio and video data streams without the requirement of perfect synchronization of these streams.

tems, etc.. While the automatic tool *Feeltrace* (Cowie et al. 2000) was developed for labelling such emotional expressions, the development of semi-supervised labelling methods remains an open issue.

As a first implementation of affect recognition in DiYSE, we chose to use the *EmoVoice* suite (Vogt et al. 2008) in combination with a voice recognition algorithms designed by UMons/Multitel. *EmoVoice* is in fact intended to be used by *non-experts*, opening further possibilities for DiY community scenarios where DiY creators can directly improve the affect recognition for an envisioned application purpose.

11.5.5 Multimodal Middleware Protocol

Multimodal approaches combining voice, image, and gesture recognition must necessarily acquire data from a *variety* of devices. The dedicated *Multimodal Middleware Protocol* (MMP) provides the low level architecture to glue different device modality components in a single user interface network. MMP's goal is to compose this network, abstracting details like underlying network protocols and the meaning of custom messages, so that all higher layer semantics and logic can relate to the composite multimodal interface. In the DIYSE concept the level above the MMP is a powerful context reasoning system, providing context-aware computing features, gathering information about users and their environment to adjust the behaviour of applications. Through the natural interfaces provided by multimodal devices such as the smart companion, context is seamlessly extended to social expressivity.

MMP interconnects devices and can store their capabilities in a central point, called a *Multimodal Hub* (MMH). Once a device modality component connects to the MMH, the MMH stores the user interfacing capabilities in terms of production and consumption of human communication events as sent by the component, and then manages the connections between components based on default or user-configured rules.

11.5.6 The Ultimate Example: Simple Smart Space Interaction with Multi-device Interfaces

Beyond the smart companion view, more heterogeneous scenarios are thus ultimately envisioned in DiYSE. Here is an illustrative example:

Peter arrives at home listening to his favourite MP3 music after an average day of work. The lights in the hallway turn on automatically as he enters and when he enters the kitchen to start making dinner the music is automatically transferred to the kitchen audio system so that he can remove his ear plugs and have his hands free. While preparing the dinner, his wife Katie arrives. She tells him with enthusiasm about the inspiring events she experiences at a work trip. She touches the screen in the kitchen with her mobile phone, which contains the pictures she has taken during her work trip. The screen comes alive and displays an overview of the pictures taken during the day. The touch screen of her phone simultaneously changes for use as a touch pad to control the cursor on the screen. She navigates to the first picture of interest and says 'Start slide show'. The screen starts to display the slideshow. When a video patch appears in the middle of the slide show, Peter's music fades out and they hear the audio track of the video. When a particularly beautiful picture comes up, Peter "steals" the picture by touching the screen with his mobile. The light slightly disturbs their viewing and Peter points at the light in the kitchen with his mobile and a personalised service view pops up. He selects a dimmed atmosphere by tapping his mobile a few times...

The implementation of scenarios like this requires the tight cooperation of all the available devices in the environment. The simultaneous use of interactive features of existing devices to operate new services constitutes to the multi-device interaction experience. Next step refinements of the DiYSE architecture will consider these aspects to yet a more complete extent.

11.6 Conclusion - Future Work of the Consortium

In this chapter we have sketched the wide variety of aspects tackled in the ongoing endeavour of enabling mass creativity in the Internet of Things, as envisioned by the DiYSE project.

As the main conclusion of the work done in the project until now, it is clear that a number of infrastructural measures and creation-supporting functions need to be in place to realise DiY application creation in the Internet of Things.

With enhanced, semantically annotated device drivers potentially auto-provisioned in a DiYSE Gateway function, a middleware for proper distributed execution across sensor network nodes, and a service framework that exposes context-awareness enabling functions and composable service building blocks towards the creation environment, a first basis for enabling such DiY application creation in the Internet of Things has been defined.

In order to realise the ultimate goal of DiY *smart spaces*, intuitively shapeable by non-technical actors, further creation-related enablers are still needed, both at the level of back-end services and tools, as well as support for *sharing* DiY experiences across large communities.

At the time of writing of this chapter, the consortium is progressing the detailed work on the DiYSE architecture according to the elements discussed, is implementing first prototypes, and is conducting interaction (co-)design user research, further fine-tuning towards the full enablement of communities sharing DiY smart space applications and smart objects.

Acknowledgements

This work is supported by the ITEA2 Eureka cluster, and the respective national funding authorities of the project partners, under the European ITEA2 project 08005, DiY Smart Experiences (DiYSE), conducted by 40 partners from 7 European countries. Beyond this book chapter, more information about the project can be found at the project's public website http://www.dyse.org.

References

Baglioni M, Macedo J, Renso C, Wachowicz M (2008) An Ontology-Based Approach for the Semantic Modelling and Reasoning on Trajectories. In: Song I-Y et al. (eds) Advances in Conceptual Modeling – Challenges and Opportunities, Springer, Berlin Heidelberg

Bastida L, Nieto FJ, Tola R (2008) Context-Aware Service Composition: A Methodology and a Case Study. SDSOA 2008 Worshop, ICSE Conference Proceedings, Leipzig, Alemania

Blum N, Dutkowski S, Magedanz T (2008) InSeRt, SEW, 32nd Annual IEEE Software Engineering Workshop

Cowie R, Douglas-Cowie E, Savvidou S, McMahon E, Sawey M, Schöder M (2000) 'Feeltrace': An Instrument for Recording Perceived Emotion in Real Time. Proc. ISCA Workshop Speech and Emotion

De Leenheer P, de Moor A, Meersman R (2007) Context Dependency Management in Ontology Engineering: a Formal Approach. J Data Semant 8:26-56

de Moor A, De Leenheer P, Meersman R (2006) DOGMA-MESS: A meaning evolution support system for inter-organizational ontology engineering. Proceedings of the 14th International Conference on Conceptual Structures (ICCS 2006)

Dormer P (1997) The Culture of Craft. Manchester University Press, Manchester, UK

Eid M, Liscano R, El Saddik A (2007) A Universal Ontology for Sensor Networks Data. Proc. IEEE International Conference on Computational Intelligence for Measurement Systems and Applications (CIMSA 2007)

ESI (2008) A3.D20. The Approach to Support Dynamic Composition. SeCSE Deliverable

Fernandez M, Gomez-Perez A, Juristo N (1997) Methontology: From Ontological Art towards Ontological Engineering. Proc. AAAI97 Spring Symposium Series on Ontological Engineering

GeSI, The Climate Group (2008) SMART 2020: Enabling the low carbon economy in the information age. Creative Commons. http://www.smart2020.org/_assets/files/02_Smart2020Report.pdf. Accessed 25 September 2010

Gruninger M, Fox M (1995) Methodology for the Design and Evaluation of Ontologies. Proc. of the Workshop on Basic Ontological Issues in Knowledge Sharing

Guinard D, Trifa V, Pham T, Liechti O (2009) Towards Physical Mashups in the Web of Things. Proceedings of INSS 2009 (IEEE Sixth International Conference on Networked Sensing Systems), Pittsburgh, USA. http://www.vs.inf.ethz.ch/publ/papers/guinardSensorMashups09.pdf. Accessed 25 September 2010

IFTF (2008) The Future of Making. http://www.iftf.org/system/files/deliverables/SR-1154%20TH%202008%20Maker%20Map.pdf. Accessed 25 September 2010

ITU-T (2008) Intelligent transport systems and CLAM. ITU-T Technology Watch Report #1

Karp P, Chaudri V, Thomere J (1999) XOL: An XML-Based Ontology Exchange Language. SRI International. http://www.ai.sri.com/pkarp/xol/xol.html. Accessed 25 September 2010

Keinonen T (2007) Immediate, product and remote design. International Association of Societies of Design and Research, Honkong

Kim J-J, Park JC (2005) Annotation of Gene Products in the Literature with Gene Ontology Terms Using Syntactic Dependencies. IJCNLP 2004. Lect Notes Comput Sci 3248:787-796

Limbourg Q, Vanderdonckt J, Michotte B, Bouillon L, López-Jaquero V (2005) USIXML: A Language Supporting Multipath Development of User Interfaces. In: Bastide R, Palanque P, Roth J (eds) Engineering Human Computer Interaction and Interactive Systems. Springer, Berlin, Heidelberg

Loiseau Y, Boughanem M, Prade H (2006) Evaluation of Term-based Queries using Possibilistic Ontologies. In: Herrera-Viedma E, Pasi G, Crestani F (eds) Soft Computing in Web Information Retrieval: Models and Applications. Springer

Luke S, Spector L, Rager D, Hendler J (1997) Ontology-based Web Agents. Proc. International Conference on Autonomous Agents (Agents97)

Machuca M, Lopez M, Marsa Maestre I, Velasco J (2005) A Contextual Ontology to Provide Location-aware Services and Interfaces in Smart Environments. Proc. IADIS International Conference on WWW/Internet

Meersman R (2010) Hybrid Ontologies in a Tri-Sortal Internet of Humans, Systems and Enterprises. Keynote talk, InterOntology'10 Conference, KEIO Tokyo

Miller E, Manola F (2004) RDF primer: W3C recommendation. World Wide Web Consortium. http://www.w3.org/TR/2004/REC-rdf-primer-20040210/. Accessed 25 September 2010

Mogensen K (2004) Creative Man. The Copenhagen Institute for Futures Studies, Denmark

Niles I, Pease A (2001) Towards a Standard Upper Ontology. Proc. 2nd International Conference on Formal Ontology in Information Systems (FOIS-2001)

Norros L, Salo L (2009) Design of joint systems - a theoretical challenge for cognitive systems engineering. Cogn Technol Work 11:43-56

Paternò F (1999) Model-based design and evaluation of interactive applications. Springer, London

Schilit B, Adams N, Want R (1994) Context-aware computing applications. Proc. IEEE Workshop on Mobile Computing Systems and Applications (WMCSA'94)

Schreiber G, Akkermans H, Anjewierden A, de Hoog R, Shabolt N, Van de Velde W, Wielenga B (1999) Knowledge Engineering and Management: The CommonKADS Methodology. MIT Press

Sennet R (2008) The Craftsman. Yale University Press, New Haven, CT

Shove E, Watson M, Hand M, Ingram J (2007) The Design of Everyday Life. Berg, London, UK

Smith B (2003) Ontology: An Introduction. In: Floridi L (ed) Blackwell Guide to the Philosophy of Computing and Information. Blackwell

Spyns P, Meersman R, Jarrar M (2002) Data modelling versus Ontology engineering. SIGMOD Rec 31:12-17

Sterling B (2005) Shaping things. MIT Press, Cambridge, MA

Strang T, Linnhoff-Popien C (2004) A context modeling survey. First International Workshop on Advanced Context Modelling, Reasoning and Management, Nottingham, England

Studer R, Benjamin R, Fensel D (1998) Knowledge Engineering: Principle and Methods. Data & Knowl Eng 25:161-197

Sundmaeker H, Guillemin P, Friess P, Woelfflé S (eds) (2010) Vision and Challenges for Realising the Internet of Things. CERP-IoT. http://www.internet-of-things-research.eu/pdf/IoT_Clusterbook_March_2010.pdf. Accessed 25 September 2010

Tang Y, Zhao G, De Baer P, Meersman R (2010) Towards Freely and Correctly Adjusted Dijkstra's Algorithm with Semantic Decision Tables for Ontology Based Data Matching. In: Mahadevan V, Zhou J (eds) Proc. of the 2nd International Conference on Computer and Automation Engineering "ICCAE 2010". IEEE, Suntec city, Singapore

Uschold M, King M (1995) Towards a methodology for building ontologies. Proc. of the Workshop on Basic Ontological Issues in Knowledge Sharing

van Harmelenand F, McGuinness D (2004) OWL web ontology language overview: W3C recommendation. World Wide Web Consortium. http://www.w3.org/TR/2004/REC-owl-features-20040210/. Accessed 25 September 2010

Vogt T, André E, Bee N (2008) EmoVoice – A Framework for Online Recognition of Emotions from Voice. Proc. Workshop on Perception and Interactive Technologies for Speech-Based Systems

Von Hippel E (2005) Democratizing Innovation. MIT Press, Cambridge, MA

Zeng Z, Pantic M, Roisman GI, Huang TS (2009) A Survey of Affect Recognition Methods: Audio, Visual, and Spontaneous Expressions. IEEE Trans Pattern Anal Mach Intell 31:39-58

12 Intelligent Cargo – Using Internet of Things Concepts to Provide High Interoperability for Logistics Systems

Jens Schumacher, Mathias Rieder, Manfred Gschweidl, Philip Masser

University of Applied Sciences Dornbirn, Austria

Abstract: The advancements in technology and increased need for streamlined business operations demand new ways in cooperation. In recent years, the Internet of Things has been recognised to be an important future technology, providing new opportunities for enhancing the exchange of information and status updates on real-time regarding business operations. Therefore, Internet of Things concepts have been adopted by several businesses to improve operations. This chapter describes the challenges that arise when implementing the ideas of the Internet of Things, regarding technology, interoperability and architecture of Internet of Things-compatible systems. Besides introducing the basic theories and concepts used to attack the mentioned challenges of Internet of Things applications, this chapter presents the EU-Framework Programme 7-funded project EURIDICE, which aims to conduct the concepts of Internet of Things to provide an open information platform for the transport sector utilising "Intelligent Cargo".

12.1 Introduction

The vision of the Internet of Things and Services is driven by technological developments like RFID and intelligent mobile devices, which can be attached to any kind of product. These gadgets close the gap between the real word and its virtual representation via, for example, seamless identification and interacting with other devices and the surrounding. This will result in a world of billions of objects, being able to report their location, identity and history by interacting with other objects or systems over wireless connections (Glover and Bhatt 2006). Regarding logistics and supply chain management, this vision will support new possibilities for business operations and business process management across enterprises and SMEs involved in the processes, but also raise new challenges regarding interoperability and data-compatibility. Although different implementations exist, mostly inter-organisational, there is still a need for open and standards-based platforms, which are necessary for the realisation of the vision of the Internet of Things and Services in logistics sector and supply chain management (SCM).

To identify the opportunities that are unleashed in the transport sector with the availability of low & high cost technology driving the ideas of the Internet of Things, such as RFID chips, the European Commission has launched the EURIDICEproject as part of the Framework Programme 7. The goal of this project is to implement a part of the Internet of Things and Services for the transport logistics domain, with the help of the technologies and services provided, which enables the long-term deployment of Intelligent Cargo for the different stakeholders involved in the transport sector. This includes e.g. customs, ports, terminals, shippers, forwarders etc. Overall, the EURIDICE project consists of more than 20 partners from different business sectors and authorities, which ensure that the project will provide an open platform. The platform enables service providers to combine different transport related services, e.g. for dangerous or high value transport in an open and freely customisable manner. For the evaluation of the project, companies and authorities from the transportation sector are integrated in the design process of the platform and pilots are installed for them, showing the applicability of the system.

The EURIDICE platform and its architecture provide new possibilities for co-operation along logistics and transportation chains by implementation of value-added cargo-centric services on item level, supporting the vision of the Internet of Things and Services and the adoption of the concepts of Intelligent Cargo (European Comission, DG INFSO 2009). EURIDICE exposes these services as Web Services in an open and standards based platform for easy integration by businesses with their back-end systems and supports the implementation of business processes across companies inside the platform using these services. The foundation of the EURIDICE platform, based on mobile agents, web services and the mediation between them, will be described later in this chapter.

Basically, three interoperability perspectives (Winters et al. 2006) must be fulfilled in order to fully provide a system that satisfies the requirements of the Internet of Things paradigm:

1. **The organisational perspective** covers the human centric aspect of interoperability in the meaning of common approaches and shared understanding of concepts, processes, beliefs and terms (Clark and Jones 1999). This also includes, for example, that an initiator of a process (e.g. functionality of an external system) is fully aware of the (semantic) consequences of a service it consumes.
2. **The IT systems perspective** handles interconnection issues between two systems. This perspective evaluates the interoperability on a physical level that builds the basis for two systems to interact with each other.
3. **The data perspective** covers data interoperability issues on the software level like the availability of the right data, data-formats and representation. Note that the data perspective considers both interoperability-domains, the semantic-

interoperability of the data, which means that the concepts and the understanding behind the data are identical in both systems (Obrst 2003), as well as the syntactical interoperability, which covers the readability of the used data-formats.

The organisational perspective is closely related to the first part of this chapter, covering aspects of business processes and the Semantic Web, e.g. automation and autonomous behavior and reconfiguration as well as a short introduction to the main technologies.

The second part describes the IT systems perspective and data perspective in more detail, split up in two sections, the first one discussing the role of ontologies in agent technologies and the second one the mediation between agent technologies and web services in the context of EURIDICE.

The last part of this chapter provides an overview about the impact of Intelligent Cargo on logistics sector and supply chain management and future developments as well as a summary and outlook.

12.2 Semantic Web

Supporting technologies and concepts for the realisation of the Internet of Things and Services are developed and defined related to the vision of the Semantic Web, sometimes also referred to as Web 3.0. This vision describes how available information can be extended with a meaning, which supports the automation of business processes using different technologies.

The fundamental idea behind the Semantic Web is to enhance available information with semantically descriptions to allow an effective implementation of information management as well as application integration. (Warren et al. 2006)

In an interview, Tim Berners-Lee formulated his vision as follows:

> "The Semantic Web is designed to smoothly interconnect personal information management, enterprise application integration, and the global sharing of commercial, scientific, and cultural data. We are talking about data here, not human documents." (Updegrove 2005)

The Semantic Web will provide access to an incredible large amount of human knowledge and additionally enhance it with machine processability. This will enable the development of different automated services, helping users and businesses to achieve their goals by accessing and processing the necessary information in a format understandable by machines. Consequently, this can lead to the implementation of distributed knowledgeable systems, including diversified reasoning services (Omelayenko et al. 2003). Therefore, the semantic integration at different levels within an organisation provides new possibilities for process integration, inter-organisational, but also across enterprises and whole supply chains. For the integration of these processes at the highest level, the organisational perspective, the basic technology is Semantic Web Services.

12.2.1 Semantic Web Services

The technologies provided by the Semantic Web are intended to transform the web to an information source interpretable by machines. This should lead to a web, where computer algorithms are able to process and reason with available information, which is currently limited to a human readable form. Services allowing access to the abilities of a company are mostly implemented by Web Services technologies and the concepts of Service Oriented Architectures (SOA), which are intended to support an environment where organisations can provide access to their abilities via the Internet. Computational capabilities are encapsulated with a Web Services interface, allowing other organisations to locate it via a registry (Universal Description, Discovery and Integration – UDDI) and interact with it using an interface description (Web Services Description Language – WSDL) (Preis 2007). These already known and widely accepted standards for the implementation of services operate at a syntactic level and, therefore, require human interaction to a great extent. For the integration of such services, it is required that developers search for appropriate Web Services in order to combine them. This limits the scalability and greatly cuts of the economic value envisioned with the advent of Web Services (Fensel and Bussler 2002). Semantic Web Services try to overcome these issues (Roman et al. 2006).

The Semantic Web Services vision (Paolucci and Sycara 2003) wants to combine the ideas behind the Semantic Web and the already available technologies for Web Services. This will enable automatic and dynamic interaction between software systems and, therefore, enable the implementation of the Internet of Things and Services. Current Web Services technology provides only support for a standard interface description, without machine-interpretable information about the software system's functionality. This issue is addressed by the concepts of Semantic Web Services by adding semantic information to Web Services. The goal is to describe them in machine-interpretable form, providing information on what the software does for a potential user and how it is done. This semantic information allows more sophisticated possibilities for discovering services than it is currently possible with UDDI. The combination of these technologies will support the development of more sophisticated applications, as services can be advertised and discovered more automatically and at high speed. Furthermore, services could also be combined to more complex services and processes, possibly automatically. Business processes should also become more robust; if a service is not available, a replacement could be rapidly found and added instead to ensure the complex service or process being still available. Semantic Web Services technology provides means for describing services and infrastructure capabilities to discover services and supporting interoperability. Semantic Web Services alone do not provide means for complex scenarios as reasoning (Preis 2007). However, Semantic Web Services technology is the enabling technology for the realisation of automatic web processes.

> "Semantic Web Services allow the semi-automatic and automatic annotation, advertisement, discovery, selection, composition, and execution of inter-organization business logic, making the Internet become a global common platform where organizations and individuals communicate among each other to carry out various commercial activities and provide value-added services." (Cardoso and Sheth 2005)

12.2.2 Semantic Web Services Processes and Lifecycle

The goal of the introduction of Web Services was the composition of loosely coupled web processes. Web processes composed of Web Services allow the representation of complex interactions among different organisations and are an evolution of existing workflow technology. Adding semantics to the web services can play an important role, as shown in the web process lifecycle (see Figure 12.1).

According to Cardoso and Sheth (2005), the semantics for the Web Services can be differentiated into:

1. **Functional semantics** are describing the aims of the services and the operations, including information about the required inputs and outputs.
2. **Data Semantics** are descriptions of the input and output data format for discovery issues and a common understanding in communication, in which data format the data has to be sent and how the receiving data can be semantically interpreted.
3. **Quality of Service (QoS) Semantics** are necessary for the selection of the most suitable service. Services can have different quality aspects, e.g. cheapest offer, fastest delivery, etc., and this provides possibilities to locate and select the service which matches best a required quality criterion in terms of QoS. This also enables monitoring of web processes based on QoS and allows evaluating alternative strategies for possible process improvements.
4. **Execution Semantics** are describing the flow of message exchange (message sequences) and provide a conversation pattern for the web service execution, flow of actions, preconditions, and effects on web service invocation, etc. They are describing the necessary interactions with a service on operational level, which is especially important for long running interactions and complex conversations involving different services from several parties.

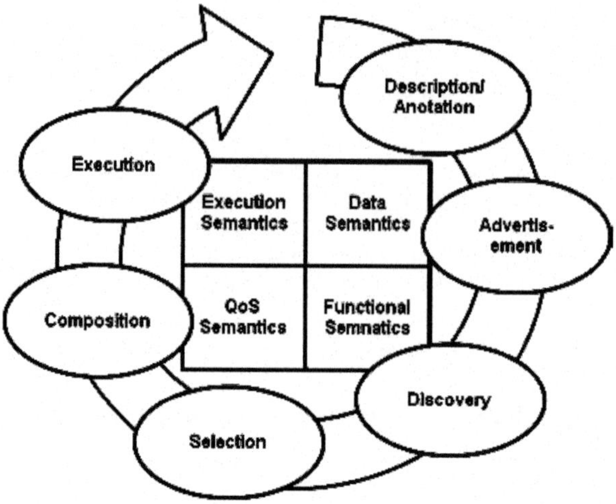

Fig. 12.1 Web Process Lifecycle and Semantics (according to Cardoso and Sheth 2005)

These semantics provide the basis for the automatic composition and realisation of semantic web processes and the underlying lifecycle. The lifecycle begins with the description / annotation, followed by advertisement, the discovery, the selection and the composition of Web Services building the web process, and, finally, the execution of the web process. For the success of the web process, these stages are all significant and should follow the next described steps (Cardoso and Sheth, 2005):

1. **Semantic Web Service Annotation**
   Description of the Web Services with the different required types of semantics for Web Services
2. **Semantic Web Service Advertisement**
   Publication of the service's capabilities on syntactical and operational level, using semantics in a repository (different technologies are available, centralised, e.g. UDDI, and even peer-to-peer). (Li et al. 2007)
3. **Semantic Web Service Discovery**
   Process of discovering appropriate services before selection, based on syntactic information, data, functional and QoS semantics
4. **Semantic Web Services Selection**
   Selection of the most suitable service matching defined QoS metrics and resulting in the best quality criteria match
5. **Semantic Process Composition**
   Automating inter-organisational processes across supply chains presents significant challenges (Stohr and Zhao 2001). Web Services are highly autonomous and heterogeneous; therefore, composition is an important stage. Semantics enhance interoperability of Web Services, and in the web process composition

(automatic, semi-automatic or manually) the functionality of the participating services (functional semantics), the data passed between the services (data semantics), the QoS of services and the process (QoS semantics) as well as the execution pattern of services and the whole process (Execution Semantics) should be considered.

6. **Execution of Web Process**
   Execution of the services and the whole process based on the execution semantics, the sequential or parallel invocations of services and operations, specified by the flow of messages and data based on the defined semantics

Using Semantic Web Services for web processes allows a higher dynamic creation and modification of the services, as the single steps can be also automated, if the involved services use the same underlying language for communication, in theses terms the same ontology.

For the implementation of Semantic Web Services a complete framework is necessary, consisting of three components (Roman et al. 2006):

1. A conceptual model (ontology)
2. A formal language to provide formal syntax and semantics for the conceptual model
3. An execution environment combining all components and carrying out the tasks for eventually service and process automation

As the idea of Semantic Web Services is relatively new, there currently exist different approaches for the realisation of Semantic Web Services, e.g. Web Service Modeling Ontology (WSMO)[140], Web Ontology Language for Web Services (OWL-S)[141] or Semantic Web Services Framework (SWSF)[142] (W3C 2004). On the *Operational Level,* Semantic Web Services and Semantic Web Processes are the enabling technologies, which could fulfill the requirements of the Internet of Things and Services.

The most promising approach for the realisation of a system of Semantic Web Services seems to be the WSMO approach, as it consists of an extensible ontology modeling language, a Web Services Modeling Language and an execution environment.

The WSMO initiative is the most important initiative in Europe regarding Semantic Web Services (SWS). It is part of the European Semantic System Initiative (ESSI)[143]-Cluster, the major initiative for standardisation and establishment of semantics for modern computer engineering. WSMO concentrates on standardisation of a unified framework for SWS and provides support for conceptual model-

---

[140] http://www.wsmo.org/

[141] http://www.w3.org/Submission/OWL-S/

[142] http://www.swsi.org/

[143] http://www.essi-cluster.org/

ing and formally representing services, and last but not least an execution framework. The three parts of the framework are (Roman et al. 2006):

- **WSMO**

This is a conceptual model for SWS providing an ontological specification for the core elements of SWS. Semantic Web technologies are intended for the transformation of the web into a world-wide system for distributed computing. Therefore, frameworks for SWS need to integrate all design principles, from web services, Semantic Web, to distributed, service-oriented computing. WSMO is based on the following design principles:

Web Compliance, Ontology-Based, Strict Decoupling, Centrality of Mediation, Ontological Role Separation and Description versus Implementation, Execution Semantics and Service versus Web service. (Roman et al. 2006)

- **Web Service Modeling Language (WSML)**

WSML[144] is a language intended for the description of ontologies, goals, web services and mediators, based on the conceptual model of WSMO. Its development is independent from existing standards for web services and the Semantic Web (Preis 2007). The major objectives for the development of the working group are (Roman et al. 2006):

1. development of a proper formalisation language for semantic web services
2. providing an adequate rule-based language for the semantic web

In its current version (1.0) WSML defines a syntax and semantics for ontology description and provides the possibility to use a Resource Description Framework Schema (RDFS)[145] and Ontology Web Language Description Logics (OWL DL), ontologies for web service description. It provides support for the dynamics of web services, choreography and orchestration.[146]

- **Web Service Execution Environment (WSMX)**

WSMX provides an execution environment enabling discovery, selection, mediation and invocation of semantic web services, and is therefore based on the conceptual model of WSMO, and also being a reference implementation. It provides support for achieving dynamic interoperability of SWS. It is available as open source and can be downloaded at the corresponding website[147] and the architecture consists of loosely coupled components, which are separately available and interchangeable (Roman et al. 2006).

---

[144] http://www.wsmo.org/wsml/index.html

[145] http://www.w3.org/TR/rdf-schema/

[146] http://www.wsmo.org/wsml/wsml-syntax

[147] http://www.wsmx.org/

In the next part of this chapter we will have a closer look at the general parts for the realisation, especially ontologies, as the different approaches differ in their implementation.

12.3 Ontology

Usually computer systems - or to speak more generally - Information Systems, work with real world concepts. If you take logistics information systems or almost every other type of information system, you will find concepts with certain properties, relationships between these concepts and vocabulary, which conceptualise real world objects and processes applied to them. “Cargo”, “vehicles”, “weight” or “delivery-deadline” are logistics terms with certain semantics behind them. Usually, the semantics of these terms, used in software, are defined by the developers via the software’s business-logic or domain-logic. The relationships between these concepts are encoded implicitly in the software guided by domain analysis methodologies and guidelines (Musen 1998).

An ontological representation of a real-world domain requires a conceptualisation of the reality (Gruber 1995). An ontology as an *“explicit, formal specification of a shared conceptualization of a domain of interest”* (Ehrig 2006) detaches the domain representation from the system’s domain implementation. Using an ontology as a basis for the system’s data representation strongly promotes truly interoperable systems. In order to understand the role of ontology for an interoperable and open Internet of Things and Services, we first have to understand the process of developing an ontology. The definition stated above (Ehrig 2006) contains two main requirements for an ontology. It must be an (a) *“explicit, formal specification”* and it must represent a (b) *“shared conceptualization of a domain”*. Let’s start here with the second request (b), since it chronologically is the first step when developing an ontology.

A conceptualisation of the real world domain involves the selection of relevant parts (e.g. concepts, properties) and the relations between those. So a conceptualisation will always produce a simplified model. The decisions which concepts and relationships become part of the conceptualised model are called *commitments*. A conceptualised model is derived from the real world domain by applying a set of commitments K to it.

The second request, which asks for an “*explicit, formal specification*” for ontology usage (a) offers big advantages when using them within computer systems. It means to use an explicit, unambiguous language (representation) for the ontology. This allows computer systems to process this information and to extract the semantic information that is encoded inside the ontology. The representation is accomplished by applying a certain *Language L* to the conceptualised model. This means to define a vocabulary for the model which means to name the elements (concepts, classes, properties, relations…) of the model.

Figure 12.2 summarises the process described above. It is important to notice that the two steps

1. Conceptualisation with the help of **Commitments K** and
2. Formalisation with the help of a **Language L**

strongly influence Internet of Things aspects like the interoperability, the openness and the potential cooperation of different systems in all three aspects (organisational, IT-System and data).

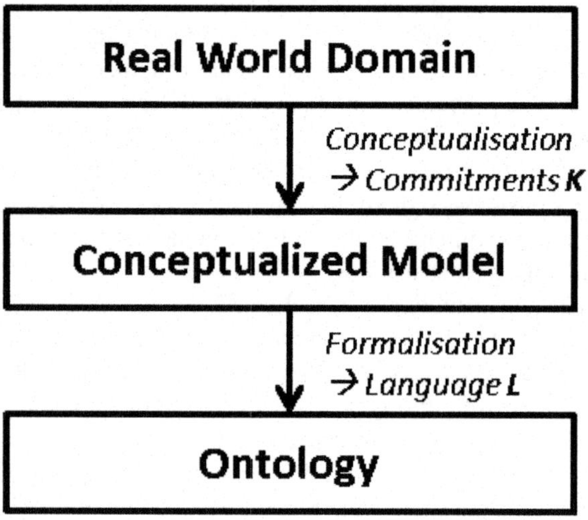

Fig. 12.2 The Process of Ontology Development as (1) Conceptualisation of the Real World Domain and (2) Formalisation of the Model to an Ontology

12.3.1 Ontology and the Organisational Perspective

When revising the problem from the *organisational perspective,* we can see that the technical implementation, or the used languages, encodings or data formats are not as crucial to this point. Applications that label themselves as compliant to the Internet of Things and Services must be aware of this organisational problem, first. There are an infinite number of possible conceptualisations of a real world domain. Two organisations can conceptualise the same real-world domain in a different way or conceptualise two different real-world domains to the same model. Both scenarios would lead to serious interoperability problems that are very often hard to detect. Interoperable systems, or Internet of Things applications that must obtain openness and interoperability to numerous (eventually yet unknown) systems, must use the concepts of ontology to state their understanding of the real-world domain or, in other words, to declare the commitments K that are used inside this application. This allows other applications or organisations to adjust their own vision in order to correctly interoperate.

12.3.2 Ontology and the IT-System Perspective

The IT-system perspective requires a more technical approach to guarantee the ability for cooperation to enable an Internet of Things and Service. Standard technologies like, for example, the SOA are able to provide open and accessible interfaces for applications. As already mentioned in the previous sections, ontology concepts already found their way into the world of SOA. Semantic Web Services are assisted by ontologies to fully semantically describe and understand the interfaces they offer and use. Computer programs can automatically discover and choose appropriate services, interfaces and data-structures to consume particular web resources.

12.3.3 Ontology and the Data Perspective

When talking about interoperability with respect to the *Data Perspective* in the meaning of the ability to cooperate and to exchange data, one must consider the interoperability of the domain models used by them. Data heterogeneity problems are well studied today (e.g., Papazoglou et al. 2000). Basically, there are the possibilities that the same representation of some data has a different intentional meaning or that there is a different representation of the same intentional meaning that cannot be translated offhand. Anyway, different representations of the same information can be an interoperability issue. In 1999 the *Mars Climate Orbiter* failed to successfully enter the orbit of Mars, due to a data interoperability issue (Isbell et al. 1999). Two software components of the orbiter, which were intended to cooperate with each other, used different intentional data models. One worked with the imperial system – where the *pounds force* is the standard unit of force – and the other one was based on the metric system – where *Newton* is the standard unit for force. The ratio between these two units is 4.45. While the used data formats of the two components where perfectly interoperable (both used decimal numbers), the different semantics behind the numbers caused the failure of a $300 million project.

Basically, there are two general ways to handle data interoperability issues (Renner et al. 1995). The first one is the data standardisation. Data interoperability issues can be avoided very effectively via defining standardised data structures in advance. A standardised data model does not only consist of the used encodings and data formats, it must also contain the used model, the interdependencies of the data, the used units, data types and precisions.

Data standardisation can be difficult and time consuming. Large scale standardisation efforts are seen to be impossible or at least inadequate, due to the large scale of the different domains (Hendler and Heflin 2000). Anyway, a lot of applications choose their data models by purpose (Renner et al. 1995). The data models are often tailored to the applications needs, regarding issues like performance, storage space or security issues. One can observe a so called "domain model evolution", especially on tailored domain models. This means that domain models naturally change from time to time (Goh et al. 1994).

Figure 12.3 shows how ontologies can be used to model an open and transparent data model. Data models, based on ontologies, provide interoperable data structures without the strict limitations of data standardisation. The data models are designed as follows (Guarino, 1997):

1. A **Top-level ontology** represents very common and general concepts like space, time, object, matter, event, action, etc., which are independent of a particular application domain.
2. The **Domain ontology** and the **Task ontology** specialise the terms defined in the top-level ontology for a certain problem domain (e.g. logistics). These ontologies differ between the concepts inside the *domain (e.g. cargo, vehicle, route...)* and the *tasks (load, unload, convey goods...)* performed there.
3. The **Application ontology** describes concepts that depend on both, the domain concepts of a domain and the tasks that are performed. Typically, it is a specialisation of both ontologies. Usually, this ontology contains domain entities playing a certain *role* while performing a certain task *(e.g. a cargo item which arrives in time at its destination)*.

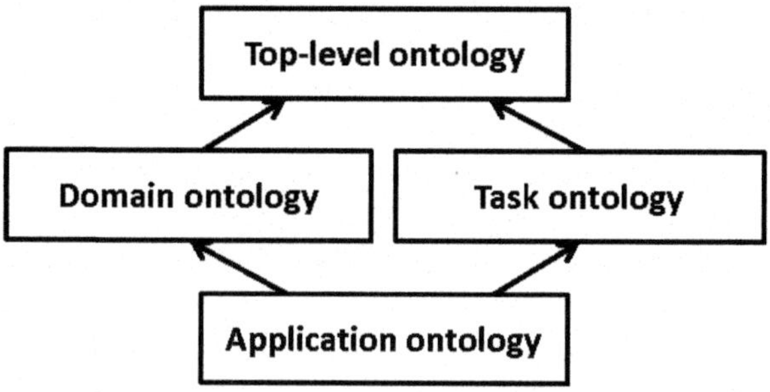

Fig. 12.3 Kinds of Ontologies, According to their Level of Dependence on a Particular Point of View (according to Guarino 1997)

12.3.4 Ontologies in Multi-agent Systems

Ontology concepts fully cover the IT-System interoperability perspective in Multi-Agent Systems. Multi-Agent Systems are "...loosely coupled networks of problem solvers that interact to solve problems that are beyond the individual capabilities or knowledge of each problem solver." (Durfee and Lesser 1989)
Multi-Agent Systems are massively distributed systems with sometimes thousands of components – so called agents. Agents are understood as software components, which are characterised by the following properties (Wooldridge and Jennings 1995):

1. **Autonomy**: agents work without any user interaction
2. **Social ability**: agents interact with other agents via agent communication languages – they offer and consume services that are beyond the capabilities of a single agent to and from each other.
3. **Reactivity**: agents perceive their environment (which may be the physical world, a user via a graphical user interface, a collection of other agents, communication channels like the Internet or perhaps all of these combined)
4. **Pro-activeness**: agents do not simply act in response to their environment; they are able to exhibit goal-directed behavior.
5. **Adaptability:** agents enjoy the ability to learn. This means that they are able to adapt their behavior according to changing surrounding parameters or events. Usually, self-adaptation is applied in order to reach the agent's design goal more efficiently.

The autonomy and the social abilities, which are requested from Multi-Agent Systems, make them very related to the ideas of the Internet of Things and Services. They must maintain a high factor of openness, interoperability and cooperativeness. Agents usually are *mediators* for users, digital resources or real world objects. So they act in the place of somebody or something and in place of the interests and the motivations of the entity they represent. These concepts enable complex systems composed of subsystems that are incompatible with each other by nature. Mediators swallow all interdependencies to the concrete entities.
Figure 12.4 shows an example of a logistics operators system made up of a multi-agent architecture. The single resource mediator agents encapsulate all concrete dependencies to, for example, the *transport vehicle*. Other agents must not deal with the detailed interaction mechanisms for a *transport vehicle* where maybe a *fleet management system* or the *concrete vehicle* itself equipped with RFID or other enhanced mobile devices sits behind the mediator.

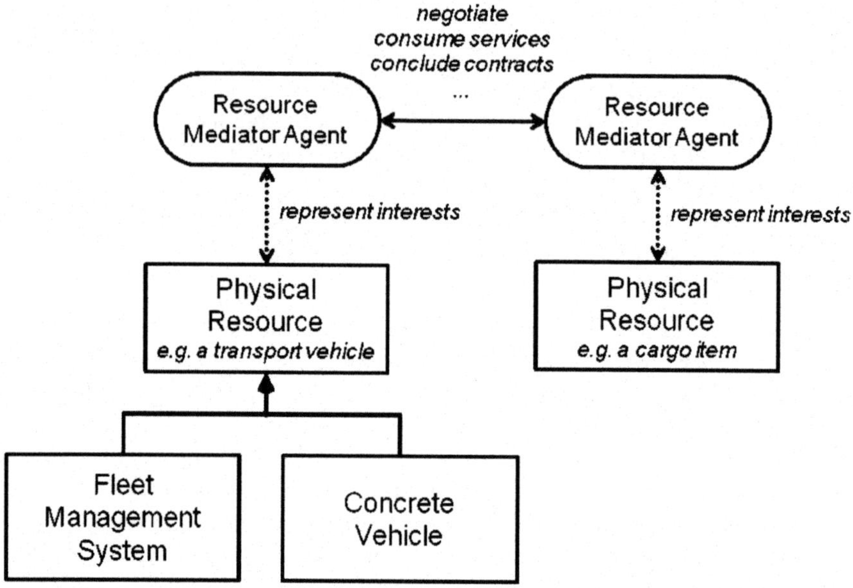

Fig. 12.4 Example Mediation Architecture in a Logistics Use Case (adapted from Maturana et al. 1999)

Since other agents, like for example the mediator of the cargo item (see Figure 12.4), only have to deal with the mediator of the transport vehicle, it is not of immediate importance to know whether the transport vehicle directly participates or is present via the Fleet Management System. Also, exchanging the subsystems (e.g. from Fleet Management System to vehicle equipment via RFID and mobile devices) will not affect the cargo item, as long as the corresponding Resource Mediator Agent still offers the same capabilities, supported languages and ontologies. On the top of the system everything is represented by an agent who produces a very coherent and flexible way of combining very different components to a smoothly working system. Those agents cooperate after the well known theories of multi-agent systems. They share a common ontology. It is the single resource mediator agent's responsibility to preserve the interoperability with its underlying system or component. This means that it must shield the rest of the system from the concrete dependencies to its resource – e.g. the resource specific ontology (see Figure 12.5) So, for example, if we think of a fully Internet of Things-based system, where a cargo item – or more precisely the corresponding cargo item agent – reaches a compromise with a transport vehicle agent on being part of its freight for tomorrows trip to Munich, the cargo item must not inform the corresponding loading staff to arrange the loading procedure. It does not matter whether there is a dedicated message necessary to put in the company's Fleet Management System or whether the freight list inside the transport vehicle's on-board unit must be updated. This implicit knowledge of how the single systems work is encapsulated by the corresponding resource mediator agents.

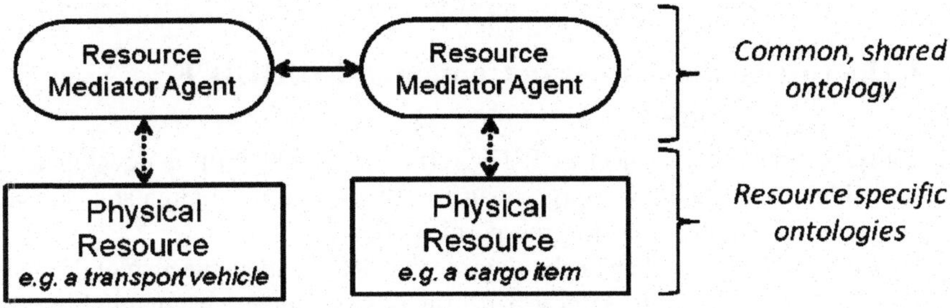

Fig.12.5 Usage of Ontologies in an Agent-mediated Architecture

12.3.5 The Role of a Top-level Ontology

In the example of the last section we showed how mediator architecture can shield the system from different issues of interoperability. In an Agent-based mediator system, a *cargo item agent* negotiates with a *transport vehicle agent* about a service – namely transportation. In order to successfully negotiate about this, the cargo item still needs some background knowledge about the services offered by the transport vehicle (e.g. transportation). We already heard that the knowledge about concepts of a domain is encoded in the domain ontology and the task ontology. These two domain-specific ontologies strongly depend on a common top-level ontology.

Usually, such top-level domains must not be developed from scratch. Different already established upper-level ontologies are available today, which can be adapted and tailored to the application's needs (Mascardi et al. 2007; Niles and Pease 2001). Nevertheless, it is important to notice that the choice of the upper level ontology should not heavily depend on the suitability to the application's domain. As shown in Figure 12.3, two ontologies are derived from the top-level ontology – the domain ontology and the task ontology, which both offer application-specific concepts to use inside the application ontology. The upper level ontology must not offer very application- specific concepts. The completeness and the ability to support inter-ontology mapping and conversion mechanisms as well as the support of standards should be emphasised instead.

As an example we take a closer look at the Cyc ontology by Cycorp, which aims to provide the most complete collection of everyday common sense knowledge (Matuszek et al. 2006). The Cyc ontology, which definitely serves as a mostly complete top-level, domain and task ontology, consists of more than 300.000 concepts, more than 3.000.000 facts and rules about 15.000 relations between the concepts. Usually, it would be of worth to think of using only a subset of this ontology in order to reduce the amount of data to process.

12.4 The Internet of Things in Context of EURIDICE

The European Integrated Project EURIDICE (Euridice 2009) aims at development and diffusion of the Intelligent Cargo, which is intended as a paradigmatic change in the field of ICT applications for transport logistics.

Despite availability of key technologies, like RFID, high-speed mobile networks and Web Services, the largest part of goods still move unsupported by information services along the route, resulting in process inefficiencies, poor communication between supply chain actors, and consequently higher societal costs in terms of environmental impact, safety and security risks. In the EURIDICE vision, Intelligent Cargo connects itself to logistics service providers, industrial users and authorities to exchange transport-related information and perform specific services whenever required along the transport chain.

The logistics sector is a field where information sharing between logistics providers, consumers, operators, authorities and other actors is essential in order to operate efficiently. EURIDICE's goal is to build a services platform centered on the individual cargo item and on its interaction with the surrounding environment and the user, allowing cargo objects and devices to perform basic interactions on their own and to involve the users' information systems if and when needed. Let us consider a case of fresh fish being transported with trucks from a supplier plant in Africa to a nearby airport, where it is palletised to be sent to Europe by plane, passing through customs after arrival, then being transported to a distribution center by train and from there to the individual customers that have originally ordered the fish. As fish is a perishable good, time plays a critical role in this case and fast handling is crucial to successfully operate, especially when the cargo is handed over from one operator to the next. However, the different companies involved in such supply chains often use proprietary Information Systems, which hinders the exchange of the required information. This scenario gives an example of the interoperability challenges faced by EURIDICE.

The EURIDICE project aims to facilitate the cooperation of various supply chain actors by providing a common platform that offers services for exchanging the transport related information. The EURIDICE platform puts the cargo item in the center of the process by equipping it with communication and computing capabilities, transforming it into *Intelligent Cargo*. The Intelligent Cargo can actively participate in the transportation process by interacting with its surrounding, consuming services, and also offering services relevant for the current situation, and thus builds an Internet of Things.

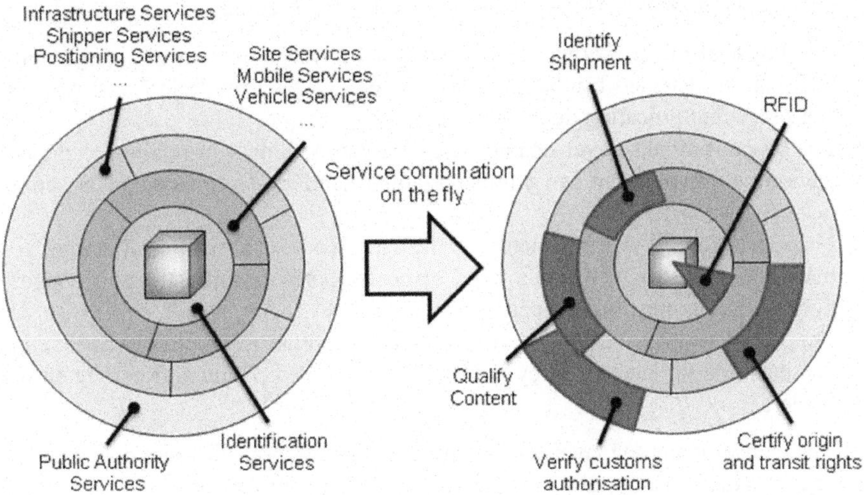

Fig.12.6 EURIDICE Cargo Centric Service Combination (Euridice 2009)

Figure 12.6 shows how the services offered by various stakeholders along the supply chain are combined on the fly to offer information relevant for the current situation, in this example for customs authorisation.

12.4.1 Interoperability in EURIDICE

The EURIDICE project has special requirements regarding interoperability and standardised data models. By using common data models and ontologies, the semantic barrier for cooperation across business domains is lowered. This has been established with a semantic framework which has two main purposes:

1. Support reusability, component-based design, and interoperability between the core services of EURIDICE and business-specific services in general, including end-user-applications and legacy systems.
2. Support context-awareness of the EURIDICE service platform, through the definition of a model describing the contextual information of the cargo and the related application domains.

The EURIDICE Context Model (ECM), shown in Figure 12.7, is a data and knowledge structure allowing access to cargo-related information and services from different domains and actors involved in Intelligent Cargo applications. The construction and definition of the ECM is organised around a core structure identifying physical cargo, vehicle and position. Around the physical kernel, the model information is organised in different layers, adding data to enrich the context definition at different levels as described in the following sections.

1. The Physical Layer (kernel) characterises the physical context of a cargo item, as it can be detected by the available technologies on the field (tags, readers, sensors, communication devices, etc.).
2. The Organisational Layer characterises the cargo context in relation to the organisations involved in cargo transportation, handling and management activities.
3. The Operational Layer characterises the cargo context in relation to cargo, vehicle and infrastructure status (e.g., cargo temperature, infrastructure congestion, etc.), ongoing and planned operations.
4. The Regulatory Layer characterises the cargo context in relation to rules, conventions and policies fixed by the different involved public and private stakeholders.

The above mentioned layers are interrelated though three different domains: the Cargo Domain, Transportation Domain and Environment Domain.

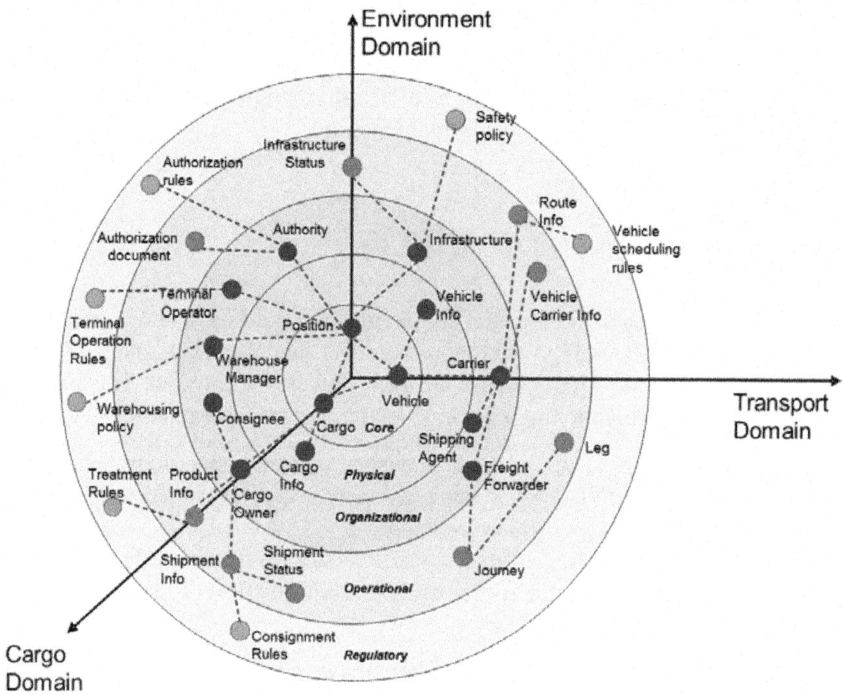

Fig. 12.7 The EURIDICE Context Model (Euridice 2009)

The EURIDICE Context Model is represented within the Cyc Knowledge Base, due to its openness, compatibility to the Cyc-ontology and interoperability characteristics concerning the IT systems perspective and the data perspective. Furthermore, the Cyc-ontology is used as a top-level ontology to build upon the EURIDICE Context Model. Due to the fact that Cyc contains a broad commonsense knowledge base, it allows reasoning and integration of live data and provides on-the-fly error detection.

The methodology for domain knowledge formalisation in Cyc uses the notion of Cyc micro theories, individuals, collection and predicates. Micro theories are used to represent thematic subsets or context of the ontology. Cyc collections are kinds of classes which instances have common attributes. The following methodology was used to integrate the EURIDICE Context Model into the Cyc Knowledge Base.

1. **Domain information identification**
   The domain experts identify the appropriate EURIDICE related domain keywords. In this module, domain relevant information, such as EURIDICE Context Model entity names and descriptions, are determined. Furthermore, dozens of transport and cargo related business documents, like for example bill of lading, consignment notes, customs documents, etc., are analysed and consolidated.
2. **Domain subset extraction**
   The extraction of the relevant domain ontology subset from a multi-domain ontology, based on the specified domain information, is taking place in the domain subset extraction module. In the beginning, the keywords are used by the upper-level domain extractor to restrict the multi-domain ontology to the specific domains of interest. Afterwards, the domain knowledge extractor uses the information about the EURIDICE Context Model entities to obtain the EURIDICE-relevant Cyc Knowledge Base subset.
3. **Domain relevant information preprocessing**
   The information from the domain information module and the extracted EURIDICE-relevant Cyc KB subset are linguistically preprocessed in the relation identification module. The preprocessing phase includes tokenisation, stop word removal and stemming.
4. **Relation Identification**
   A ranked list of the relevant Cyc concepts and possible new relationships is composed. During relation identification, the EURIDICE Context Model takes entities and descriptions as well as a relevant subset of the Cyc Knowledge Base and, creates a ranked list of similar concepts for each EURIDICE Context Model entity.

The EURIDICE Context Model is used to capture the knowledge of the transportation domain. This knowledge is based on a broad basis of standardised and domain-specific knowledge. The EURIDICE architecture offers two main interfaces to the information gathered inside the EURIDICE Context Model, as shown in Figure 12.8.

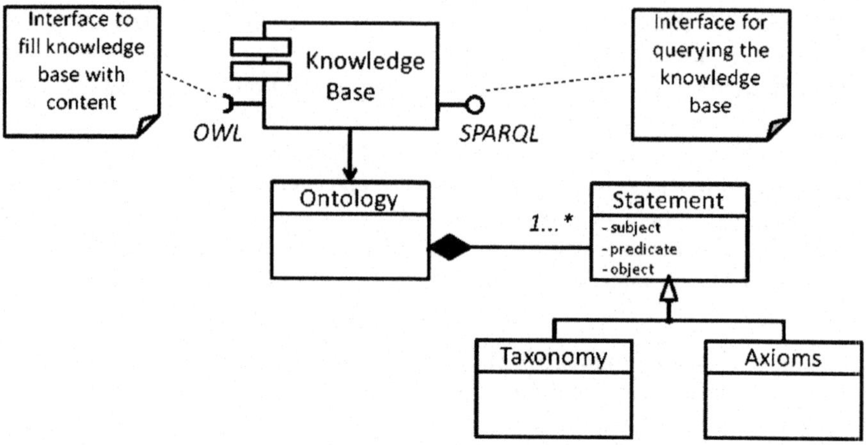

Fig. 12.8 The EURIDICE Knowledge Base Conceptual Model

The Ontology Web Language is chosen because it allows the modification and reuse of concepts definition and relations through well defined and sufficiently expressive semantics. Ontology-based models provide logic characterisation for the interpretation of objects, classes and relationships, allowing semantically consistent inferences, and also assuring one shared and reusable representation of the contextual information among service platforms and applications.

SPARQL is a Resource Description Framework (RDF) query language. RDF is a standard model for data interchange on the Web. RDF extends the linking structure of the Web to use Uniform Resource Identifiers (URIs) to name the relationship between things as well as the two ends of the link (this is usually referred to as a —triplet||). Using this simple model allows structured and semi-structured data to be mixed, exposed, and shared across different applications.

A statement is a triplet that contains a subject, a predicate and an object. The subject identifies the kind of information; the predicate can be used to specify the relation between subject and object, where the object represents a value or the concepts to be linked with. Taxonomy is a hierarchical system of concepts and axioms are rules, principles, or constraints among the concepts.

The challenge for external users of the EURIDICE platform is to convert the information they need to provide it to the platform in OWL format, and, vice versa, convert the OWL data back to the data format of their choice (e.g. the data format of the used Transportation Management System). The formalised ontology and the open and standardised technologies used, strongly support the conversation process as described in Section 12.3.

12.4.2 The EURIDICE Architecture

The EURIDCE project does not intend to build a new supply chain management tool that is to be used by all participating actors, replacing existing systems, but rather encourages them to share relevant information and services in a common, trusted, open platform. The platform itself is built upon open standards like SOA, Web Services, Foundation for Intelligent Physical Agents (FIPA) standards, Business Process Execution Language (BPEL) and ontologies. It offers services to interface with the cargo- and transportation-relevant information, appropriate for the various stakeholders that participate in the transportation chain, but also enables providers to extend the available services by custom services representing supply chain specific needs, and to develop applications that interact with existing legacy systems. On top of these EURIDICE- and application-specific services lies an orchestration layer that enables the definition of inter-organisational business processes along the supply chain.

As mentioned before, the Intelligent Cargo items are integral components of the EURIDICE platform, and are also equipped with computing and reasoning capabilities. To be able to effectively manage the myriad of cargo items that will eventually be participating in the EURIDICE applications, an MAS is used. Since the cargo services are of interest not only for other business services and applications in the back end, but as well for the devices and staff, physically interacting with the actual cargo, each Intelligent Cargo item is represented by an agent pair in the EURIDICE architecture. This pair splits the EURIDICE platform into two sub-platforms, the *fixed platform,* consisting of the back end with its services, and the *mobile platform* with the devices and services accompanying the cargo in the field. The corresponding Agents representing an Intelligent Cargo item are:

1. The **Assisting Cargo Agent (ACA)**. The ACA is the representation of the cargo in the virtual world. It offers services to other back-end services and applications via standard Web Service interfaces. It also serves as a single point of access to interact with the actual cargo item in the field, and acts on its behalf when it needs to execute back-end services. Therefore, it conceptually *is* the cargo item itself in the EURIDCIE fixed platform.
2. The **Operational Cargo Agent (OCA)**. The OCA supports the physical cargo item in the real world by connecting and interacting with its surrounding and watching over the cargo's conditions, like temperature constraints. It detects the local context of the cargo and invokes or offers the appropriate local services, e.g. customs clearance documents. Through the connection with the corresponding ACA it can also invoke global services in the back-end, e.g. for getting traffic information.

Together, these agents build the Intelligent Cargo Network of the overall EURIDICE infrastructure as depicted in Figure 12.9, surrounded by other components on either platform of the architecture.

On the fixed platform, there are components located that offer services concerning multiple items. The identification and discovery component provides services for uniquely identifying an item, and finding its corresponding ACA based on that identifier. Hierarchical positioning allows the reduction of communication acts, as information about one item can be categorically transferred to other items in the same hierarchy. For example, the position of all cargo items transported in the same truck is virtually identical.

The Semantic Framework contains the EURIDICE Context Model and ontologies as described in Section 12.4.1. This model is used as a basis for the context detection and cargo intelligence components. The former provides additional context information about cargo items for other services, based on sensor measurements reported by the OCA, and its business process state within the overall process. The latter provides global reasoning capabilities, like trend and anomaly detection in business process.

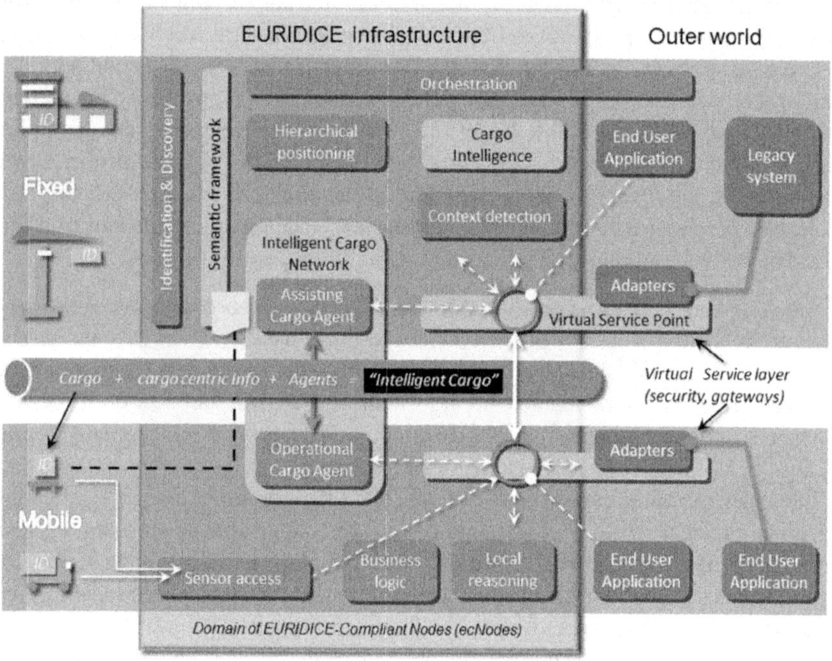

Fig. 12.9 EURIDICE Architecture Overview (Schumacher et al. 2009)

The mobile platform components support a single cargo item on its way through the transportation process. The sensor access component enables it to search for and access sensors in its local surrounding. Based on the measurements, the local reasoning and business logic components build a local context and identify violations of the cargo's mission, like delays or temperature limit exceeding.

Both platforms contain a virtual service point, which acts as a gateway to the EURIDICE components, enabling to implement custom applications and services.

12.4.3 Integration

To benefit from the services offered by the EURIDICE infrastructure, new applications have to be developed based upon it, and, more importantly, existing legacy systems, like ERP systems, have to be integrated and made interoperable between corporations across the supply chain. The EPC Information Services (EPCIS) is a well established and widely used EPCglobal standard for data sharing across enterprises (EPCglobal 2009). It consists of services for capturing, querying and subscribing to events related to business objects. The EURIDICE Infrastructure builds upon this standard and extends it to be applicable to all items handled within it. Enterprises that already make use of the EPCIS services can directly access and interoperate with the Intelligent Cargo items, otherwise adapters converting data from the EURIDICE Context Model into a system specific representation can be attached to the VSP.

The orchestration component of the EURIDICE architecture can build upon these and other EURIDICE services, and build and execute cross-organisational business processes, as shown in Figure 12.10.

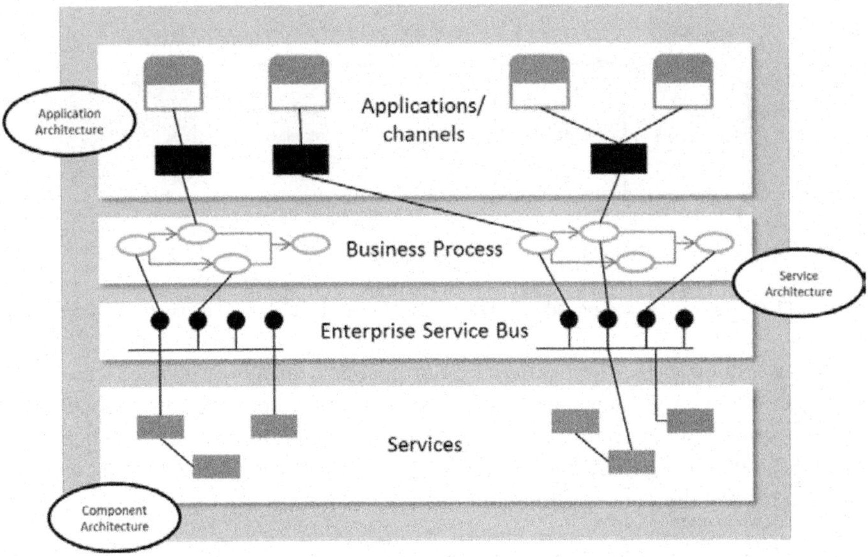

Fig. 12.10 Combination of EURIDICE Services to Business Processes (Schumacher et al. 2009)

Identification of the right service provider for a given cargo item is crucial to successfully operate these business processes, as the architecture is highly distributed among the participating enterprises. The ACA representing the cargo item is the prime information source and directly coupled with the item; it can be found through the discovery service. It can provide additional information sources, like the EPCIS services responsible for this specific item.

12.4.4 Deployment

The EURIDICE architecture is highly distributed among the participating enterprises, as its intention is not a central platform, but to use existing systems of the various enterprises and make them interoperable with each other.

For the fixed platform a service provider infrastructure is required. This will be provided by the Object Recognition and Positioning Hosted European Service (ORPHEUS) and made accessible for the EURIDICE users (Euridice 2009).

The mobile platform is more delicate, as a myriad of devices is required for supporting the cargo items in the field. Furthermore, the types of devices used vary between use cases and operators, ranging from on-board units fixed on trucks, over smart phones, to active RFID tags. Therefore, the mobile platform has to be flexible. This is achieved by the ACA-OCA pair representing a single cargo item, where the responsibilities can be located on either side, or distributed among them, as shown in Figure 12.11.

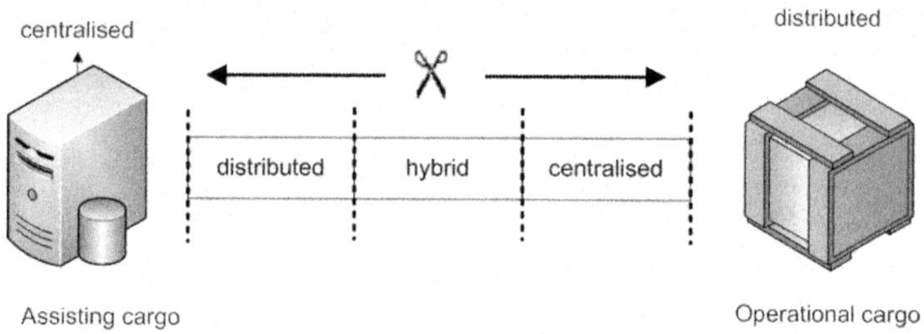

Fig. 12.11 Distribution of Responsibilities between ACA and OCA (Schumacher et al. 2009)

As an example, if a box is scanned at customs, in the distributed case (e.g. smart phone) a cargo item may derive from the local context, which documents are relevant for the current user, and directly transfer it to the user's device. In a centralised case (e.g. RFID tag), the user's device will have to contact the cargo's ACA to request the same information.

12.4.5 Project Evaluation

The project results within EURIDICE are evaluated using several pilot cases. These pilot cases consist of a set of representative use cases for a certain transportation domain. The eight business cases are hosted by different European companies of various types, including representatives of consignors, producers or distributors, infrastructure managers, logistics operators, authorities and consigners.

These scenarios have been selected to test the EURIDICE infrastructure and technologies on real cases, with the aim of demonstrating the Intelligent Cargo concept and its advantages. Various modes of transportation are covered including:

1. Road transport
2. Rail transport
3. Maritime transport
4. Air transport
5. Multimodal combinations

Each scenario refers to a precise business context and problem to be solved. The aim is certainly not to cover all the possible activities in a generic transport process, but rather to map different relevant situations where Intelligent Cargo can be put into practice to the benefit of different stakeholders. Typical current problems of logistics processes include:

1. Missing information (e.g. arrival / departure times) or low quality of information
2. Inefficient utilisation of resources
3. Delays of information distribution
4. etc.

These indicators are measured qualitatively as well as quantitatively. The project strongly expects that the applied concepts of the Internet of Things considerably improve the above mentioned indicators. The evaluation phase of EURIDICE is scheduled for the second half of 2010. The results of the evaluation phase will be published at the end of the EURIDICE project (2011).

12.4.6 EURIDICE and the Internet of Things

The capabilities and services offered by the EURIDICE project present a first step to build an Internet of Things for the logistics sector. In the virtual world, the cargo items can be seamlessly integrated in the web of services, and, thus, the cargo item can be incorporated in high level business processes. So, events that occur during the transportation process can be used to trigger business processes, like the detection of a delay in the cargo transport can lead to a re-scheduling in subsequent transportation trips. Through such real time reporting of abnormalities, the barrier between physical execution of the transport process and the virtual representation of the same process in the back-end system vanishes.

In the mobile world the cargo item discovers and interconnects itself with other devices to share information and services. Consequently, a device with certain capabilities, e.g. sensing the temperature, could make this information available for other devices in its surrounding, like other cargo items transported in the same truck. Furthermore, the cargo item is able to present only the relevant information, based on the current context, to human operators interacting with the item. So, the cargo item could automatically detect which kind of documents are of interest for the current user and present either a customs clearing document, a transportation schedule, or an invoice.

The EURIDICE platform also supports the realisation of the three interoperability perspectives needed to build an Internet of Things. Through the modeling of all exchanged information – the business documents presented to human operators, the mission documents defining the cargo agents' behaviors as well as the information exchanged between devices – a formalised model with clear syntax and semantics is used. This is a key requirement to interoperate in the Data Perspective. Moreover, the communication between agents as defined in the FIPA standards is based on ontologies (FIPA 2006).

The IT Systems perspective is supported by the usage of SOA standard to connect the various services with each other. On the higher orchestration layer, on top of the services defined by EURIDICE, and the specific custom application, BPEL is used to compose business processes. This layer can also be used to overcome the organisational interoperability perspective issues.

12.5 Business Impact

The previously described technologies and concepts are the base technologies for the implementation of the vision of the *Internet of Things and Services*. Logistics and supply chain management are important parts of this vision. Single goods are flowing through an interconnected and interoperable information system with a virtual representation consuming offered services for their own purposes (mainly transportation services), enabling an automatic transportation of its real word representation. Therefore, this virtual representation needs some intelligence for operation, which is often referred to as cargo intelligence (also Intelligent Cargo) in the logistics sector, implemented by already mentioned technologies and concepts and shown as practical realisation in the context of the EURIDICE platform for logistics. These already available implementations for parts of an Internet of Things and Services do not only drive technological advancements in business operations, they are also causing a huge impact on current and future business operations from the management perspective. These business operations are:

1. Business models
2. Business opportunities
3. Management concepts
4. Logistics and supply chain management
5. Supply Chain Networks
6. Freight exchange platforms
7. Auctions models
8. Regulations
9. Sustainability

The step to cargo intelligence is going to change the way logistics operations are done. Cargo intelligence represents an important step form a functional process oriented view towards a real object oriented view. Nowadays, logistics operators think mainly about their operating processes, e.g. the trucks, planes, hubs, terminals etc. – the focus of their work is clearly process oriented. The object oriented view differs dramatically from this perspective. Here, the focus is on the individual cargo item, meaning that the underlying processes are treated as services that are offered to the cargo item.

This vision can be achieved by representing each individual cargo item as an intelligent agent. Thus, the physical item is accompanied by a virtual representative, which is able to identify and use services, which are described as parts of the logistics system architecture (e.g. EURIDICE). The intelligence is represented by an agent on e.g. a mobile device, which can connect to the internet.

Allowing cargo intelligence to take place in the logistics sector will have a profound impact on the underlying business processes. As in other areas, this would mean a de-tangling of the service providers and a new role for transport chain integrators. While a principal planning would be still necessary in order to achieve a "mission description" for a cargo item, the services used could vary dramatically, depending on the real-time environment and circumstances a cargo item could get into during real shipping operations. Thus, long-term transport contracts would have to be adapted to this new situation. However, while the process view is replaced by an object oriented view, there is a good chance that the utilisation of the transport vehicles will start to rise, as now the space in transport vehicles can be easily identified and allocated with a common open, standardised and interoperable platform in place, like the EURIDICE platform discussed in more detail in this chapter. A better utilisation will then, in return, transfer into cheaper and, at the same time, more environmental friendly transports. Currently, the transparency about the services and the transported cargo is missing. Logistics providers keep their consignments hidden, the fear of losing a customer if this customer is served by competitors (even if only temporarily) is understandable, however, a fundamental change has to take place here, like thinking of a marketplace for logistics.

Over the last decade, Europe has deployed a unique communication infrastructure, which allows communication everywhere throughout Europe. And with Galileo just around the corner, even positioning will enhance accuracy substantially, paving the way for new logistics services and views.

The newly gained real-time transparency in logistics chains requires new logistics processes, e.g. transport providers need to be exchangeable, as required in order to achieve the logistics goals identified before. As long as one third of trucks are running empty through the streets in Europe, it is clear that there is a good opportunity to save money but, maybe even more important, a good opportunity to contribute to de-carbonisation, supporting therefore economic and environmental sustainability developments.

However, a focused approach is required to address these problems, and there is no question that also political actions have to be taken, in order to sustainably regulate the transport in Europe.

One approach to deal with this is by regulating the flow of goods in Europe. As was shown in France some years back, the operation of freight exchange platforms can be successful, if they are accompanied by the right regulations. The transparency gained with such systems can clearly contribute to reducing the carbon footprint of transport operations. If logistics providers are e.g. forced to publish one third of their consignments on a dedicated freight exchange platform and free bidding could take place onto these consignments, it would be possible that bidders could better use synergies with their existing transport operations. These synergies could be used in all modes of transportation and are not restricted to road transport. With an unsurpassed accuracy, the information about logistics operations can be passed tailor-made to the different stakeholders involved in transport or logistics operations, leading to the much needed trust in the service provision in this sector. This transparency and openness in logistics and supply chain (supply networks) operations enables the logistics sector to be part of the Internet of Things and Services vision.

12.6 Future Developments

In the long-term view, it is obviously that the information and services covered by the agents can be used for further services definitions in other sectors. The applications are almost unlimited with the advent of avatars representing humans in virtual worlds; there is an even bigger demand for product avatars that represent products and their requirements for services in the virtual world. While a human being can make intelligent decisions autonomously, such decisions are even more important for products, since they need an "intelligent" support to achieve their possible different kind of goals and objectives.

Based on the view of a product avatar, a whole platform of services around specific products is conceivable. A product centric approach to maintenance, upgrade and other services, could lead to a re-vitalisation in the area of ERP and CRM systems. The concept of the product avatar allows the establishment of a new customer channel that has a value for the customer as well as the producer of a product or a related service provider. The software industry has been working with these kinds of intelligent products for some years with great success, but here the products themselves are virtually, meaning that the step to provide intelligence in this setting doesn't require a substantial re-thinking in the way products and services are understood. The big challenge in the next years will be to transfer these experiences into the world of real products and real "non-virtual" processes.

12.7 Conclusion

Although the vision of the Internet of Things and Services exists for some years now, the real applications for it are still missing. It proposes a future of value added services for businesses and consumers. Advancements in technology are following this vision, but, as always with visions and their implementations, research and developments need to be done to provide the base for achieving the goals and objectives. It is no question that it could be done, but standards and open interoperable platforms are necessary for supporting the successful implementation, based on already available ICT infrastructures. Tools for the implementation of virtual business processes exist; the next step is the implementation of business processes interconnected with objects from the real world. Therefore, solutions are necessary for allowing these objects to exist in a virtual world. The technical base is available as well as the concept of agents and associated technologies. But for the wide adoption and application of the Internet of Things vision, it is necessary to provide open standards based platforms for easy implementation and participation of enterprises in an interconnected virtual and real world. EURIDICE tries to close the gap between the virtual and real world for the logistics sector. It provides a common language for all three perspectives, business, IT-systems and data, adopts the concepts of Multi-Agent Systems and allows the integration with existing and future systems by using an SOA.

Such systems support the shift from traditional, centralised planning and decision systems to distributed, intelligent, cooperative systems, putting the individual objects and their interaction with their environment in the center of processes. For the future, it can be predicted that the Internet of Things and Services will become reality, if the necessary technologies become available and applicable. It is no fiction anymore, it is a near future.

Acknowledgments

This work was supported by the Seventh Framework Programme of the EUROPEAN Commission through the EURIDICE Integrated Project contract number ICT 2007-216271

References

Cardoso J, Sheth A (2005) Semantic Web Services and Web Process Composition. First International Workshop - SWSWPC 2004 -Revised Selected Papers. Springer, Berlin, Heidelberg

Clark T, Jones R (1999) Organisational interoperability maturity model for C2. Proceedings of the 1999 Command and Control Research and Technology Symposium. Newport, USA

Durfee EH, Lesser VR (1989) Negotiating task decomposition and allocation using partial global planning. In: Huhns M, Gasser L (eds) Distributed Artificial Intelligence Vol 2. Morgan Kaufmann Publishers Inc, San Francisco

Ehrig M (2006) Ontology Alignment: Bridging the Semantic Web Gap. Springer Verlag, Berlin

EPCglobal (2009), The EPCglobal Architecture Framework. http://www.epcglobalinc.org/standards/architecture/architecture_1_3-framework-20090319.pdf. Accessed 30 June 2010

Euridice (2009) EURIDICE Whitepaper. http://www.euridice-project.eu. Accessed 30 June 2010

European Comission, DG INFSO (2009) Intelligent Cargo Systems study. http://ec.europa.eu/information_society/activities/esafety/doc/2009/intelligent_cargo_study_final.pdf. Accessed 30 June 2010

Fensel D, Bussler C (2002) The web service modeling framework WSMF. Electron Commer Res Appl 1: 113-137

FIPA (2006) Foundation for Intelligent Physical Agents. FIPA Abstract Architecture Specification. http://www.fipa.org/repository/architecturespecs.php3. Accessed 30 June 2010

Glover B, Bhatt H (2006) RFID Essentials - Theory in Practice. O'Reilly Media, USA

Goh C, Madnick S, Siegel M (1994) Context interchange: overcoming the challenges of large-scale interoperable database systems in a dynamic environment. Proceedings of the third international conference on Information and knowledge management. ACM Press, New York

Gruber TR (1995) Toward Principles for the Design of Ontologies Used for Knowledge Sharing. Int J Hum Comp Stud 43: 907-928

Guarino N (1997) Semantic Matching: Formal Ontological Distinctions for Information Organization, Extraction, and Integration. In: Pazienza MT (ed) Information Extraction: A Multidisciplinary Approach to an Emerging Information Technology. Springer, Heidelberg

Hendler J, Heflin J (2000) Semantic Interoperability on the Web, Proceedings of the Extreme Markup Languages 2000, Graphic Communications Assoc., Montreal, Canada

Isbell D, Hardin M, Underwood J (1999) Mars Climate Orbiter Team Finds Likely Cause of Loss. http://www.spaceref.com/news/viewpr.html?pid=2937. Accessed 30 June 2010

Li H, Du X, Tian X (2007) Towards semantic web services discovery with QoS support using specific ontologies. Proceedings of the Third International Conference on Semantics, Knowledge and Grid (SKG 2007). IEEE Digital Library, Xi'an, China

Mascardi V, Cordì V, Rosso P (2007) A comparison of upper ontologies. Proceedings of the WOA07. Genoa, Italy

Maturana F, Shen W, Norrie DH (1999) MetaMorph: An Adaptive Agent-Based Architecture for Intelligent Manufacturing. Int J of Prod Res 37: 2159 – 2173. doi: 10.1080/002075499190699

Matuszek C, Cabral J, Witbrock M, DeOliveira J (2006) An Introduction to the Syntax and Content of Cyc. Proceedings of the 2006 AAAI Spring Symposium on Formalizing and Compiling Background Knowledge and Its Applications to Knowledge Representation and Question Answering. Stanford

Musen MA (1998) Domain ontologies in software engineering: use of protege with the EON architecture. Methods Inf Med 37: 540-550

Niles I, Pease A (2001) Towards a standard upper ontology. Proceedings of the international conference on Formal Ontology in Information Systems Vol 2001. ACM Press

Obrst, L (2003) Ontologies for semantically interoperable systems. Information and knowledge management, ACM, New York

Omelayenko B, Crubzy M, Fensel D, Benjamins R, Wielinga B, Motta E, Musen M, Ding Y (2003) UPML: The Language and Tool Support for Making the Semantic Web Alive, chapter Ontologies and Knowledge Bases. Towards a Terminological Clarification. The MIT Press, Cambridge

Paolucci M, Sycara K (2003) Autonomous semantic web services. IEEE Internet Comp 7: 34-41

Papazoglou MP, Tari Z, Spaccapietra S (2000) Advances in Object-Oriented Data Modeling. MIT Press, Cambridge

Preis C (2007) Semantic Web Services: Concepts, Technologies, and Applications, chapter Goals and Vision. Springer, Berlin

Renner SA, Rosenthalm AS, Scarano JG (1995) Data Interoperability: Standardization or Mediation. Proceedings of the IEEE Metadata Workshop. Silver Spring, USA

Roman D, de Bruin J, Mocan A, Toma I, Lausen H, Kopecky J, Bussler C, Fensel D, Domingue J, Galizia S, Cabral L (2006) Semantic Web Technologies - trends and research in ontology-based systems. John Wiley & Sons Ltd., Chichester

Schumacher J, Rieder M, Gschweidl M (2009) EURIDICE – An enabler for intelligent cargo. The 2nd International Multi-Conference on Engineering and Technological Innovation (IMETI). Florida, USA

Stohr EA, Zhao JL (2001) Workflow automation: Overview and research issues. Inf Syst Front 3: 281-296. doi: 10.1023/A:1011457324641

Updegrove A (2005) The future of the web: An interview with Tim Berners-Lee. http://www.consortiuminfo.org/bulletins/jun05.php. Accessed 30 June 2010

W3C (2004) OWL-S: Semantic Markup for Web Services. http://www.w3.org/Submission/OWL-S/. Accessed 30 June 2010

Warren P, Studer R, Davies J (2006) Semantic Web Technologies - trends and research in ontology-based systems. John Wiley & Sons Ltd. Chichester

Winters LS, Gorman MM, Tolk A (2006) Next Generation Data Interoperability: It's all About the Metadata. Proceedings of Fall Simulation Interoperability Workshop. Orlando, USA

Wooldridge M, Jennings NR (1995) Intelligent agents: Theory and practice. Knowl Eng Rev 10: 115-152

Index

About the Editors

Dieter Uckelmann is managing director of the LogDynamics Lab at the University of Bremen since July 2005. His main research area includes the synchronisation of material, information, and financial flows under the influence of Auto-ID and billing technologies. Dieter Uckelmann is co-founder and president of the Global RF Lab Alliance – an international network of RFID-focused research organisations with representatives in Europe, USA and Asia and associate editor of the International Journal of RF Technologies: Research and Applications.

Mark Harrison is the Director of the Cambridge Auto-ID Lab, providing expertise in information architectures and technologies. He is deeply involved in standardisation activities at EPCglobal and has co-chaired the Tag Data Translation work group and Data Discovery Joint Requirements Group and currently co-chairs the Discovery Services work group, as well as participating in the EPCglobal Architecture Review Committee and GS1 Architecture Group. Mark is also a co-founder of the Fosstrak open source project and has contributed to the Tag Data Translation modules.

Florian Michahelles is a Project Manager and Associate Director of the Auto-ID Labs at ETH Zurich. His research interests comprise the design of an infrastructure of an Internet of Things for involving consumers, Human Computer Interaction (HCI) aspects of consumer RFID systems and the development of novel Internet of Things applications tailored towards consumers' needs. Michahelles has published 50+ papers in international journals (e.g. IEEE Pervasive Computing, Computer & Graphics) and at major conferences (IoT, Ubicomp, Pervasive, MobileHCI).

Many of the initial developments towards the Internet of Things have focused on the combination of Auto-ID and networked infrastructures in business-to-business logistics and product lifecycle applications. However, the Internet of Things is more than a business tool for managing business processes more efficiently and more effectively – it will also enable a more convenient way of life.

Since the term *Internet of Things* first came to attention when the Auto-ID Center launched their initial vision for the EPC network for automatically identifying and tracing the flow of goods within supply-chains, increasing numbers of researchers and practitioners have further developed this vision.

The authors in this book provide a research perspective on current and future developments in the Internet of Things. The different chapters cover a broad range of topics from system design aspects and core architectural approaches to end-user participation, business perspectives and applications.

Printed by Printforce, the Netherlands